Emergence, Complexity and Computation 1

For further volumes:
http://www.springer.com/series/10624

Andrew Adamatzky

Reaction-Diffusion Automata: Phenomenology, Localisations, Computation

Springer

Author
Prof. Andrew Adamatzky
Unconventional Computing Centre
and Department of Computer Science
University of the West of England
Bristol
UK

ISSN 2194-7287 e-ISSN 2194-7295
ISBN 978-3-642-31077-5 e-ISBN 978-3-642-31078-2
DOI 10.1007/978-3-642-31078-2
Springer Heidelberg New York Dordrecht London

Library of Congress Control Number: 2012940855

Printed on acid-free paper

Springer is part of Springer Science+Business Media (www.springer.com)

To those who made my life good

Preface

Cellular automata are regular uniform networks of locally-connected finite-state machines, or cells. Cells take discrete states and update their states simultaneously in discrete time. Each cell chooses its next state depending on states of its closest neighbours. The cell-state transition rules are very simple and intuitive yet allows for coding a non-trivial space-time dynamics. Thus the cellular automata is an ideal tool for a fast-prototyping of non-linear media models, massive-parallel computers and mathematical machines. Using cellular-automaton models of reaction-diffusion and excitable systems we analyse phenomenology spatial dynamics and show how to implement computation in these automaton models of a non-linear media. We complement cellular-automaton models with automata on planar proximity graphs.

The book consists of three parts. In the first part we introduce reaction-diffusion and excitable cellular automata and automata on proximity graphs, study phenomenology of propagating patterns, and represent a 'zoo' of travelling and stationary localizations. Reaction-diffusion is represented in terms of inter-species interactions in automaton models of populations in the second part. There we discuss dynamics and complexity of inter-species interactions and analyse mutualistic relationships. Computation in reaction-diffusion and excitable automata is overviewed in the third part. There we describe automaton networks which execute a space tessellation, we demonstrate how adders and multipliers are implemented by colliding gliders in excitable medium, we show how operate binary strings in reaction-diffusion automata and overview minimalistic models of memristive networks.

The book is self-consistent and does not require any special knowledge to apprehend. All models are intuitive and can be implemented with minimal knowledge of programming. Abundant illustrations help to appreciate expressive power of dynamical systems on cellular automata and planar automaton networks. Ideas, implementations and analysis offered will attract readers from all walks of life: everyone intrigued by sophisticated behaviour of cellular automata, non-linear media and mathematical machines.

Andrew Adamatzky
Bristol

Acknowledgements

I am thankful and grateful to

- Genaro Martinez for contributing to Chapter 2 by discussing phenomenological classification of binary-state reaction-diffusion automata and co-authoring paper [16].
- Emmanuel Sapin for evolving cell-state transitions matrices of reaction-diffusion automata discussed in Chapter 8 and selecting the rules rich in travelling localisations [20].
- Martin Grube for adding biological flavour to Chapters 9 and 10.
- Liang Zhang for advancing my ideas on collision-based based binary arithmetics and original implementation of 2^+-medium one bit half-adder; most designs of the chapter are based on papers co-authored with Liang [281, 282].
- Andy Wuensche for telling me about his beehive rule, which pushed me to discover the spiral rule automata [17], and for contributing to Chapter 13
- Leon Chua for introducing me to memristors which inspired some automaton constructs in Chapters 14 and 15.
- Thomas Ditzinger, Engineering Editorial, Springer-Verlag, for being so supportive and helpful.

Contents

Part III Computation with Excitation

Chapter 1
Introduction

Cellular automata are regular uniform networks of locally-connected finite-state machines, called cells. A cell takes a finite number of states. Cells are locally connected: every cell updates its state depending on states of its geographically closest neighbours. All cells update their states simultaneously in discrete time steps. All cells employe the same rule to calculate their states. Cellular automata are discrete systems with non-trivial behaviour. They are mathematical models of computation and computer models of natural systems. The cellular automata forms theoretical background and, at the same time simulation tools and implementation substrates, of mathematical machines with unbounded memory, discrete theoretical structures, digital physics and modelling of spatially extended non-linear systems; massive-parallel computing, language acceptance, and computability; reversibility of computation, graph-theoretic analysis and logic; chaos and undecidability; evolution, learning and cryptography. It is almost impossible to find a field of natural and technical sciences, where cellular automata are not used. For those not familiar with cellular automata we recommend to have a look in few classical titles. You can start with a Toffoli-Margolus's bestseller [242] and then spoil yourself with lavishly illustrated atlas by Wuensche [266] and thought-provoking cellular automaton treatise by Wolfram [262]. Plenty of interesting state transition rules and useful hints and tips on can be found in Ilachinski's compendium of cellular-automaton universe [144]. Conway's Game of Life is the most popular cellular automaton, we recommend the collection of chapters as a treatise [25] of the Game- of Life investigations, approaches and findings. Comprehensive specialised texts on modelling space-time dynamics of natural processes in cellular automata are authored by Boccara [56], Chopard and Droz [77], Weimar [258] and Deutsch and Dormann [95].

Since their inception in [122], cellular automaton models of excitation became a usual tool for studying complex phenomena of excitation wave dynamics and chemical reaction-diffusionactivities in physical, chemical and biological systems [77, 144]. They are now essential instruments in computational analysis of non-linear systems, and exhaustive search for non-trivial functions in cell-state transition rule spaces [9], [16]. Cellular-automaton models of reaction-diffusion and excitable systems are of particular importance because by using them we can —

A. Adamatzky: Reaction-Diffusion Automata, ECC 1, pp. 1–18.
springerlink.com

without too much effort — map already established architectures of massively parallel computing devices onto novel material base of chemical systems, and design non-classical and nature-inspired computing architectures [14]. For example, over ten years ago a range of mobile localizations was discovered in two-dimensional automaton models of excitable medium [5] and these localizations are demonstrated to be indispensable in implementing architecture-less computing schemes. The automaton models gave us indication that similar mobile self-localizations should exist in chemical excitable system, and in after few years of laboratory experiments such mobile localizations were discovered in real-world chemical media [219].

Why have we chosen cellular automata to study computation in reaction-diffusion media? Because cellular automata can provide just the right fast prototypes of reaction-diffusion models. The examples of 'best practice' include models of Belousov-Zhabotinsky reactions and other excitable systems [116, 187], chemical systems exhibiting Turing patterns [273, 275, 278], precipitating systems [14], calcium wave dynamics [274], and chemical turbulence [131].

Cellular automata are proved to be computationally efficient, accurate and user-friendly tools for simulation of huge varieties of spatially extended systems [77, 144]. They are essential in computational analysis of non-linear systems, and exhaustive search for non-trivial functions in cell-state transition rule spaces [9, 16]. Cellular-automaton representation of reaction-diffusion and excitable systems is of particular importance because this allows us to effortlessly map already established massively-parallel architectures onto novel material base of chemical systems, and also design non-classical and nature-inspired computing architectures [14]. We therefore consider it reasonable to interpret the cell-state transition rules we have discovered in terms of reaction-diffusion chemical systems. We envisage that this interpretation will provide the basis for experimental chemical laboratory designs of reaction-diffusion computers, allowing stationary localisations to be used as memory units [10].

We study automaton models of reaction-diffusion and excitable systems, analyse phenomenology and show how to implement computation in these automaton models of non-linear media. Apart of cellular automata we also consider automata defined on proximity planar graphs. The automata on proximity graphs are not cellular automata however share with cellular automata such basic features as local connectivity, discrete number of states, uniformity of state-transition rules and parallel updates of node states.

Cellular automata are very often used as fast-prototyping tool for developing novel algorithms of wave-based computing (see e.g. [9]). A distinctive feature of the prototyping is that it is made on an intuitive, we can say interpretative rather then implementative, level, where states are interpreted as chemical species and cell-state rules as quasi-chemical reactions [15]. That is we do not have to follow reaction-diffusion dynamic to simulate it in automata [241] but map models of cellular automata onto a space of quasi-chemical reactions. There are a well known variety of evolution rules that support behaviour similar to reaction and diffusion, see e.g. [207] and [182], however so far no one undertook a systematic analysis of 'reaction-diffusion' rules, particularly in terms of constructing parametric space, establishing conditions of pattern formation.

Every cell of a cellular automaton is a finite state machine, which updates its states in discrete time depending on states of its local neighbours. In all cellular automata studied in the book cells calculate their next states depending on sums of states of their closest neighbours. Diagrams of cell-state transitions are shown in Fig. 1.1.

The simplest yet powerful state transition diagram is shown in Fig. 1.1a. It represents cell-state transition for a minimalist model of a reaction-diffusion. This is two-dimensional cellular automaton with binary cells states: 0 and 1. Transitions between states 0 to 1 and 1 to 0 are determined by numbers of neighbours in state 1. In Chapter 2 we show that such an automaton imitates a quasi-chemical system with a substrate and a reagents. Quasi-chemical reactions are represented by semi-totalistic transition rules. Every cell switches from state 0 to state 1 depending on if a sum of the cell's neighbours in state '1' belongs to some specified interval (Fig. 1.1a). We investigate space-time dynamics of 1296 automata. We establish morphology-bases classification of the rules and explore precipitating and excitatory families of the automaton reaction-diffusion medium.

Diagram Fig. 1.1b is a classical state transition diagram for an excitable automaton: transition from resting state 0 to excited state + are determined by excited neighbours and transitions from + to refractory state – and from – to 0, are unconditional, i.e. happen independently on neighbours' states. This type of cell-state transition is employed in computational experiments with sub-excitable automata, Chapter 12, excitable proximity graphs, Chapters 6 and 7, and structurally-dynamic excitable automata, Chapter 15. Three modifications of the classical diagram are shown in Figs. 1.1cde. They are state transition diagrams where excitation is determined by excited and refractory neighbours (Figs. 1.1c), Chapter 4 and diagrams where a cell can stay in excited state (Figs. 1.1d), Chapter 3, and refractory state (Figs. 1.1e), Chapter 14.

The fully connected diagram (Figs. 1.1f) is employed in hexagonal reaction-diffusion automata studied in Chapters 8 and 13. Diagram (Figs. 1.1g) represents state transitions for automata imitating dynamics of two-species population in mutualistic relationships, where species are 1 and 2 and substrate is 0, Chapter 10. There is no predating in the system and thus no arrows between 1 and 2. And finally, a hybrid of excitation and precipitation shown in (Figs. 1.1h) is employed in Voronoi automata, which are not only based on Voronoi diagram but also approximate Voronoi diagrams, Chapter 11.

1.1 Weird Mods of Excitable Automata

In a classical, *a la* Greenberg-Hasting [122], automaton models of excitation a cell takes three states — reseting, excited and refractory. A resting cell becomes excited if number of excited neighbours exceeds a certain threshold, an excited cell becomes

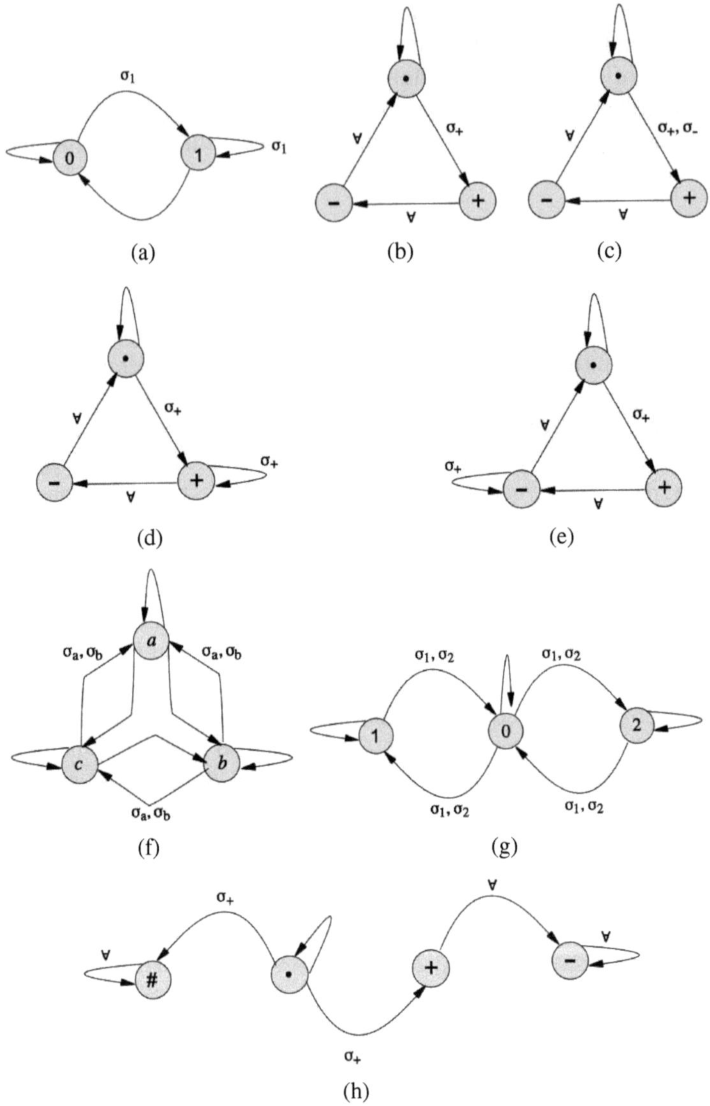

Fig. 1.1 Cell-state transition diagrams of automata studied in the book. (a) Binary state reaction-diffusion automata, Chapter 2, and minimal automaton model of population dynamics, Chapter 9. (b) Excitable automata, Chapters 6, 7, 12, 15. (c) Excitable automata with mutualistic excitations, Chapter 4. (d) Excitable automata with retained excitation, Chapter 3. (e) Excitable automata with retained refractoriness, Chapter 14. (f) Three-state reaction-diffusion, Chapters 8 and 13. (g) Automata model of mutualistic populations, Chapter 10. (h) Voronoi automata, Chapter 11. Symbols in grey discs symbolise cell-states and arrows are cell-state transitions. If an arrow from state s_1 to state s_2 is labelled with symbol σ_+, σ_-, σ_1, σ_2, σ_a or σ_b then a cell being in state s_1 takes state s_2 depending on a number of neighbours in state $+, =, 1, 2, a$ or b, respectively. Arrows labelled with $\forall$ symbolise unconditional transitions, and unlabelled arrows are 'otherwise' transitions, i.e. transitions which can take place when neither of the labelled conditions occurs.

refractory, and a refractory cell returns to its original resting state. In [9] we demonstrated that extension of the model to so-called interval excitation — a resting cell excites if number of its neighbours belong to a certain interval — allows to uncover a rich phenomenology of space-time excitation dynamics. We illustrate two further modifications of excitable automata: an excited cell is allowed to remain in its excited state (Fig. 1.1d), Chapter 3, and a resting cell excites depending on a number of its excited and refractory neighbours (Fig. 1.1c), Chapter 4 .

The retained excitation model, Chapter 3, allows an excited cell can remain in the excited state if a number of excited neighbours belong to some interval; the cell takes refractory state otherwise (Fig. 1.1d). Every resting cell excites if a number of excited cells in its neighbourhood belong to some other interval. Cell-state transition from refractory to resting state is unconditional. We classify over thousand rules of retained excitation based on how dynamics of excitable lattices develop after initial stimulation. Several modes of space-time activity dynamics are discovered. They include not growing but persistent domains of activity, domains with rectangular, octagonal and almost circular growth, amoeba-like growing patterns, travelling and stationary localizations.

Chapter 4 provides one more deviation from 'classical' excitation. A resting cell excites depending on numbers of excited and also refractory neighbours (Fig. 1.1c). We call this the mutualistic excitation model because excited and refractory states benefit from each other. We made exhaustive study of spatio-temporal excitation dynamics for all rules of this type and selected several classes of rules with non-trivial behaviour. Classes supporting self-localizations are particularly interesting. We uncover basic types of mobile (gliders) and stationary localizations, and characterise their morphology and dynamics.

Retained excitation and mutualistic excitation models exhibit stationary — still or oscillating — localisations. These can be interpreted as standing waves. Standing excitation waves in natural systems are not unknown of. Thus, externally induced standing excitation waves are observed in myocardium [121]. The standing waves are also found in oscillatory reaction of carbon monoxide on a platinum surface [149] and electronic excitations in carbon nanotubes [168]. As demonstrated in [98], standing and traveling waves can, in principle, coexist in the same reaction-diffusion medium. Some possible mechanisms of emergence of standing waves in general type excitable systems are discussed in [238]. Also it is numerically demonstrated in [84] that standing waves in excitable medium may emerge due to competition between diffusive propagation of perturbations, and pulse propagation, transitions from excitation to refractoriness do also contribute to birth of the standing waves.

Localisations — compact and long-living local disturbances of a medium's characteristics — are becoming hot topic of interdisciplinary non-linear sciences [65, 228, 248]. They can be found in almost any type of spatially extended non-linear systems, from liquid crystals to monomolecular arrays to reaction-diffusion chemical media [10]. From a computer science point of view, localizations are ideal candidates for elementary processing units in 'free-space', or collision-based computing devices [10, 50, 109, 184]. This why we pay so particular to rules supporting the localisations.

Cells in automata studied in Chapters 2–4 undergo state-transitions based on whether number of their neighbours in certain state belongs to some fixed specified interval. The interval based local transitions rules exhibits rich dynamics of discrete reaction-diffusion and excitation patterns. Can we make the excitation dynamics even richer by allowing the boundaries of the intervals to change dynamically? In Chapter 5 we introduce an excitable cellular automaton where boundaries of the excitation interval of every cell are updated at every step of the automaton development depending on the sums of excited and refractory neighbours in the neighbourhood of the cell. The automaton exhibits intriguing phenomena in its space-time dynamics.

1.2 Excitation on Proximity Graphs

A planar graph consists of nodes which are points of Euclidean plane and edges which are straight segments connecting the points, edges intersect only in the points/nodes. A planar proximity graph is a planar graph where two points are connected by an edge if they are close in some sense. Usually a pair of points is assigned certain neighbourhood, and points of the pair are connected by an edge if their neighbourhood is empty. Delaunay triangulation [92], relative neighbourhood graph [150] and Gabriel graph [191], and indeed spanning tree, are most known examples of proximity graphs. Proximity graphs found their applications in fields of science and engineerings: geographical variational analysis [112, 191, 226], evolutionary biology [183], spatial analysis in biology [81, 82, 152, 170], simulation of epidemics [245]. Proximity graphs are used in physics to study percolation [52] and analysis of magnetic field [230]. Engineering applications of proximity graphs are in message routing in ad hoc wireless networks, see e.g. [172, 196, 221, 227, 254], and visualisation [218]. Road network analysis is yet another field where proximity graphs are invaluable. Road networks are well matched by relative neighborhbood graphs, see e.g. study of Tsukuba central district [256, 257]. Biological transport networks also bear remarkable similarity to certain proximity graphs. Foraging trails of ants [11] and protoplasmic networks of slime mold *Physarum polycephalum* [19, 23] are most striking examples.

Structure of proximity graphs represents so wide range of natural systems that it is important to uncover basic mechanism of activity propagation on the graphs, which could be applied in future studies of particular natural systems. This is why in Chapters 6, and 7 we undertook computational experiments with two families of proximity graphs — β-skeletons and Delaunay triangulations.

Given a finite set of planar points Delaunay triangulation is a planar proximity graph which subdivides the space onto triangles with nodes in the given set such that the circumcircle of any triangle contains no points of the given set other than the triangle's vertices [92]. Delaunay triangulation is a graph-theoretic dual of Voronoi diagrams [253]. It represents connectivity of Voronoi cells. Voronoi diagram and its dual Delaunay triangulation are widely used in studies related to filling a space with connected structural units. Voronoi diagram and Delaunay triangulation are used to approximate arrangements of discs [117], sphere packing [107, 177, 179], to make structural analysis of liquids and gases [37], and protein structure [210], and to model dense gels [280] and inter-atomic bonds [138].

We are interested in studying excitable Delaunay triangulation because they may provide a good alternative to existing approaches of modelling unstructured unconventional computers [239]. Experimental research in novel and emerging computing paradigms and materials shows a great progress in designing laboratory prototypes of spatially extended computing devices. In these devices computation is implemented by excitation waves and localisations in reaction-diffusion chemical media [14], geometrically constrained and compartmentalised excitable substrates [90, 126, 127, 154], organic molecular assemblies [47], and gas-discharge systems [40]. These unconventional computing substrate can be formally represented by Delaunay triangulations with excitable nodes. Thus it is important to uncover most common types of excitation dynamics on the Delaunay diagrams. Despite being a ubiquitous graph representation of wide range of natural phenomena the Delaunay triangulation was not studied from automaton point of view. Will excitable Delaunay triangulation behave as a conventional excitable cellular automata or there will be some unusual phenomena?

We answer the question by slightly modifying classical Greenberg-Hasting model [122] (Fig. 1.1b) and considering not only a threshold of excitation but also a ratio of excited neighbours as an essential factor of nodes' activation. In an excitable Delaunay automaton every node takes three states (resting, excited and refractory) and updates its state in discrete time depending on a ratio of excited neighbours. All nodes update their states in parallel. By varying excitability of nodes of the Delaunay automata, in Chapter 6, we produce a range of phenomena, including reflection of excitation wave from edge of triangulation, backfire of excitation, branching clusters of excitation and localised excitation domains.

The proximity graphs β-skeletons, proposed in [160], form a unique family of proximity graphs monotonously parameterised by parameter β. A β-skeleton, $\beta \geq 1$, is a planar proximity undirected graph of an Euclidean point set where nodes are connected by an edge if their lune-based neighbourhood contains no other points of the given set. Parameter β determines size and shape of the nodes' neighbourhoods. In an excitable β-skeleton every node takes three states — resting, excited and refractory, and updates its state in discrete time depending on states of its neighbours (Fig. 1.1b). In Chapter 7 we design families of β-skeletons with absolute and relative thresholds of excitability and demonstrate that several distinct classes of space-time excitation dynamics can be selected using β. The classes include spiral and target waves of excitation, branching domains of excitation and oscillating localizations.

1.3 Automated Search for Localisations

So far we have been selecting localisation-rich rules by exhaustive search, essentially visual observations. In our previous works we have designed and studied a range of hexagonal cellular-automaton models of reaction-diffusion excitable systems, particularly those with concentration dependent inhibition of the activator [15, 17]. We have analysed three-state totalistic cellular automata on a two-dimensional lattice with hexagonal tiling, and discovered a set of specific rules that support a variety of mobile (gliders) and stationary (eaters) localisations, and generators of localisations (glider guns). In [17] we demonstrated that rich spatio-temporal dynamics of interacting localizations and generators of localizations can be used in implementing purposeful computation, including signal routing, multiple-valued logical operations and finite state machines.

Despite success of preliminary studies, and some techniques developed [270] to pinpoint 'best' rules supporting localisations, we remained somewhat puzzled and uncertain on whether rules manually selected are good representatives of a set of localisation-supporting cell-state transition rules. We therefore applied the full power of evolutionary computation methods to evolve and select all possible rules that support mobile localizations in two-dimensional hexagonal ternary state cellular automata. The automated approach is illustrated in Chapter 2.

There we consider hexagonal cellular automata with immediate cell neighbourhood and three cell-states. Every cell calculates its next state depending on the integral representation of states in its neighbourhood, i.e. how many neighbours are in each one state (Fig. 1.1f). We employ evolutionary algorithms to breed local transition functions that support mobile localizations, or gliders. We characterise sets of the functions selected in terms of quasi-chemical systems. Analysis of the set of functions evolved allows to speculate that mobile localizations are likely to emerge in the quasi-chemical systems with limited diffusion of one reagent, where a small number of molecules is required to amplify travelling localizations. Techniques developed can be applied in cascading signals in nature-inspired spatially extended computing devices, and phenomenological studies and classification of non-linear discrete systems.

1.4 Reaction-Diffusion, Automata and Populations

Reaction-diffusion equations is a classical tool in mathematical modelling of population dynamics [75, 118, 197, 214]. Populations dynamics itself is indeed akin to diffusion of species and reaction between the species. For example, diffusion can be written as

$$\text{Substrate} + \text{Species A} \rightarrow 2\ \text{Species A}$$

and reaction between say prey species A and predator species B as follows:

$$\text{Species A} + \text{Species B} \rightarrow 2\ \text{Species B}$$

Cellular automata have been used to simulate population dynamics for over twenty years. A 'mass-usage' of automata to imitate space-time dynamics of species has started with the popular article by Dewdney [96], supported by scientific publications on lattice-gas automata [63] and automata models of host-parasite interaction [71, 132]. See an overview in [104] and discussion on advantages of cellular automata models of population dynamics in [83]. Cellular automata models are nowadays uncontested models of pattern formation in population dynamics [95], spatial ecology [55, 97, 124, 234, 240], and the geomorphology-ecology interface [69]. The automata are handy in imitating and simulating propagation of species [73], developments of plant populations [45], prediction of epidemics dynamics in spatially heterogeneous environments [100], competitive interactions [66], hierarchical ecological systems [265], stochastic species invasion [74, 161], predation chains [166] and competition in complex landscapes [64], and prey-predator systems [67].

1.5 Minimal Models of Population Dynamics

Most cellular automata models of population dynamics aimed to simulate real-world phenomena. They are tied to particular species, landscapes or development scenarios and loaded with substantial number of parameters and characteristics. Any attempt to bring a model closer to reality blurs skeletal features of the model, and prevents, due to complicated designs, complete classification of space-time dynamics. Therefore we decided to strip automata models of inter-species interactions back to their bare bones and consider the most primitive model of two-species population without resources. In Chapter 9 we analyse space-time dynamics of basic interactions between two species using a minimal hexagonal cellular automaton model. A cell of a hexagonal lattice takes two states and updates its state depending on how many of its neighbours are in each particular state (Fig. 1.1a). We design and study threshold-based cell-state transition rules imitating mutualism, commensalism, parasitism and predation, amensalism and competition.

Given two species a and b we can depict their relationships by tuples (γ_a, γ_b), where γ_a (γ_b) shows how much species a (b) benefits or suffers from interaction with species b (a) [203]. The following interactions are considered in the Chapter 9:

- Mutualism $(++)$: Both species benefit from inter-species interaction. Previously indirect mutualism was discussed in [1], in models where increase of one species in numbers leads to saturation of predators and thus indirectly allows another species (also preyed by the same predator) to prosper.
- Commensalism $(+0)$: One species benefits while another is not affected, e.g. mixed cultures of milk-fermenting bacteria [44, 119], relationships between hermit crab and its associated species [260], commensalism between predators [125, 135, 235].
- Parasitism (predation, herbivory) $(+-)$: One species benefits while another species suffers. Positive and negative features of parasitism have been studied

via cellular automaton simulation in [164], cellular automaton models of starfish predation in corral reef developed in [165], cellular automata equivalents of Lotka-Volterra model discussed in [99, 134], automaton models reflecting distribution of individual characteristics of predating species and influence of predation on mimicry are studied in [142] and [157]. Recent results of non-automaton modelling can be found in [155, 180, 279]. Models of predatory relationships are well known for their complex dynamics [70, 101, 129, 158, 233].

- Amensalim (0−): One species is not affected while other species suffers. Examples include mixed cultures of easts [209], relationships between surface and subsurface feeding polechaeta, where ragworms took advantage from the absence of lugworms [252], and 'apparent' amensalism: wildcats in nature parks of central Spain are negatively affected by red deers and wild boards (due to direct competition between these hoofed animals and rodents which are preys of the wildcats) [178].
- Neutralism (00): None of the species is affected by interaction. In two-state cellular automaton neutralism-rule means that cells never change their states and any initial configuration becomes fixed one immediately. Therefore we will not discuss neutralism further.
- Competition (−−): Both species are badly affected by the interaction. Competition was simulated in cellular automata in the context of spatially explicit models of competition [93], analysis of propagation in population of competing species (as a function of the species' competing abilities) [38], emergence of rare species [113], and competition between plants [73, 189].

These six types of interactions are clearly idealistic and hardly realised in this clarity in nature. Naturally occurring interaction rather locate a continuum between beneficial (mutualistic) and pathogenic (parasitic) poles. In certain cases, the context of ecological conditions, such as habitat conditions can modify the behaviour of interacting species [156]. Moreover, some classic cases are hard to place in these anthropo-centrical categories.

We provide a morphological classification of inter-species interactions, supported by integral measures, and find that mutualism is the most complex interaction type. Spatial mutualistic interactions produce sophisticated patterns of stationary, oscillating and propagating species. Being intrigued by mutualism we decided to study it details and undertook exhaustive analysis of automaton rules imitating mutualism in Chapter 10.

In two-species interactions, mutualism is an interaction of species of organisms that benefits both [202]. Most classical symbioses were initially described as two-species interactions, but modern research shows that many of these cases comprise more interacting species and a higher level of complexity than previously thought [110, 175]. Mutualism is the most intriguing and still not well understood type of inter-species interactions [60, 231]. Even simplest models of mutualism exhibits higher behavioural complexity than predatoring, parasitism, amensalism and commensalism.

Mathematical and computer studies of mutualism are growing extensively last years. Most analytical models are based on modifications of Lotka-Volterra model with positive inter-species interactions [59, 115], stabilised with feedbacks of limited resources [139], tuned by diffusiveness and transport effects [94], and other relations between interacting species [76, 198]. Other mathematical models are based on limit per capita growth [120] and feedback delays [176].

Almost no results are obtained in spatial evolution of mutualistic populations. Spatially extended prey-predator systems produce characteristic wave patterns, what are the patterns emerging in mutualistic systems? So far only existence of stationary localised domains, or patches, of species is known. Their existence is demonstrated by two different techniques: lattice Lotka-Volterra model [236] and reaction-diffusion model of population dynamics [195]. Apart of particular case of interacting lattices [61] we are unaware of any published results concerting pure cellular automata models of mutualistic systems. In Chapter 10 we study a two-dimensional hexagonal three-state cellular automaton model of a two-species mutualistic system (Fig. 1.1g). The simple model is characterised by four parameters of propagation and survival dependencies between the species. We map the parametric set onto the basic types of space-time structures emerged in the mutualistic population dynamic. The structures discovered include propagating quasi-one dimensional patterns, slowly growing clusters, still and oscillating stationary localizations.

1.6 Towards Computing in Reaction-Diffusion Automata

Reaction-diffusion chemical systems are widely known for their ability to perform various types of computation, from image processing and computational geometry to the control of robot navigation and the implementation of logical circuits. In a reaction-diffusion computing medium, data are represented by the spatial configuration of the medium (e.g. local drastic changes of reagent concentrations or excitations), information is transferred by spreading diffusion or excitation waves and patterns, computation is implemented by interactions between spreading patterns, and the results of computations are represented by the final concentration profile or the dynamic structure of excitations. Numerous examples of simulated and chemical laboratory computers can be found in [14]. In the remaining chapters of the book we demonstrate several computing schemes implemented in reaction-diffusion and excitable cellular automata.

Excitable chemical media are amongst most promising 'wet' computing devices capable for solving a wide range of tasks from optimisation, computational geometry, image processing and robot control. The excitable media can also implement functionally complete sets of logical gates thus qualifying as (logically) universal computing systems. See theoretical background and details of laboratory implementations of reaction-diffusion computers in [14].

Most chemical processors designed in experimental laboratories are specialised, or a task oriented. They only can solve only one computational problem or a family of similar problems. In Chapter 11 we design an automaton model of such a specialised, or a task oriented, reaction-diffusion chemical processor. It solves the problem of bisecting planar data points and shapes. When logical universality is concerned the experimental implementations do usually deal with one or two logical gates, see e.g. [12, 18, 89, 246]. To move from abstract computational universality to general-purpose machine we must firstly demonstrate how we can cascade simple logical gates in more complicated circuits able to execute sensible computational tasks. An implementation of arithmetical chip in reaction-diffusion medium would be enough to convince laymen that excitable chemical computing devices are not just a matter of curiosity but viable candidates for a role of future non-silicon computers. Intriguing examples of collision-based arithmetical circuits and binary string transformers are presented in Chapters 12 and 13.

In Chapter 11 we explore automaton model of ensembles of compartmentalised excitable and precipitating chemical system. The automaton model is inspired by arrangements of vesicles filled with excitable chemical mixture [26, 32, 141, 141, 159]. When vesicles are in close, at least in a diffusion terms, contact with each other, via tiny pores, excitation waves can pass from one vesicle to its close neighbour. Excitation wave-fragments keep their shape, more or less constant, inside each Belousov-Zhabotinsky vesicle. A wave-fragment passing from one vesicle to another it contracts, due to the restricted size of the connecting pore. When two or more wave-fragments collide inside a vesicle they can annihilate, deviate, or multiply. When interpreting presence/absence of wave-fragments in any given as a value of Boolean variable we can implement all basic operations of a Boolean logic via collisions between wave-fragments in a vesicle. In computer experiments we designed a binary adder in a hexagonal array of vesicles filled with excitable chemical mixture [26], built polymorphic logical gates (switching between XNOR and NOR) by using illumination to control outcomes of inter-fragment collisions [27] and geometry-modulated complex arithmetical circuits [140]. Our theoretical ideas and results got experimental chemical laboratory back up — results on information transfer between Belousov-Zhabotinsky mixture enclosed in lipid membrane are successful [200].

While in computer models regular arrangement of uniform vesicles is effortless real-life experiments bring nasty surprises. Usually vesicles are different sizes, they do not form a hexagonal lattice as a rule, and they may be unstable, coalescence transforms fine-grained networks of elementary vesicle-processors into a coarse-grained assembles of monstrous vesicular structures. What kind of computation can be done on an irregular arrangement of non-uniform vesicles? We address the question by representing the vesicle assembles by automata networks and studying how planar subdivision, Voronoi diagram, problem can be solved in such vesicle-automata.

We abstract vesicle assembles as planar Voronoi diagrams of planar sets, points of which are centres of the vesicles. Voronoi diagram is routinely as approximation of arrangements of discs [117] and sphere packing [107, 177, 179]. The diagram is also used in structural analysis of liquids and gases [37], and protein structure [210], and to model dense gels [280] and inter-atomic bonds [138].

A planar Voronoi diagram of a point set **P** is a partition of the plane into such regions, that for any point p of **P**, a region corresponding to p contains all those points of the plane which are closer to p than to any other node of **P**. In Chapter 11 we define finite-state machines on a planar Voronoi diagram. Every Voronoi cell takes four states: resting, excited, refractory and precipitate. A diagram of cell-state transitions is shown in (Fig. 1.1h). A resting cell excites if it has at least one excited neighbour. The cell precipitates if a ratio of excited cells in its neighbourhood to its number of neighbours exceed certain threshold. To approximate a Voronoi diagram on Voronoi automata we project a planar set onto automaton lattice, thus cells corresponding to data-points are excited. Excitation waves propagate across the Voronoi automaton, interact with each other and form precipitate in result of the interaction. Configuration of precipitate represents edges of approximated Voronoi diagram. We discover relation between quality of Voronoi diagram approximation and precipitation threshold, and demonstrate feasibility of our model in approximation Voronoi diagram of arbitrary-shaped objects and a skeleton of a planar shape.

We assume that every cell of a Voronoi diagram is a finite-state that takes four states and updates its states depending on states of its first and second order neighbours. We design a cell-state transition function which combines generalised, and highly-abstracted, properties of both excitable and precipitating chemical media: a local disturbance gives birth to quasi-circular waves of excitation while collisions between the waves lead to precipitation.

The problem solved by Voronoi automats is the approximation of Voronoi diagram. Approximated Voronoi diagram is much more coarse-grained than Voronoi diagram on which excitable-precipitating automaton is built. To approximate a Voronoi diagram on Voronoi automata we project a planar set onto automaton lattice, thus cells corresponding to data-points are excited. Excitation waves propagate across the Voronoi automaton, interact with each other and form precipitate in result of the interaction. Configuration of precipitate represents edges of approximated Voronoi diagram. In our model precipitation depends on a local density of excitation — precipitation threshold. For low precipitation threshold the medium becomes cluttered with meaningless clusters of precipitate, for high threshold few domain of precipitation is formed.

1.7 Collision-Based Computing in Excitable Automata

Many physical, chemical and biological spatially extended non-linear systems exhibit a wide range of stationary and mobile localizations: solitons, kinks, breathers, excitons, defects and wave-fragments. The localizations can be used to transmit and transform information, and ultimately to perform computation [10]. A unit of information, such as the value of a Boolean variable, is decoded into presence (logical truth) or absence (logical false) of a localisation in some specified site of space at a specified moment of time. When two localizations (representing the values of two logical variables) collide, they change their trajectories (or annihilate, reproduce, or change their shape). The new trajectories of the localizations encode the values of some logical function over the two variables. This is how most collision-based computing devices work [50], [109], [10, 89, 184, 247, 281].

The collision-based, or free-space, computing devices typically do not have wires and — in principle – are not supposed to use any other stationary components to perform computation. Any point of the computing media can act as a wire, a trajectory of a traveling localisation can be seen as a momentary wire. Any site where two or more localizations collide is a logical gate. Thus space can be used efficiently and nothing is wasted. However, there is a price to pay. Initial positions and launch time of the traveling localizations should be precisely specified: one wrong time step destroys the whole computing scheme. We envisage that novel arithmetic chips, to be built in an excitable medium will be based on principles of collision-based computing. They will be based on logical schemes of computation in Conway's Game-of-Life [50], Fredkin-Toffoli's conservative logic [109] and Margolus's physics of computation [184].

A sub-excitable Belousov-Zhabotinsky medium [219] is an ideal substrate to build collision-based arithmetical chips in chemical systems. In a normal, excitable, mode the Belousov-Zhabotinsky medium responds to local perturbations by forming target or spiral waves, which propagate in all possible directions away from perturbation site. In a sub-excitable mode, we observe generation of a localised excitations, or wave-fragments that preserve their shape and travel like dissipative solitons [57] in one pre-determined direction for a substantial amount of time. We have demonstrated [12, 18, 89, 246] that it is possible to implement logical gates by colliding excitation wave-fragments.

In Chapter 12 we use a cellular-automaton sub-excitable lattice [2, 3, 8] to simulate a sub-excitable chemical medium. We integrate together our previous results [281, 282] on arithmetical operations in collision-based media. We construct adders and multipliers using a two-dimensional three-state cellular-automaton model of an excitable medium – the 2^+-medium, originally introduced in [3, 8]. The 2^+-medium consists of an orthogonal array of finite automata, where every automaton, called a cell, takes three states: resting, excited and refractory and updates its state depending on the states of its eight neighbours. All cells update their states simultaneously and in discrete time, using the same rule. A resting cell becomes excited if it has exactly two excited neighbours (so the name 2^+). The transitions from excited state to refractory state, and from refractory state to resting state are unconditional (Fig. 1.1b). The cellular automaton exhibits compact localised excitations, which can be arranged to implement schemes of collision-based computing. In Chapter 12 we expand the collision-based design of a one-bit binary half-adder [13] into more complicated circuits of full adders and multipliers.

1.8 Spiral Rule Automata

Using localised wave-fragments in experimental and simulated reaction-diffusion systems we could implement functionally complete sets of logical gates and varieties of binary logic circuits [14]. The functionality of these constructions, however, lasts for a markedly brief time because the unstructured reaction-diffusion excitable devices lack stationary localisations (which could be used as memory units) and stationary generators of mobile localisations (which are essential for implementing negation). In our search for real-life chemical systems exhibiting both mobile and stationary localisations we discovered a cellular-automaton model [15, 17] of an abstract reaction-diffusion system, which ideally fits the framework of the collision-based computing paradigm and reaction-diffusion computing. We introduce this automaton in Chapter 13.

We design a hexagonal ternary-state two-dimensional cellular automaton which imitates an activator-inhibitor reaction-diffusion system (Fig. 1.1f). The activator is self-inhibited in particular concentrations and the inhibitor dissociates in the absence of the activator. The automaton exhibits both stationary and mobile localizations (eaters and gliders), and generators of mobile localizations (glider-guns). A remarkable feature of the automaton is the existence of a spiral glider-gun, a discrete analogue of a spiral wave that splits into localised wave-fragments (gliders) at some distance from the spiral tip. This unique glider-gun gave name to the Spiral Rule automaton.

We demonstrate how spatio-temporal dynamics of interacting traveling localizations and their generators can be used to implement computation: manipulation with signals, binary logical operations, multiple-value operations, and finite-state machines. Constructing logical gates is a pre-requisite for the demonstrating the computational universality of a system. However, to build working prototypes we need to have more detailed techniques: for manipulating signals, memorising the intermediary results of a computation, and feeding data into the computing device, to name but a few. This is why we mainly concentrate on these 'auxiliary' means of computation with spiral rule automata in Chapter 13.

The substrate-sites between inhibitor-sites of the eater can be switched to an inhibitor-state by a colliding glider, or even a glider just brushing past..Can we do without perfect timing? Asynchronous cellular-automaton based computers do pretty well [147, 169, 237] by using predetermined wires and valves. We choose the reaction-diffusion spiral rule cellular automaton [15, 17, 268, 271] as a testbed for ideas of asynchronous collision-based computing. In the present chapter we are trying to combine pure collision-based computing ideas (gliders only) with stationary architectures (breather-like localisation) to implement computing schemes with relaxed timing. We demonstrate how to do transformations of 2-, 4- and 6-bit numbers in interactions between traveling and stationary localizations in the spiral rule cellular automaton. Mechanics of the computation is based on interactions between gliders and eaters. When a glider brushes an eater the eater may slightly change its configuration, which is updated once more every next hit. We encode binary strings in the states of eaters and sequences of gliders. We show what types of binary compositions of binary strings are implementable by sequences of gliders brushing an eater.

What is our rationale behind selecting the spiral rule automaton to study novel concepts of the collision-based computing? The spiral rule cellular automaton [17] plays a unique role in unconventional computing. On the one hand, this is a simple ternary state hexagonal automaton with Conway's Game of Life type of behaviour: it has gliders, still lives and eaters, and glider. Therefore it is very suitable for experimenting with collision-based computing schemes. On the other hand, the spiral rule automaton is a unique discrete model of a non-linear reaction-diffusion chemical system with an activator and inhibitor. The gliders and glider guns in the spiral rule automaton are analogues of excitation wave-fragments and generators of wave-fragments in a light-sensitive sub-excitable Belousov-Zhabotinsky medium [90]. This means that prototypes of computing schemes designed in the spiral rule automaton can then be almost straightforwardly implemented in chemical laboratory prototypes of reaction-diffusion computers.

1.9 Memristors in Cellular Automata

The memristor — a passive resistor with memory — is a device whose resistance changes depending on the polarity and magnitude of a voltage applied to the device's terminals and the duration of this voltage's application. The memristor's existence was theoretically postulated by Leon Chua in 1971 based on symmetry in integral variations of Ohms laws [78–80]. The memristor is characterised by a non-linear relationship between the charge and the flux. This relationship can be generalised to any two-terminal device in which resistance depends on the internal state of the system [79]. The memristor cannot be implemented using the three other passive circuit elements — resistor, capacitor and inductor – therefore the memristor is an atomic element of electronic circuitry [78–80]. Using memristors one can achieve circuit functionalities that it is not possible to establish with resistors, capacitors and inductors, therefore the memristor is of great pragmatic usefulness. The first experimental prototypes of memristors are reported in [105, 261, 276]. Potential unique applications of memristors are in spintronic devices, ultra-dense information storage, neuromorphic circuits, and programmable electronics [232].

Despite explosive growth of results in memristor studies there is still a few (if any!) findings on phenomenology of spatially extended non-linear media with hundreds of thousands of locally connected memristors. We attempt to fill the gap and develop a minimalistic model of a discrete memristive medium. There are two way to introduce memristance in cellular automata: via cell-states [30] and via links between neighbouring cells [29, 148]. The first approach is explored in Chapter 14 and the second in Chapter 15.

In Chapter 14 we consider a three-state cellular automaton model of an excitable medium. A resting cell excites if number of excited neighbours lies in a certain interval. An excited cell becomes refractory independently on states of its

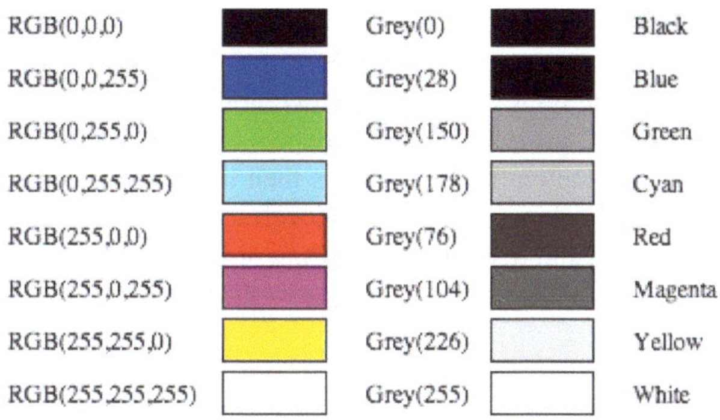

Fig. 1.2 Samples of grey-values of colours used in the illustrations

neighbours. A refractory cell returns to a resting state only if the number of excited neighbours belong to some recovery interval (Fig. 1.1e). The model is an excitable cellular automaton abstraction of a spatially extended semi-memristive medium where a cell's resting state symbolises low-resistance and refractory state high-resistance. The medium is semi-memristive because only transition from high- to low-resistance is controlled by density of local excitation. We discuss phenomenological classification of the automata behaviour for all possible excitation intervals and recovery intervals. We describe eleven classes of cellular automata with retained refractoriness based on criteria of space-filling ratio, morphological and generative diversity, and types of travelling localizations.

A cellular automaton is structurally-dynamic (also called topological) when links between cells can be removed and reinstated depending on states of cells these links connect [130, 143]. Structurally-dynamic automata are now proven tools to simulate physical and chemical discrete spaces [33, 133, 137, 213, 215] and graph-rewriting media [244]; see overview in [145]. We design a minimalistic model of a regular network of memristors using structurally-dynamic cellular automata in Chapter 15.

Each cell of the automaton gets info about states of its closest neighbours via incoming links. A link can be in 'conductive' or 'non-conductive' states. States of every link are updated depending on states of the cells the link connects. Every cell of a memristive automaton takes three states: resting, excited (analog of positive polarity) and refractory (analog of negative polarity). A cell updates its state depending on states of its closest neighbours which are connected to the cell via 'conductive' links (Fig. 1.1b). The chapter analysis behaviour of memristive automata in response to point-wise and spatially extended perturbations, discusses structure of localised excitations coupled with topological defects and interfacial mobile excitations and growth of information pathways.

1.10 On Colors

Most illustrations of the book are originally in colour. The book is printed in black and white. The Tab. 1.2 gives us a rough samples of basic represented in levels of grey. If a cellular automaton has three states then usually we represent excited or activator state with red colour or grey level 76, refractory state or inhibitor with blue colour or grey level 28, and resting state is white or light-grey.

Part I
Phenomenology and Localisations

Chapter 2
Reaction-Diffusion Binary-State Automata

Here we consider a simplest version of reaction-diffusion automaton: a two-dimensional cellular automaton, where a cell has eight closest neighbours and takes two states, 0 and 1. We earmark state 0 as a substrate and state 1 as a reagent. To imitate diffusion and reaction we allow state 1 to spread: a cell in state 0 takes state 1 – under certain conditions – if it has some neighbours in state 1. We consider three types of automaton quasi-chemical systems (Fig. 2.1):

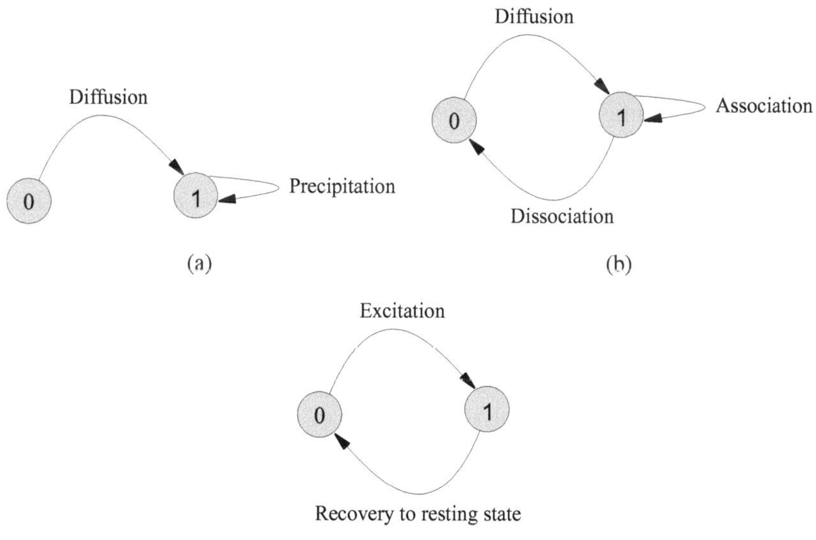

Fig. 2.1 Cell-state transition diagrams. (a) Precipitating automaton. (b) Diffusion-association automaton. (c) Excitable automaton without refractory state.

A. Adamatzky: Reaction-Diffusion Automata, ECC 1, pp. 21–35.
springerlink.com

- Precipitating automata: If state 1 is absorbing, i.e. a cell in state 1 always stays in state 1, the automaton imitate precipitating system: transition $1 \to 1$ is a precipitation.
- Diffusion-association automata: If cell in state 1 can take – under certain conditions – state 0 we assume the automaton imitates association-dissociation system. A transition $1 \to 1$ is association and transition $1 \to 0$ is dissociation.
- Excitable automata without refractory state: Cell in state 1 takes state 0 independently on states of its neighbours. In classical, e.g. Greenberg-Hasting [122] excitable automata, there is a dedicated refractory state. In the binary excitable automata there are only excited state 1 and resting state 0.

2.1 Precipitating Automata

Let σ_x^t be a sum of cells in state 1 in neighbourhood $u(x)$ of cell x at time step t. Every cell x updates its state by the rule:

$$x^{t+1} = \begin{cases} 1, \text{ if } (x^t = 0 \text{ and } \sigma_x^t \in [\theta_1, \theta_2]) \text{ or } (x^t = 1) \\ 0, \text{ otherwise} \end{cases} \tag{2.1}$$

A cell in state 0 takes state 1 if number of its neighbours in state 1 belong to interval $[\theta_1, \theta_2]$, $1 \leq \theta_1 \leq \theta_2 \leq 8$; once in state 1 a cell remains in this state forever. The model represents a quasi-chemical precipitating system, where 0 is a substrate and 1 is a reagent, when reagent diffuses onto the substrate it is bound to substrate, a kind of precipitation occurs. We can call interval $[\theta_1, \theta_2]$ a diffusion-precipitation interval. Examples of configurations generated by such cell-transition rules are shown in Fig. 2.2.

Finding 1. *Branching patterns are formed when diffusion and precipitation occur at low concentration of reactant.*

For narrow diffusion-precipitation intervals with small lower boundaries, $[1,1]$, $[1,2]$, $[2,2]$, structures formed by precipitate resemble intersection and overlapping branching trees (see configurations marked $[1,1]$, $[1,2]$ and $[2,2]$ in Fig. 2.2). The branching structures are produced because every cell in state 1 or two neighbouring cells in state 1 generate a multiply-branching trees of precipitate, see Fig. 2.3a–e. Low values of θ_1 and θ_2 are interpreted as limited diffusion and precipitation.

Finding 2. *Precipitating automata approximate planar Voronoi diagram when diffusion and precipitation occurs at medium concentration of reactant.*

Further increase of upper boundary, $[1,3]$ and $[1,4]$, and lower boundary, $[2,3]$, $[2,4]$, $[3,3]$ and $[3,4]$ leads to formation of Voronoi-like domains around each cell that was initially in state 1 (see definition of Voronoi diagram in Chapter 11) more details about cellular automaton generation of Voronoi diagrams in [9]). Edges of Voronoi

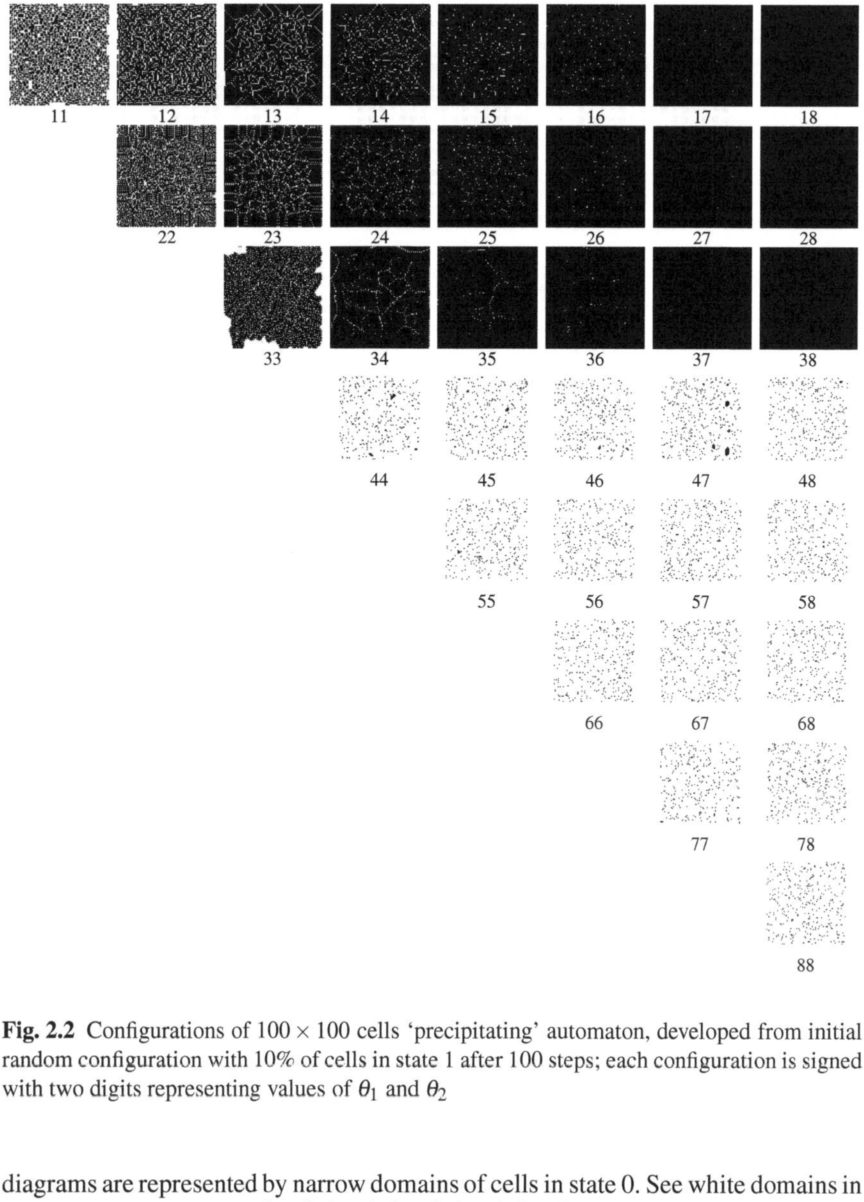

Fig. 2.2 Configurations of 100×100 cells 'precipitating' automaton, developed from initial random configuration with 10% of cells in state 1 after 100 steps; each configuration is signed with two digits representing values of θ_1 and θ_2

diagrams are represented by narrow domains of cells in state 0. See white domains in configurations signed by $[1,3]$, $[1,4]$, $[2,3]$, $[2,4]$, $[3,3]$, $[3,4]$ in Fig. 2.2, especially configuration $[3,4]$. This happens because at sites where fronts of precipitation (originating from different sources) approach each other, cells in state 0, trapped between propagating fronts of state 1, have number of 1-state neighbours exceeding θ_2 and so these cells do not take state 1 (Fig. 2.4).

When lower boundary of 'precipitation' interval $[\theta_1, \theta_2]$ becomes more then 3, typical configuration of automaton developing from random initial configuration is

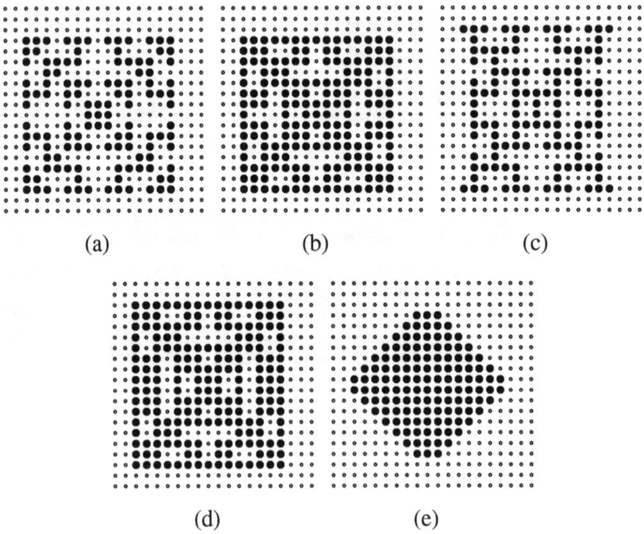

Fig. 2.3 Configuration of 20 × 20 cells 'precipitating' automaton developed from initial configuration with a single cell in state 1. Configuration displayed is recorded after seven iterations. Cells in state 1 are shown by solid discs and cells in state 0 by dots. Initially all cells but those indicated below are in state 0. (a) $\theta_1 = \theta_2 = 1$, $x^0_{n/2,n/2} = 1$, $n = 20$, (b) $\theta_1 = 1, \theta_2 = 2$, $x^0_{n/2,n/2} = 1$, (c) $\theta_1 = \theta_2 = 1$, $x^0_{n/2,n/2} = 1$ and $x^0_{n/2,n/2+1} = 1$, (d) $\theta_1 = 1, \theta_2 = 2$, $x^0_{n/2,n/2} = 1$ and $x^0_{n/2,n/2+1} = 1$, (f) $\theta_1 = 2, \theta_2 = 3$, $x^0_{n/2,n/2} = 1$ and $x^0_{n/2,n/2+1} = 1$.

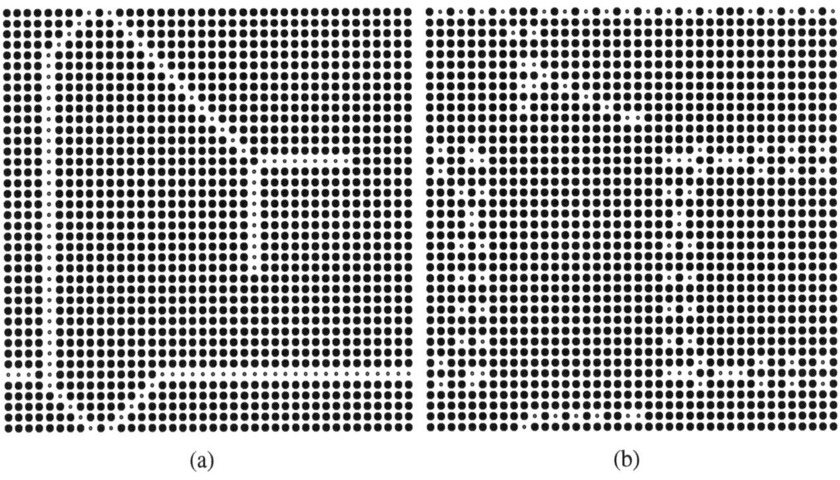

Fig. 2.4 Configuration of 40 × 40 cellular automaton recorded at step $t = 40$, the automata started their development in configuration where almost all cells were in state 0 but cells with coordinates $(n/2, n/2)$, $(n/2, n/2+1)$, $(n/2+n/4, n/2)$, $(n/2+n/4, n/2+1)$, $(n/2+n/4, n/2+1+n/4)$ and $(n/2+n/4, n/2+1+n/4+1)$ were in state 1. (a) $\theta_1 = 1$ and $\theta_2 = 3$, (b) $\theta_1 = 2$ and $\theta_2 = 3$.

a population of mostly sparsely distributed states 1 (black pixels in Fig. 2.2). This is because in sparse random initial configuration a cell in state 1 is usually surrounded by neighbours in state 0 and therefore initial random configuration remains in general unchanged during automaton development.

For every value of intervals' boundaries and any initial condition the automaton eventually got trapped in fixed point of evolution, where no cell changes its state (because 1 is an 'absorbing' cell state, or a precipitate). To make things more attractive dynamically we can allow for 'dissociation' of reagent '1' from substrate '0', the model is discussed in following section.

2.2 Diffusion-Association Automaton

Let σ_x^t be a sum a cells in state 1 in neighbourhood $u(x)$ of cell x at time step t, and $0 \leq \theta_1 \leq \theta_2 \leq 8$ and $0 \leq \delta_1 \leq \delta_2 \leq 8$. Every cell x updates its state x^t by the following rule:

$$x^{t+1} = \begin{cases} 1, \text{ if } (x^t = 0 \text{ and } \sigma_x^t \in [\theta_1, \theta_2]) \text{ or } (x^t = 1 \text{ and } \sigma_x^t \in [\delta_1, \delta_2]) \\ 0, \text{ otherwise} \end{cases} \tag{2.2}$$

Selecting particular values of intervals' boundaries we change cell state transition rule, and subsequently development of space-time dynamics in the automaton. We write rules in the form $R(\delta_1, \delta_2, \theta_1, \theta_2)$. The rule can be interpreted as a simple discrete model of a quasi-chemical system with substrate '0' and reagent '1', and $[\theta_1, \theta_2]$ is analogous to diffusion rate, or association between substrate and reagent, and $[\delta_1, \delta_2]$ is analogous to a reaction rate, or in other terms degree of affinity between substrate and reagent.

Finding 3. *Diffusion-association automaton rules can be split into five main classes. They are the rules evolving an initially random configuration to an empty configuration (**E**-class), almost not transforming a random configuration (**S**-class), developing labyrinthine structures (**L**-class), evolving towards string-like disorganised patterns (**M**-class), and evolving towards porous structures, sometimes with regular sub-structures (**P**-class).*

We analysed patterns produced by each of 1296 rules, from initially random configuration (the same random configurations were used for all rules) where every cell was assigned state 1 with probability 0.3, cell state transitions Eq. (2.2) for all possible values of intervals' boundaries. Several morphology-based classes of rules are discovered.

The rules which transform random initial configuration to a uniform configuration of cells in state 0 are grouped in the **E**-class. For example, rules $R(33ab)$, $a \geq 5, b \geq 5$ (it is assumed that $a \leq b$, $a, b \leq 8$; $R(44ab)$, $a, b \geq 4$; $R(4cab)$, $a, b \geq 4$, $c = 4, 5$. The interval boundaries of the rules are such that initial 1-states do not

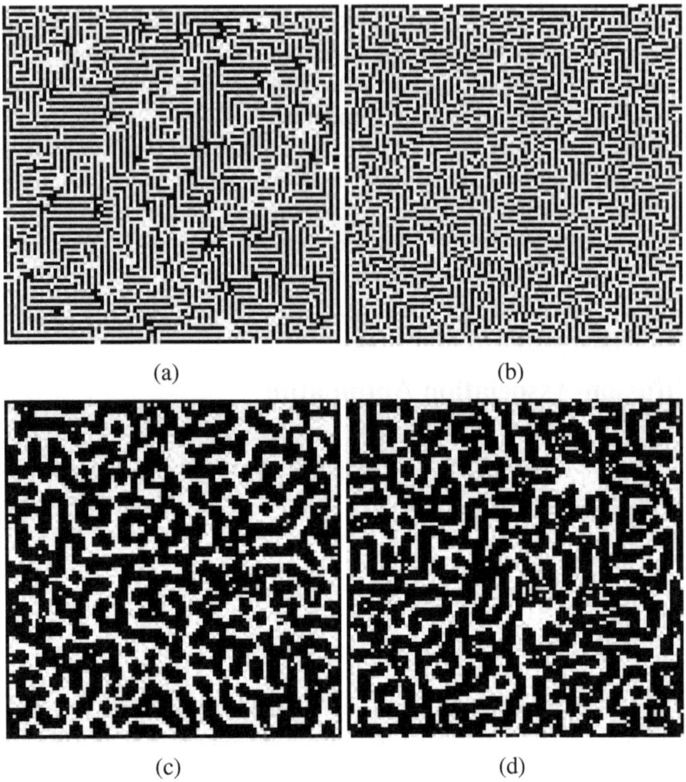

Fig. 2.5 Examples of configurations generated by rules of **L**-class. (a) $R(1312)$, (b) $R(1322)$, (c) $R(1616)$, (d) $R(4617)$

spread to neighbouring 0-state sites, and cells in state 1 could not be in this state for longer then few time steps of development. The class **E** represents a quasi-chemical system with sub-threshold diffusion and reaction rates.

The second class, **S**, is comprised of rules which almost did not change initially random configurations. Small clusters of 1-states are formed but never spread for more then few cells, e.g. configurations generated by rules $R(11ab)$, $a,b \geq 3$ or $R(22ab)$, $a,b \geq 4$. Diffusion rate is still low for substantial patterns to form but reaction rate is high enough to keep formed patterns stable.

The third simple class, **D**, comprised of rules that generate solid (all almost solid) configurations, where the whole lattice is filled with 1-states. For example, the class include rules $R(d8a8)$, $1 \leq a \leq 4$, $d = 2,3$ and $R(d8a8)$, $1 \leq a \leq 3$, $d = 4,5$. The rules imitate high reaction and diffusion rates.

L-class includes rules that generate fine labyrinths of 1-states. Some exemplar configurations are shown in Fig 2.5. There are at least two distinctive types of structures: $\mathscr{L}_1$-labyrinths with walls one cell sick (see e.g. Fig 2.5ab), and $\mathscr{L}_2$-labyrinths with walls few cell sick (see e.g. Fig. 2.5cd). $\mathscr{L}_1$ labyrinths are formed because

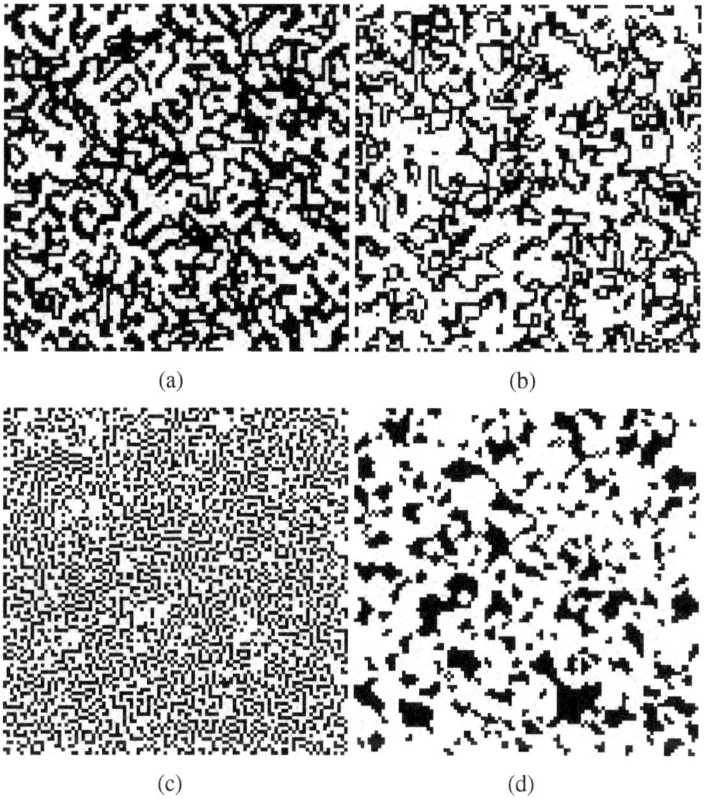

Fig. 2.6 Examples of configurations generated by rules of **M**-class. (a) $R(1315)$, (b) $R(1425)$, (c) $R(2213)$, (d) $R(2858)$.

diffusion rate, interval $[\theta_1, \theta_2]$ is narrow enough to allow only stripes of 1-states to grow and join each other, and the one cell wide stripes are stabilised by reaction rate, interval $[\delta_1, \delta_2]$ allowing for only cells surrounded by, e.g., one, two or three neighbours in 1-state to persist. Decreasing reaction interval, e.g. going from rule $R(1312)$ to rule $R(1322)$, dramatically reduces amount of free hanging ends of stripes (where cell in state 1 has just one neighbour, as in Fig. 2.5a) and therefore development favours continuous (but possibly frequently turning) stripes (see Fig. 2.5b). Increasing upper boundary of reaction interval $[\delta_1, \delta_2]$ leads to formation of sick stripes, $\mathscr{L}_2$-labyrinths, see e.g. Fig. 2.5cd. Formation of $\mathscr{L}_2$-labyrinths may be attributed to high degree of diffusion and relative high reaction rate comparing to that in rules producing $\mathscr{L}_1$-labyrinths.

With increase of both lower and upper boundaries of diffusion interval $[\theta_1, \theta_2]$ we see formation of irregularly oriented and often branching stripe-like domains of 1-states (Fig. 2.6). We group the rules generating such patterns to **M**-class. The strips become thinner with decrease of diffusion rate, see transitions between configurations (a) $\rightarrow$ (b) $\rightarrow$ (c) $\rightarrow$ in Fig. 2.6. The strips are transformed to interconnected

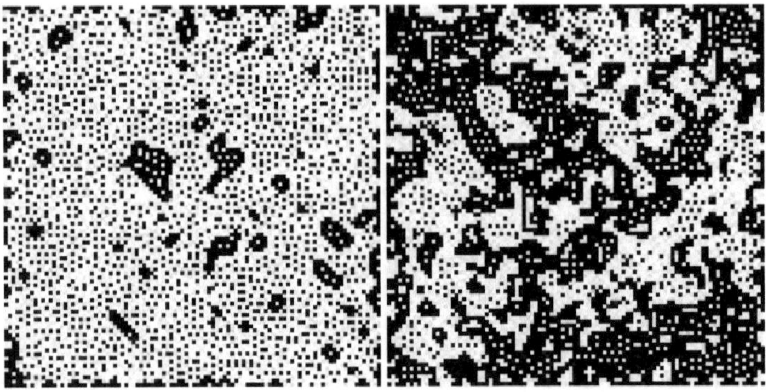

(a) (b)

Fig. 2.7 Examples of configurations generated in **P**-class. (a) $R(2318)$, (b) $R(2418)$

(a) (b)

Fig. 2.8 Examples of configurations generated by rules of **O**-class. (a) $R(3511)$, (b) $R(3811)$

domains with increase of diffusion rate and increase of lower boundary of reaction interval (Fig. 2.6d).

Rules representing high diffusion rate — wide interval $[\theta_1, \theta_2]$ — and low reaction rate — narrow $[\delta_1, \delta_2]$ — generate configurations where domains of mostly 0-states with scattered 1-states compete with domains of mostly 1-states with scattered 0-states (Fig. 2.7). The rules are grouped in **P**-class.

Configurations of irregularly distributed spots (domains of fews cells in 1-state) are typically generated by rule in **O**-class. The spots may be connected by thin filaments of 1-states. The rules represent very low rate of diffusion (self-inhibiting diffusion) and high reaction rate.

2.3 Functional Classification

Finding 4. *Diffusion-association automata can implement collision-based logical circuits, approximate Voronoi diagram and a discrete convex hull.*

Three more classes can be specified based on computational functionality of developing configurations. Here we follow a paradigm of reaction-diffusion computing, where data are represented by initial concentrations of reagents, computation is implemented by spreading and interacting patters (e.g. diffusion or excitation waves), and result is given by final (either stationary or oscillating) distribution of reagents (see details in [9]). In automaton classes studied in this chapter, computationally functional rules are morphologically indistinguishable from 'computationally useless' rules.

The first, computationally functional class **G** consist of rules $R(2c22)$, where $2 \leq c \leq 8$. The rules support mobile localizations, or gliders, in automaton development. By colliding gliders we can implement any kind of logical operations [10]. Other examples of functionally complete collision-based binary-state automata include Conway's Game of Life [22, 50].

Set of rules grouped in class **V** allows for partial computation or approximation of a Voronoi diagram — given a set of planar points compute such domain for each point p of data set that any point in the domain is closer to p then to any other point of given data set. Boundaries of the domains constitute Voronoi diagram. See details of constructing Voronoi diagram in chemical reaction-diffusion systems in [9, 14]. As you can see in Fig. 2.9 edges of Voronoi diagram can be represented either by loci without reagent, i.e. domains of cells in state 0 (Fig. 2.9cd), or by different types of tiling (Fig. 2.9b).

Rules grouped in class **C** approximate discrete convex hulls [4] of connected subsets of cells being initially in 1-state, so an initially random configuration of 1-state cells is transformed by the rules of **G** to a set of discrete convex hulls of various size (Fig. 2.10). The class include rules $R(4747)$, $R(4847)$ and $R(4848)$.

Convex hulls generated by automata rule $R(4747)$ are most amazing because they remind us on formation of living cellular structures (at a very high level of abstraction indeed0. A typical pattern (Fig. 2.10c) is comprised of a stationary boundary and dynamical interior. Inside the convex hull on can observe formation and collapse of 'bubbles', propagating waves of activity, and porous still domains (groups of cells in '0' and '1' states).

2.4 Excitable Automata without Refractory State

Binary state reaction-diffusion automata have also a family of excitable cellular automata without a refractory state. In such automata $1 \leq \theta_1 \leq \theta_2 \leq 8$ but $\delta_1 = \delta_2 = 9$. There are only eight neighbours in cell's neighbourhood, therefore value nine is impossible to achieve. Thus the condition $\delta_1 = \delta_2 = 9$ means that a cell

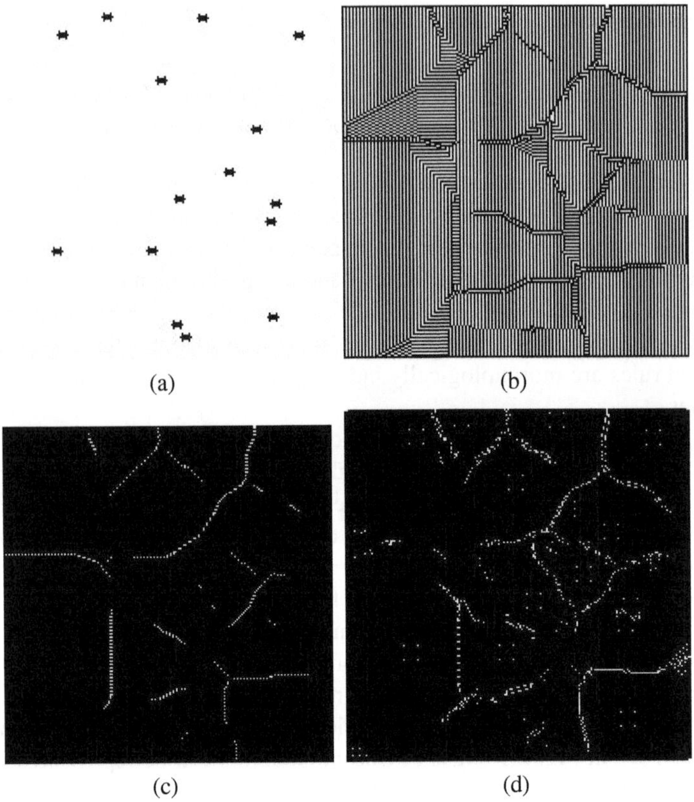

Fig. 2.9 Partial construction of Voronoi diagram in cellular automaton governed by rules (b) $R(1634)$, (c) $R(1834)$, (d) $R(4834)$. Initial data set represented by domains of six cells in 1-state is shown in (a).

in state 1 will always (unconditionally, independently on states of its neighbours) takes value 0. This alike an excitable automata but without a refractory state, 1 is an excited state, 0 is a resting state.

Finding 5. *With increase of upper boundary θ_2 of the excitation interval $[\theta_1, \theta_2]$ space-time dynamics of excitation changes from disorganised patterns of string like excitations to target and spiral waves.*

For $\theta_1 \geq 3$ a random initial configuration (with low ratio of 1-state cells) evolves to a configuration that is empty or contains just few breathing or blinking domains.

For $\theta_1 \leq 3$ dynamics of automata resembles 'classical' excitation dynamics of two-dimensional discrete media with interval excitation [9]. Example configurations generated from a random configuration are shown in Fig. 2.11.

For $\theta_1 = 1$ and $\theta_2 = 1$ automaton exhibits filaments of excitation, e.g. configuration 11 in Fig. 2.11, which are 'transformed' toconventional target waves, e.g. configuration 18 in Fig. 2.11, or spiral, e.g. configuration 28 in Fig. 2.11, waves.

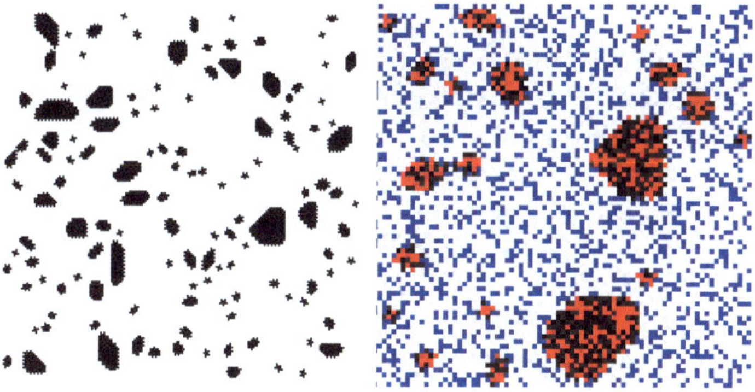

(a) (b)

(c)

Fig. 2.10 Examples of configurations constructed by rules $R(4848)$ and $R(4747)$. (ab) Constructing convex hull by cellular automaton rule $R(4848)$. (a) A typical final configuration for initial random configuration where every cell took state 1 with probability 0.3. (b) Superposition of initial random configuration, grey pixels are cells in state 1, and black domains are final configurations. (c) Configuration of a cellular automaton rule $R(4747)$.

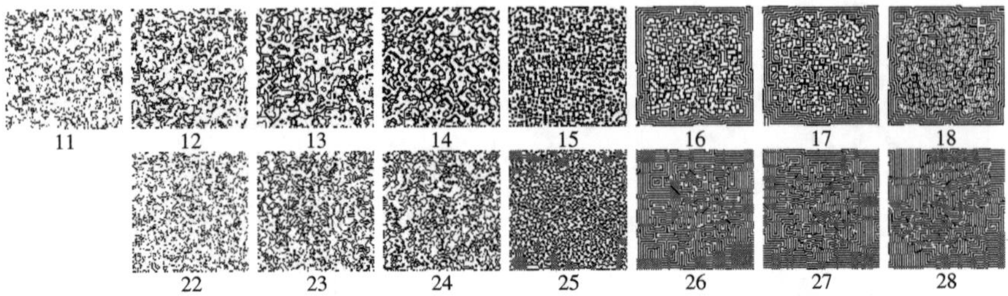

Fig. 2.11 Configurations of 100 × 100 cell excitable automaton, developed from an initial random configuration with 10% of cells in state 1 after 100 steps. Each configuration is signed with two digits representing values of θ_1 and θ_2.

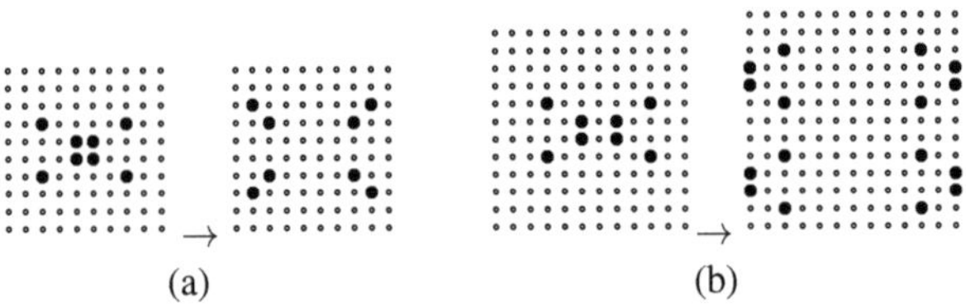

Fig. 2.12 Outcomes of 'head-on' collisions of gliders, initial position of gliders is shown on left configuration of each sub-figure; resultant configurations – on the right of each sub-figure. (a) Even distance between gliders, nil side-shift. (b) Odd distance between gliders, nil side-shift.

Cell state rule determined by interval $[\theta_1, \theta_2] = [22]$ allows for existence of mobile self-localisations, or gliders. Minimal glider consists of four cells in state 1, and can move in four directions. For example, a glider travelling west has the following configuration:

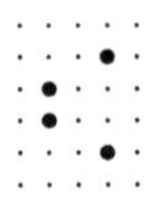

A minimal stationary localisation is a blinker of two cells in state 1:

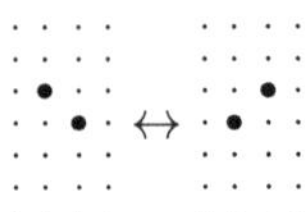

A minimal stimulus of 'excitation' is a couple of two neighbouring cell in state 1.

Head-on collisions between gliders have various outcomes depending on odd-even distance, or 'phase-difference' of, between colliding gliders and their side-shifts relative to each other. Two gliders colliding head on at even distance annihilate and produce quadruple of blinkers (Fig. 2.12a), while those colliding at odd distance are multiplied (Fig. 2.12b).

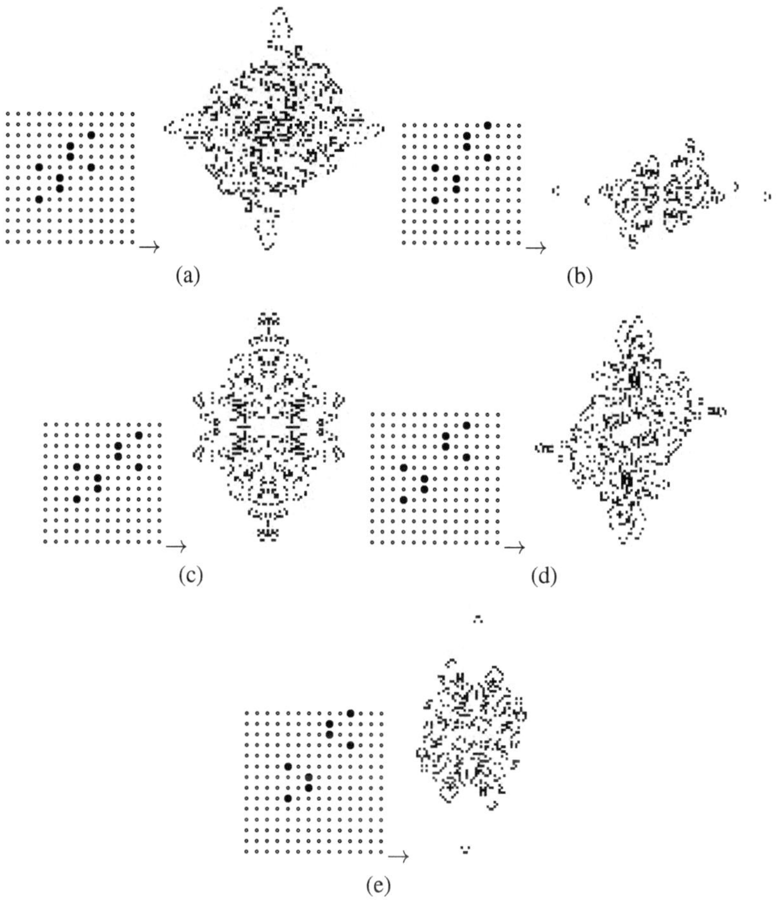

Fig. 2.13 Outcomes of 'head-on' collisions of gliders, initial position of gliders is shown on left configuration of each sub-figure; resultant configurations — on the right of each sub-figure — are recorded at step 50th, lattice size is 120×120. (a) Odd distance, side-shift 3. (b) Odd distance, side-shift 4. (c) Even distance, side-shift 3. (d) Even distance, side-shift 4. (e) Even distance, side-shift 5. Thanks to Genaro Martinez [16].

Situations with head-on shift-shifts collisions, which do not result in just annihilation of gliders, are demonstrated in Fig. 2.13; for side-shifts more then 4 (for odd distance) or more then 5 (for even distance) gliders do not interact. When gliders are shifted relatively to each other, they produce a spreading patterns, led, in most cases by gliders with extensive 'tail-waves' attached. For situations (b) and (e) in Fig. 2.13, speed of growing patterns is lower then speed of glider traveling, for other situation both gliders and the growing patterns have the same speed [16].

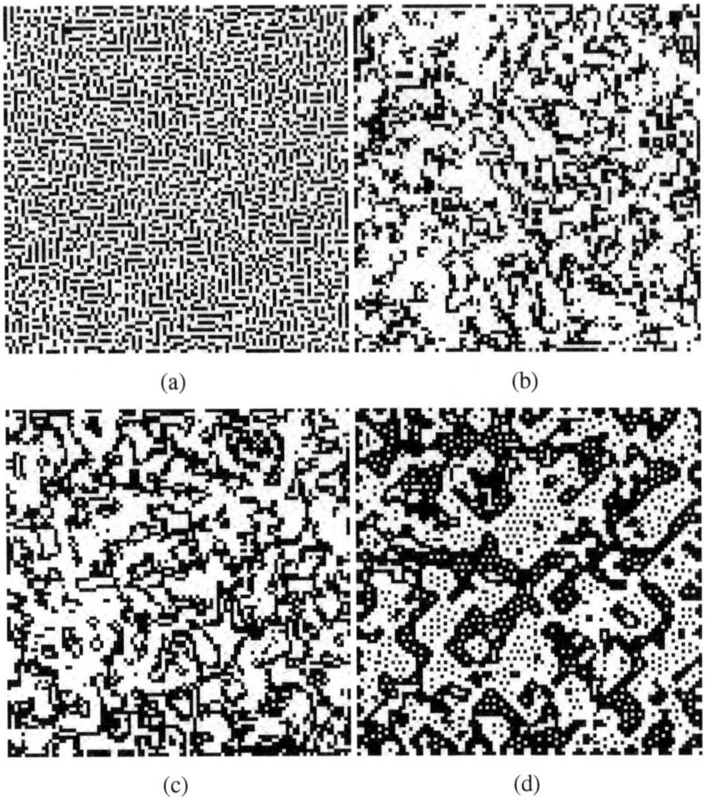

Fig. 2.14 Configurations generated by rules (a) $R(1311)$, (b) $R(1212)$, (c) $R(1313)$, (*d*) $R(1318)$.

2.5 Summary

A cellular automata with a minimal possible number of states demonstrate very rich spectra of reaction-diffusion and excitation patterns. We demonstrate this on a two-dimensional cellular automaton with binary cell states and interval-based cell-state transition rules. In one of the models considered a cell in state 0 takes state 1 if number of neighbours in state 1 belongs to interval $[\theta_1, \theta_2]$, and cell in state 1 remains in state 1 if number of neighbours in state 1 belongs to interval $[\delta_1, \delta_2]$, $1 \le \theta_1 \le \theta_2 \le 8$, $1 \le \delta_1 \le \delta_2 \le 8$. The model can be seen as reaction-diffusion quasi-chemical system with substrate 0 and reagent 1, where $[\theta_1, \theta_2]$ is interval of diffusion and $[\delta_1, \delta_2]$ is interval of reaction. Exhaustive analysis of configurations generated for all 1296 rules allowed us to draft a morphological classification of the rules, and show that increasing upper boundary of diffusion interval usually lead cellular automaton to make a transition from complex ordered spatio-temporal behaviour to disordered behaviour. An example is given in Fig. 2.14: a narrow intervals of reaction (Fig. 2.14ab) lead to formation of ordered patterns, while wider intervals of reaction, (Fig. 2.14cd) generate disordered patterns. In the precipitating automaton increasing interval of diffusion causes transition from disordered to labyrinthine to uniform domains while increasing lower boundary of reaction interval causes transition from disordered and labyrinth/ordered to sparse patterns. The automata also have rules which support mobile localisations, gliders, resembling travelling wave-fragments in sub-excitable Belousov-Zhabotinsky medium [12, 89].

Chapter 3
Retained Excitation

In classical models of excitation an excited cells returns to refractory states either immediately or in few steps but in any case independently on a state if its neighbourhood. How space-time dynamics of excitation will change if we allow excited cells to stay excited for certain configurations of excitation in their neighbourhoods? Let us consider excitation rules which allow excitation to persist by enabling excited cells 'support' their already excited neighbours. Every cell of the automaton studied in this chapter takes three states: resting ·, excited •, refractory ∘. Cells update their states in parallel by the same rule, depending on states of their immediate neighbours. A resting cell excites if a number of excited neighbours belongs to interval $[\theta_1, \theta_2]$, $1 \leq \theta_1 \leq \theta_2 \leq 8$. An excited cell remains excited if a number of excited neighbours belongs to interval $[\delta_1, \delta_2]$, $1 \leq \delta_1 \leq \delta_2 \leq 8$; the excited cell takes refractory state otherwise (Fig. 3.1).

This is a principal different of the rule adopted with 'classical' excitation rule, where cell-state transition from excited to refractory state is unconditional. A refractory cell becomes resting unconditionally similarly to all 'classical' discrete models of excitation.

A resting and excited cells x 'calculate' number of excited neighbours

$$\sigma_a^t(x) = \sum_{y \in u(x)} \{y : y = a\}, \text{ where } u(x) = \{y : |x - y| <= 1\}.$$

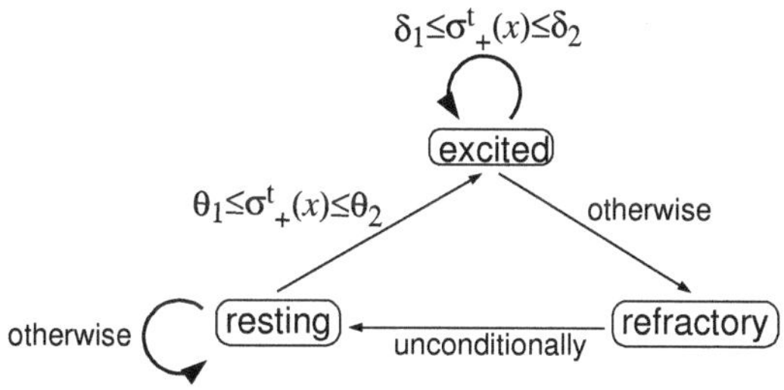

Fig. 3.1 Cell state transition diagram of excitable cellular automaton with retained excitation

A. Adamatzky: Reaction-Diffusion Automata, ECC 1, pp. 37–66.
springerlink.com

Every cell updates its state by the following rule:

$$x^{t+1} = \begin{cases} \bullet, \text{ if } (x^t = \cdot \text{ and } \sigma^t_+(x) \in [\theta_1, \theta_2]) \text{ or } (x^t = \bullet \text{ and } \sigma^t_+(x) \in [\delta_1, \delta_2]) \\ \circ, \text{ if } x^t = \bullet \text{ and } \sigma^t_+(x) \notin [\delta_1, \delta_2] \\ \cdot, \text{ otherwise} \end{cases}$$

We denote rules by tuples $R(\theta_1 \theta_2 \delta_1 \delta_2)$, and call $[\theta_1, \theta_2]$ interval of excitation, and $[\delta_1, \delta_2]$ interval of excitation retention.

To classify the rule we used the following experimental scheme. We stimulated cellular-automaton lattice by closed domain of random configuration and observed how domain of excitation spreads on the lattice and what kinds of configurations, or phases, emerge inside the growing domains. In some cases we also constructed a refractory skeleton: every cell that takes refractory state at least once in automaton development corresponds to the black pixel in the refractory skeleton picture.

3.1 Rectangularly Growing Domains

In the section we study rules which cause initial stimulation field to be transformed to a growing rectangle-shaped domain of excitation and refractoriness.

Almost Everywhere Excited Rectangular Domains

In family of rules $R(1ab8)$, $a = 3, \cdots, 8$ and $b = 1, \cdots, 3$, excitation dynamics is characterised by domination of vertical and horizontal wave-fronts. Initial domain of random stimulation is transformed to a growing domain of almost everywhere excited states, no refractory states are present, and resting states can be found just occasionally at the segments corresponding to singularities of the wave-fronts. The excitation domains are almost invariant to the shape of the initial stimulation domain (Fig. 3.2). Both rectangular stimulation (Fig. 3.2a) and discoidal stimulation (Fig. 3.2b) lead to almost the same growing shape, with the only difference that rectangular domains emerged from discoidal stimulation field may have rounded corners. With increase of excitation retention (size of interval $[\delta_1, \delta_2]$ and its upper boundary δ_2) number of resting states in a locus corresponding to domain of initial stimulation decreases till the whole configuration becomes a linearly growing solid domain of excitation.

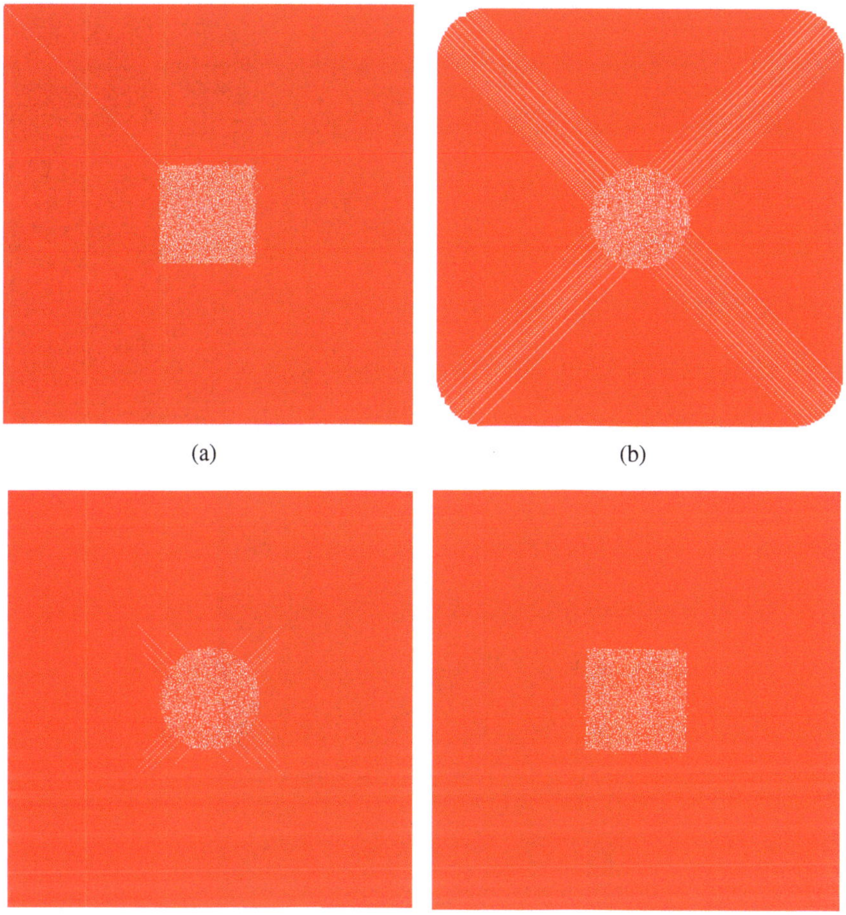

(a) (b)

(c) (d)

Fig. 3.2 Snapshots of excitation dynamics in rules $R(1318)$ (a) and (b) and $R(1428)$ (c) and (d), the lattice of 500×500 cells was stimulated by (a) centred rectangle of 100×100 cells, snapshot made at time step 188, and (b) centred disc of radius 50 cells, snapshot made at time step 199. (c) domain of initial excitation is rectangle, (d) domain of initial excitation is disc. Sizes of growing patterns look different cause they are taken at different steps of development. Every cell in the stimulation domain takes non-resting state with probability 0.2.

Mixed-State Rectangular Domains

Rules $R(1112)$ and $R(1113)$ exhibits rectangular labyrinthine-like growing structure. The labyrinthine configuration inside the domains has no refractory states, but refractory states are present at the fronts of the growing structure. In configurations generated by rules $R(1114), \cdots, R(1116)$ we see a disordered core surrounded

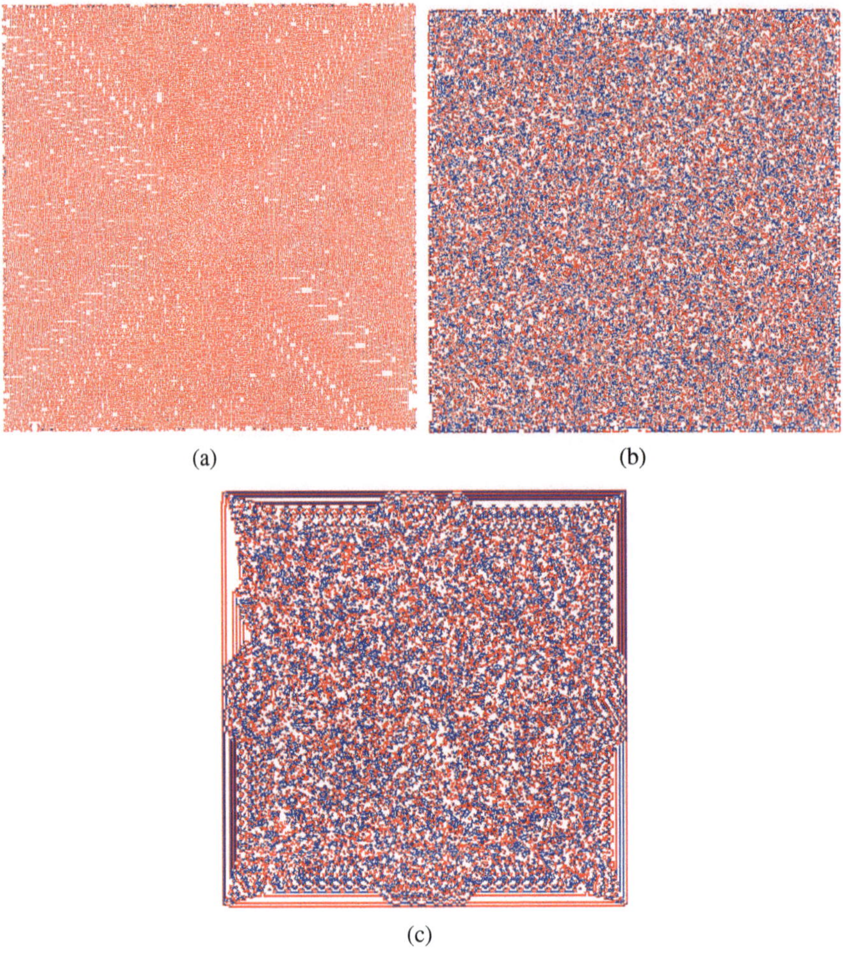

Fig. 3.3 Snapshots of patterns growing in automata governed by rules (a) $R(1114)$, (b) $R(1212)$, (c) $R(1312)$. Initially stimulated domain is a 100×100 cell rectangle centred on the lattice, every cell inside the square initially takes non-resting state with probability 0.45.

by ordered pattern looking like Pascal triangles, see Fig. 3.3a. Similarly behaving rules are $R(1213), \cdots, R(1218)$, there patterns become more ordered till complete segregation is reached in patterns exhibited in configurations of automata ruled by $R(1218)$.

Starting from rule $R(1114)$ no refractory states are visible and in rules $R(1117)$ and $R(1118)$ only excitable states present in the patterns. In rule family $R(1122)$, $\cdots, R(1212)$ disordered patterns combining excited and refractory states, and small domains of excited and refractory states are observed (Fig. 3.3b). Such distribution of patterns between rules continues till rule $R(1288)$, assuming lexicographic order. In rules $R(1312)$ and $R(1313)$ sub-domain of disordered patterns penetrates through

(a) (b)

Fig. 3.4 Snapshots of patterns growing in automata governed by rules (a) $R(1314)$, (b) $R(1317)$, stimulated domain is a disc radius 50 cells centred on the lattice, and every cell inside the disc initially takes non-resting state with probability 0.45

wave fronts, Fig 3.3c; rules $R(1412)$, $\cdots$, $R(1418)$ and $R(1612)$, $\cdots$, $R(1617)$; $R(1812), \cdots, R(1817)$ behave similarly.

Rules in family $R(1314)$, $\cdots$, $R(1317)$ show that the labyrinthine structure typically occupying domain of initial stimulation becomes destroyed with increase of upper boundary of excitation retention interval (Fig. 3.4).

In rules $R(1322)$, $\cdots$, $R(1327)$ we can observe similar 'evolution of matter'. A 'pre-matter' of patterns of excitation dynamics is formed in rule $R(1322)$ then a labyrinthine structure emerges (which can symbolise the 'matter'). Automata governed by rules of the family $R(1a2b)$, $4 \leq a \leq 8$, $2 \leq b \leq 7$, behave similarly.

Excitation dynamics in series $R(1333), \cdots, R(1337)$ is similar to above discussed dynamics but no labyrinths are formed, just slightly ordered structure.

Rules in family $R(1a4b)$, $4 \leq a \leq 8$, $4 \leq b \leq 7$, demonstrate a transition from disordered internal core to an ordered structure (Fig. 3.5) similarly to that observed in family $R(2344)$, $\cdots$, $R(2348)$. The same can be said about space-time dynamics of automata governed by rules $R(1512)$, $\cdots$, $R(1518)$ and $R(1712), \cdots, R(1717)$.

In rules $R(1433)$, $\cdots$, $R(1438)$ domain of initial stimulations adopts 'homogeneous' pattern of excitation and refractoriness, same applies to rules $R(1633)$, $\cdots$, $R(1637)$; $R(1833)$, $\cdots$, $R(1837)$.

Disordered patterns of excitation and refractoriness in the initial stimulation domain and 'classical' excitation wave-fronts elsewhere are typical for narrow interval of excitation retention, with the low upper boundary δ_2 or high low boundary δ_1. The families of rules $R(1a11)$, $a = 3, \cdots 8$, and $R(1abc)$, $a = 3, \cdots 8$, and $5 \leq b \leq c \leq 8$ exhibit homogeneous configuration inside the stimulation core, then regular excitation wave-fronts form that spread outwards (Fig 3.6). Wave fronts keep rectangular

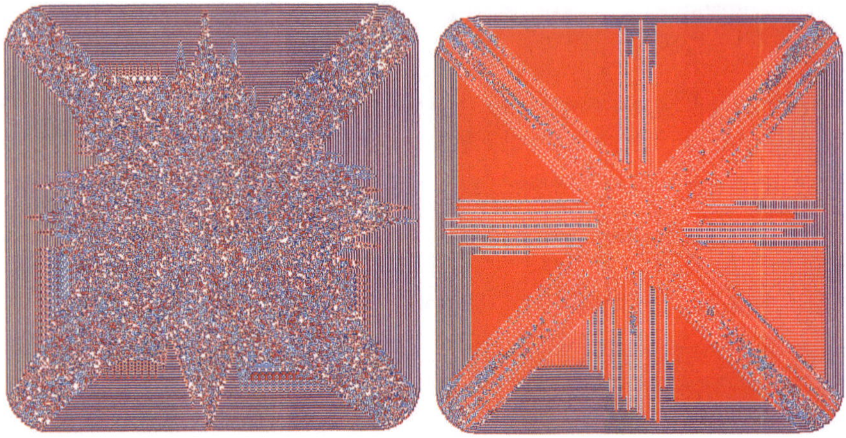

(a) (b)

Fig. 3.5 Snapshots of patterns growing in automata governed by rules (a) $R(1344)$ and (b) $R(1348)$ stimulated domain is a disc with radius 50 cells centred on the lattice, and every cell inside the disc initially takes a non-resting state with probability 0.45

shape independently on shape of the stimulation domain. Waves along the diagonals of the lattice are generated with less frequency than waves traveling along rows and columns of the lattice.

3.2 Diamond-Shaped Growing Domains

Rules discussed in this section transform a pattern of initial stimulation to diamond-shaped growing domain of excitation and refractoriness.

Almost Everywhere Excited Diamond-Shaped Domains

When automaton lattice governed by family of rules $R(2ab8)$, $a = 3, \cdots, 8$ and $b = 1, 2, 3$, is stimulated by rectangular or circular pattern of random states a diamond-shaped growing pattern is formed, there are no or almost no refractory states, and resting states are only present in traces of singularities. Propagating wave-front of the growing domain consists of six segments, four *hv*-segments are relatively oriented along vertical or horizontal axes (they do corresponding to edges of initially stimulated domain), and four segments, *d*-segments, are parallel to diagonals, (they link *hv*-segments together). Initially the growing pattern looks like octagon (Fig. 3.7a), however while lengths of *hv*-segments remain constant lengths of *d*-segments increases proportionally with time (Fig. 3.7b). Therefore, in case of unlimited development, the excitation domain takes diamond shape. Formation of

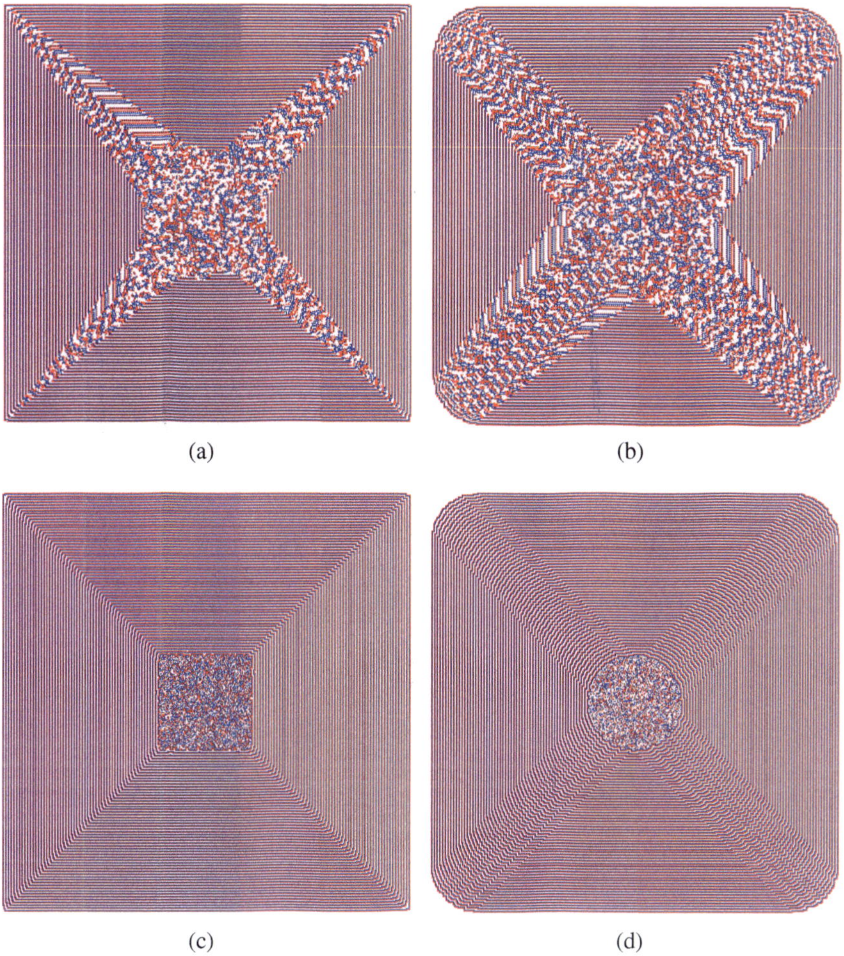

Fig. 3.6 Snapshots of patterns growing in automata governed by rules $R(1311)$ (a) square start, (b) disc start, and $R(1888)$ (a) square, (b) disc. Stimulated domain is a disc with radius 50 cells or square 100×100 cells centred on the lattice, and every cell inside the domain initially takes non-resting state with probability 0.45.

the diamond shape is invariant to initial shape of the stimulated domain, compare Figs. 3.7a and Figs. 3.7c.

With increase of upper boundary θ_2 of excitation, density of excitation in the central core increases, until solid excitation domain remains in rule $R(2838)$ (Fig. 3.8). But with increase of degrees of excitation and retaining of excitation, mobile localizations emerge. The localizations are 'attached' to the rest of the growing pattern, so can be thought of as 'guiding' loci of the growing domain. We trace movement of localizations in Fig. 3.8b by constructing refractory skeleton.

Fig. 3.7 Snapshots of patterns growing in automata, 500×500 cells, governed by rule $R(2318)$: (a) configuration at time step $t = 111$, (b) at time step $t = 464$. Configurations are displayed disproportionally, excitation pattern in (a) is 288 cells along horizontal, and patter in (b) is 842 cells along horizontal. Configuration shown in (c) is stimulated by discoidal domain, probability of non-resting state is 0.2.

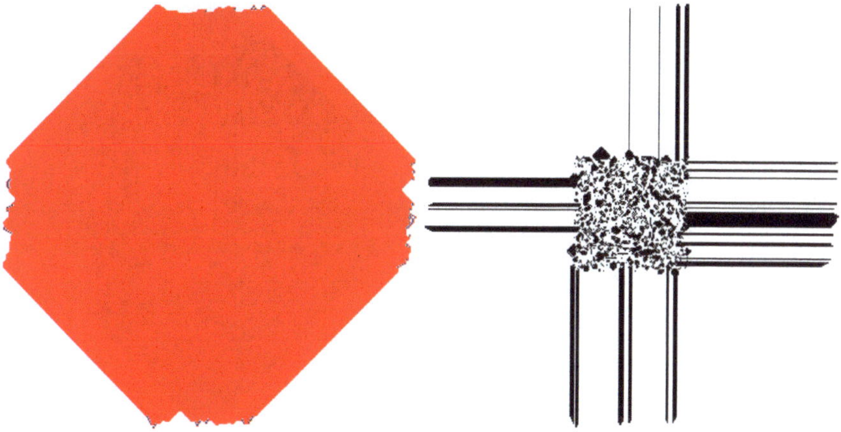

Fig. 3.8 Snapshots of patterns growing in automaton lattice governed by rule $R(2838)$: (a) snapshot of excitation dynamics at time step $t = 148$, (b) refractory skeleton. Probability of excited and refractory states are 0.2 in the domain of initial stimulation.

Mixed State Diamond-Shaped Domains

The rules discussed in this section cause formation of diamond-shaped growing domains filled with a mixture of all three states. The shape of the domains is invariant to shape of initially stimulated domain. Disordered patterns inside the domains are typical for rules $R(3413), \cdot, R(3613)$ (Fig. 3.9a).

Automata governed by $R(3447)$ are quite specific — even with high concentration of excited cells in domain of initial stimulation — dynamics is a subject of a chance, either excitation remains contained (just small quasi-stationary domains persist) or, on rare occasions, the domains merge and a slowly growing global domain emerges, which eventually takes diamond shape.

In rules $R(221a)$, $1 \leq a \leq 8$, domain of initial stimulation is preserved more when upper boundary of excitation retaining interval is higher.

In series of rules $R(222a)$, increase of a from 2 to 8 transform growing domains from quasi-disordered to ordered, subdivided onto several sub-domains of coherent excitations, Fig. 3.10ab. Automaton governed by rule $R(2222)$ exhibits a lot of localised mobile excitations, or gliders, however most collisions between gliders lead to production of new gliders, so internal space of the lattice becomes filled with newly generated patterns and localised excitations. Gliders can also be found in dynamics of rule $R(2228)$, there they only present as leading parts of the horizontal and vertical wave-fronts. The similar dynamics observed in the rules $R(2433), \cdots, R(2437)$, and $R(2422), \cdots, R(2427)$. The same applies — subject to variety of guiding gliders and refractory states in central domain — to the rules $R(2733), \cdots, R(2737)$ and $R(2833), \cdots, R(2837)$

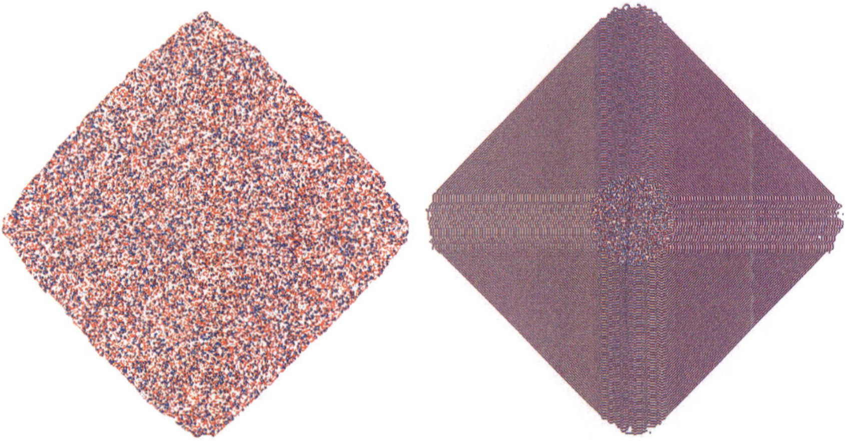

(a) (b)

Fig. 3.9 Snapshots of patterns growing in automaton lattice governed by rule (a) $R(3413)$, (b) $R(2455)$. Initial stimulation disc, radius 50 cells, centred, probability of non-resting state inside the disc was 0.45.

(Fig. 3.10cd); $R(3613),\cdots,R(3617)$; $R(3725),\cdots,R(3727)$, $R(2533),\cdots,R(2538)$; and, $R(2633),\cdots,R(2637)$.

In automata governed by rules $R(2344),\cdots,R(2348)$ growing patterns consist of centrally located astroid-shaped domain of irregular excitation and refractoriness, lines of vertical and horizontal growth, and wave-fronts normal to diagonals; central excitation core develops when upped boundary of excitation retention interval increases, as shown in Fig. 3.11ab. Increase of upper boundary of excitation interval (Fig. 3.11cd) leads to disappearance of central irregularity domain, and production of 'classical' wave-fronts, rules $R(2444),\cdots,R(2448)$.

This type of dynamics is also typical for rules $R(2355),\cdots,R(2358)$, $R(2366),\cdots,R(2368)$, $R(2377),\cdots,R(2378)$, $R(2388)$, $R(2411),\cdots,R(2417)$. There are also certain particulars of the dynamics, like exact shape of central domain of irregular states, and presence or absence of horizontal and vertical growth lines, but the general structure of the dynamics is the same. Such development reaches its climax in rules $R(2544),\cdots,R(2548)$, when increase in excitation retention interval leads to formation of central solid domains of standing excitation (Fig. 3.12). This behaviour is exhibited by rules $R(2644),\cdots,R(2648)$, $R(2655),\cdots,R(2658)$, $R(2744),\cdots,R(2747)$.

In rules $R(2455),\cdots,R(2458)$ we observe domains of initial stimulation exhibit quasi-chaotic dynamics, the rest of the growing pattern is filled with of covered by 'classical' wave-fronts of target waves (Fig. 3.9b). This applies to dynamics ruled by family $R(2bcd)$, $4\ leqb \leq 7$, $5 \leq c,d \leq 8$,

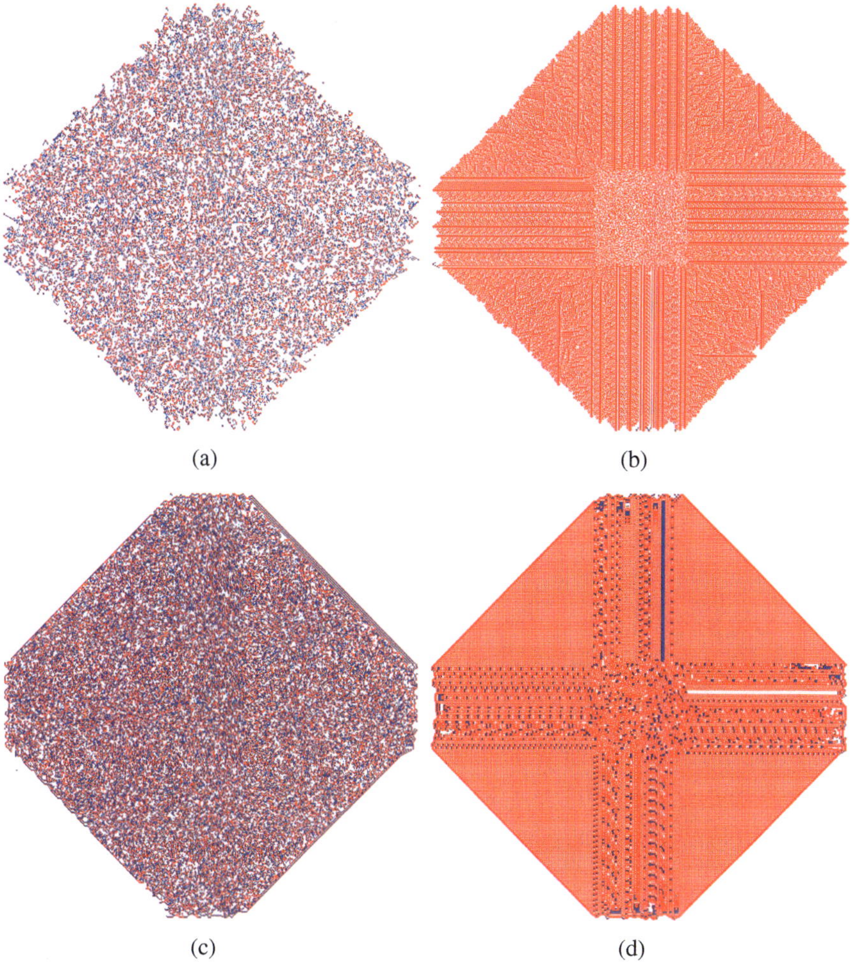

Fig. 3.10 Examples of morphological transitions caused by increase of excitation retention interval, (a) rule $R(2222)$, (b) rule $R(2228)$, (c) rule $R(2733)$, (d) rule $R(2737)$. Each automaton was stimulated by 100×100 cells centred rectangle, where probability of non-resting state was 0.45.

Transition from rule $R(2511)$ to rule $R(2527)$ leads to transformation of the central pattern, shaped as Cassinian curve, of disordered excitation to pattern with localised irregular domains, corresponding to initial excitation, and standing excitation pattern (Fig. 3.12cd); the same happens in families of rules $R(2522),\cdots,R(2527)$, $R(2611),\cdots,R(2617)$, $R(2622),\cdots,R(2627)$, $R(2711),\cdots,$ $R(2718)$, $R(2722),\cdots,R(2728)$, $R(3815)$, $R(3816)$, $R(3817)$.

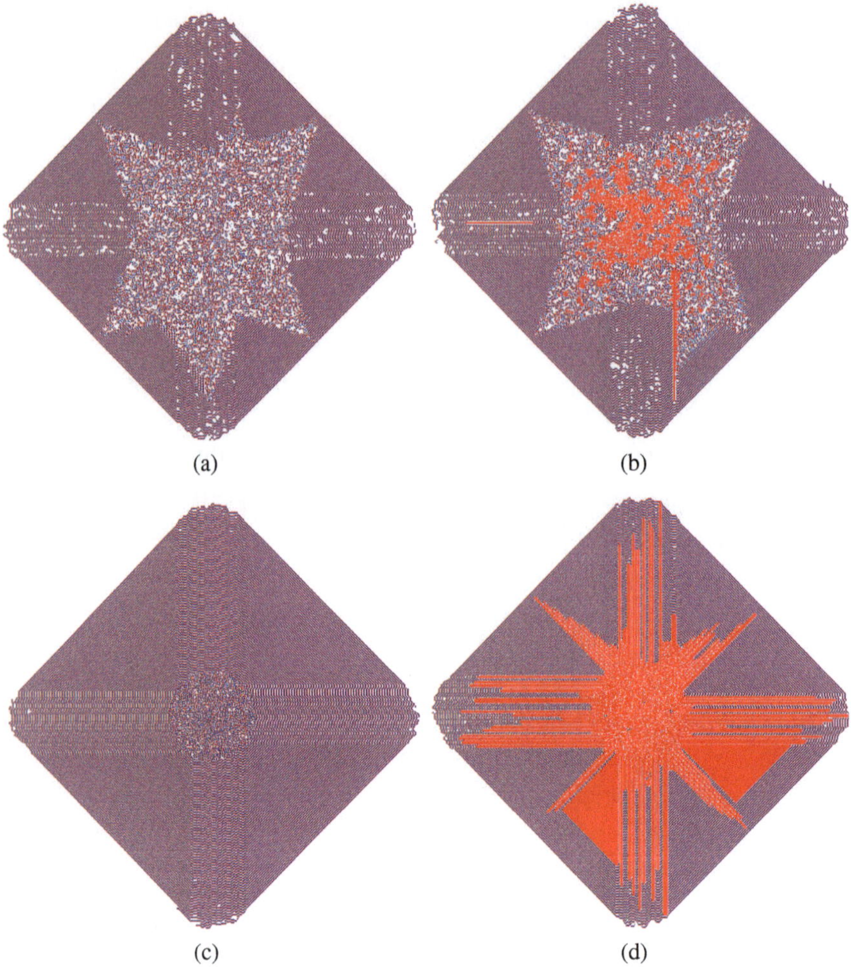

Fig. 3.11 Examples of morphological transitions caused by increase of boundaries of excitation and excitation retention intervals, (a) rule $R(2344)$, (b) rule $R(2348)$, (c) rule $R(2444)$, (d) rule $R(2448)$. Initial centred disc of radius 50 cells was stimulated, probability of non-resting state inside the stimulation disc was 0.45.

3.3 Octagonally Growing Domains

Almost Everywhere Excited Octagonal Domains

A necessary condition for emergence of solid domains of excitation is that upper boundary of excitation retention interval equals 8. In this case every excited cell surrounded by excited neighbours will stay excited. Shape of the growing domain

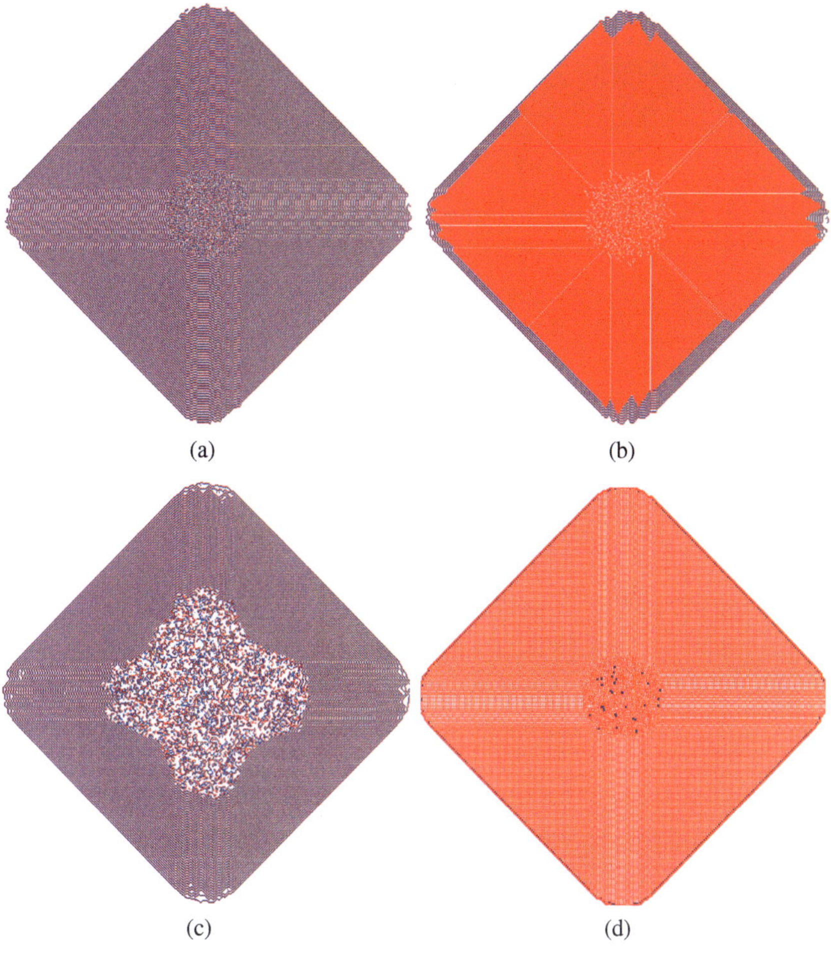

Fig. 3.12 Examples of morphological transitions caused by increase of excitation retention interval: (a) rule $R(2544)$, (b) rule $R(2548)$, (c) rule $R(2511)$, (d) rule $R(2517)$. Initially cells in centred disc of radius 50 cells are stimulated, probability of non-resting state inside the stimulation disc was 0.45.

is determined by boundaries of the excitation interval. Thus rules $R(1ab8)$, exhibit rectangularly growing patterns, the rules $R(2ab8)$ diamond-shaped patterns, for the same values and a and b: $3 \le a \le 8$ and $1 \le b \le 3$. Further increase of the lower boundary θ_1 of excitation interval leads to the formation of octagonally growing domains. Namely, such domains are typical for rule $R(3418)$ (Fig. 3.13. The domains are invariant to shape of the initial stimulation domain — discoidal and rectangular stimulations of the same size produce identical growing domains (Fig. 3.13ac). Their growth is sensitive to symmetry of initial stimulation domain, compare Fig. 3.13a and 3.13b.

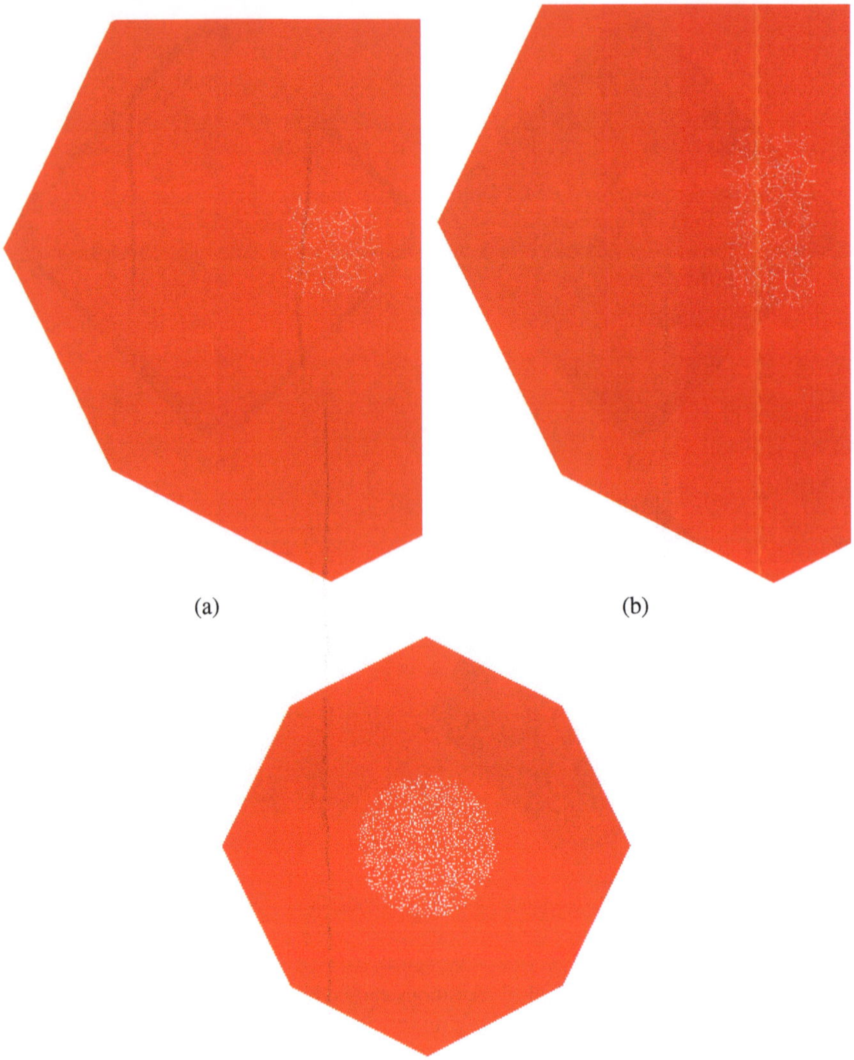

(a) (b)

(c)

Fig. 3.13 Snapshots of patterns growing in automata governed by rule $R(3418)$ (a) centred domain of 100×100 cells is stimulated, (b) centred domain of 100 (horizontal) $\times$ 200 (vertical); and (c) rule $R(3418)$, stimulated domain is a disc radius 50 cells centred on the lattice. In stimulated domains every cell takes non-resting state with probability 0.3.

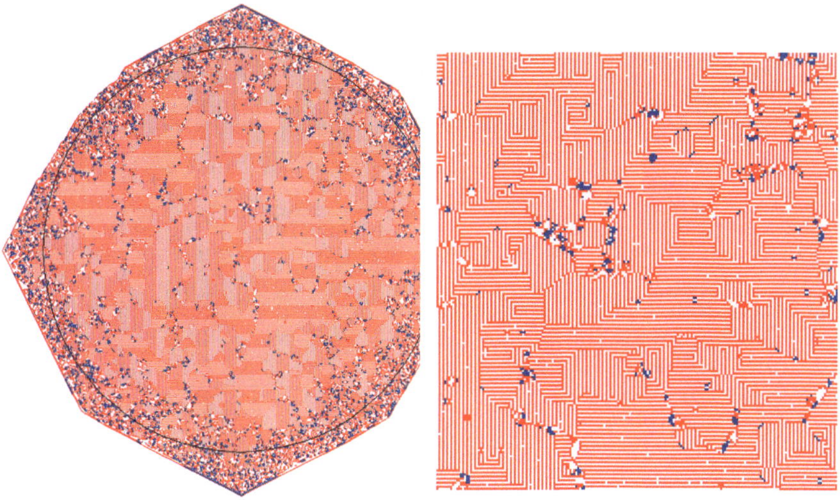

(a) (b)

Fig. 3.14 (a) Snapshot of patterns growing in automaton governed by rule $R(3414)$, stimulated domain is a disc radius 50 centred on the lattice, every cell inside the disc initially takes one of three state with probability 0.3, (b) zoomed fragment of configuration (a)

Mixed State Octagonal Domains

Families of rules discussed are responsible for various configurations of excited and refractory states inside the growing domain, however all of the domains have octagonal shape. In automata governed by rule $R(3414)$ growing domain consists of labyrinthine sub-domains (Fig. 3.14a) and also sub-domains of disordered dynamics (Fig. 3.14b).

Internal core of labyrinthine patterns surrounded by domains of co-oriented stripes of excitation (Fig. 3.15ab) are typical for rules $R(3415), \cdots, R(3417)$, $R(3426)$, $R(3437)$, $R(3425)$, $R(3516)$, $R(3526)$, $R(3527)$, $R(3537)$, $R(3616)$, $R(3617)$, $R(3626)$, $R(3627)$, $R(3637)$.

In Fig. 3.15b we can see traces of traveling singularities (straight segments leading to corners of the octagonal shape), and also less regular trajectories of mobile defects and generators of defects, distant analogs of glider guns.

Disordered configurations of growing domains (Fig. 3.15c) are characteristic for rules $R(3524)$, $R(3515), \cdots, R(3526)$, $R(3724)$, $R(3824)$, $R(3614)$, $R(3615)$, $R(3624)$, $R(3635)$, $R(3636)$, $R(3736)$, $R(3737)$, $R(3813)$, $R(3814)$, $R(3836)$, $R(3837)$, and $R(3825), \cdots, R(3827)$. In later case, increase of the upper boundary of the excitation retention interval causes formation of sub-domains of ordered excitation in the disordered excitation domain (Fig. 3.15d).

Rules $R(3446)$ and $R(3437)$ exhibits visibly uniformly growing domains, which are subdivided onto domains of co-oriented stripes of excitation (Fig. 3.16).

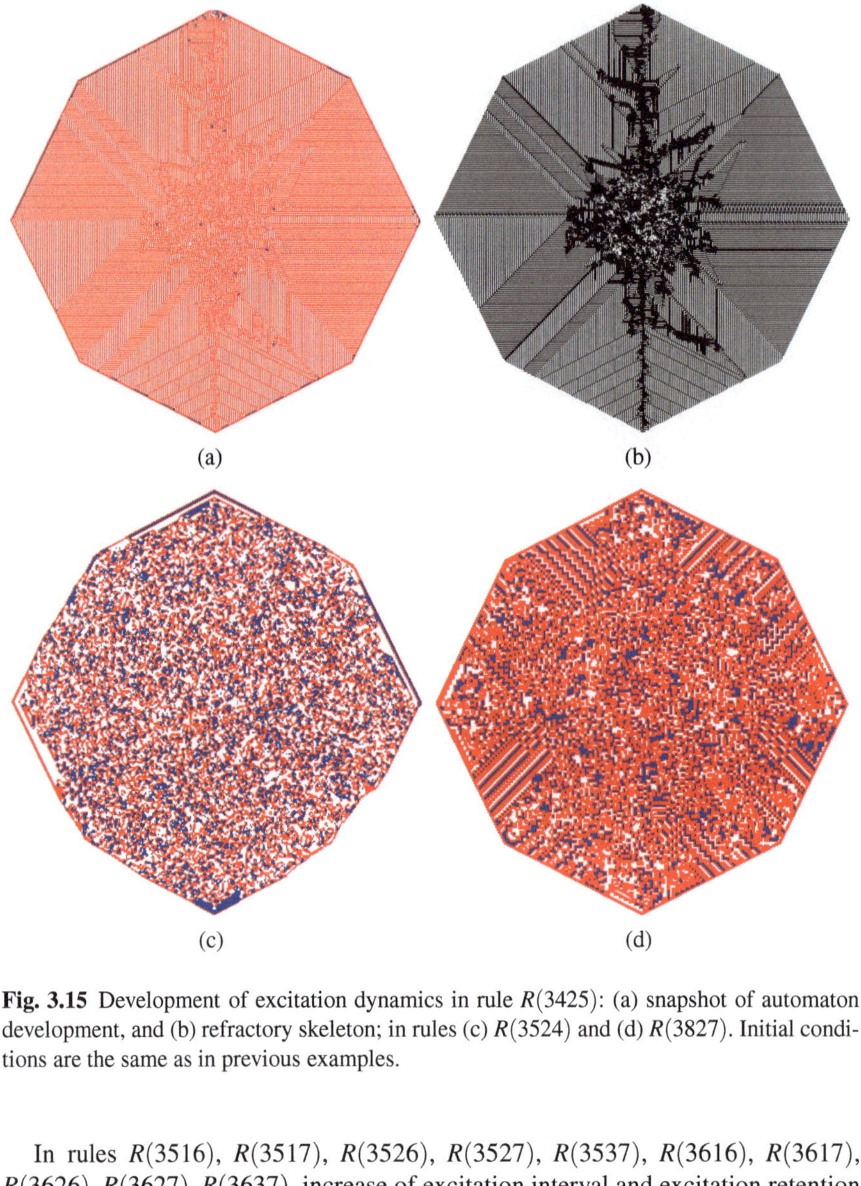

Fig. 3.15 Development of excitation dynamics in rule $R(3425)$: (a) snapshot of automaton development, and (b) refractory skeleton; in rules (c) $R(3524)$ and (d) $R(3827)$. Initial conditions are the same as in previous examples.

In rules $R(3516)$, $R(3517)$, $R(3526)$, $R(3527)$, $R(3537)$, $R(3616)$, $R(3617)$, $R(3626)$, $R(3627)$, $R(3637)$, increase of excitation interval and excitation retention interval leads to 'homogenisation' of the growing excitation pattern (Fig. 3.17).

3.4 Almost Disc-Shaped Growing Domains

Automata governed by certain rules of excitation retention show growing domains of excitation and refractoriness which are very close to or have nearly discoidal

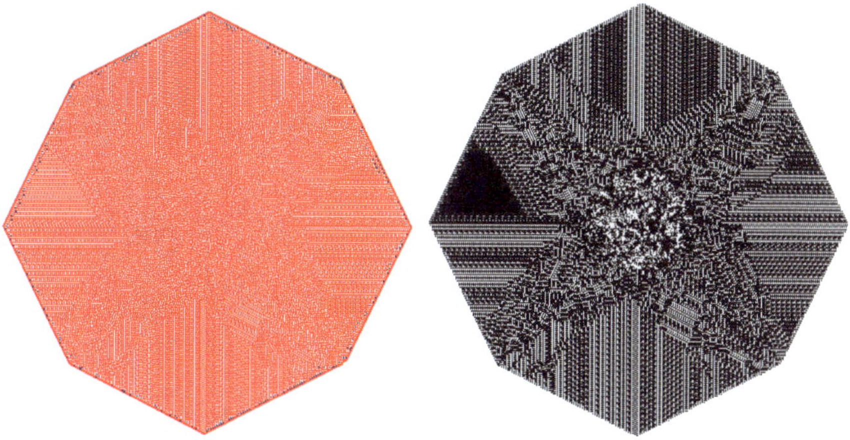

(a) (b)

Fig. 3.16 Snapshot of rule $R(3436)$ (a) and its refractory skeleton (b). Initial stimulation with disc-shaped random configuration as in previous examples.

shape. Most of them exhibit sub-linear rate of growth, radius of the domain at any given time step t is less then t.

In rule $R(3313)$ we observe ordered domains of labyrinthine structures 'immersed' into disordered configuration (Fig. 3.18a), the domain grows with approximate rate $\frac{1}{4}t$. In long term the growing shape of excitation in automata governed by rule $R(3435)$ tends to be more octagon-looking, as shown in Fig. 3.18b, the same applied to $R(3535), \cdots, R(3835)$. Automata governed by rules $R(3314)$ and $R(3523)$ exhibit discoidal growing domain (Fig. 3.18c) filled with labyrinthine structure of excited cells (Fig. 3.18d).

3.5 Amoeboid Growth of Mixed-State Patterns

In this group rules — $R(33ab)$, $1 \leq a \leq 3$, $5 \leq b \leq 8$ — exhibit amoeboid growth of excitation: domain of initial stimulation is transformed to amoeba-shaped pattern which irregularly spreads on the lattice by extending its pseudopodia-like structures (Fig. 3.19). There are no refractory states inside the growing body, but cells in refractory states are present on the front of the growing pattern; essentially, refractory state may be thought of as guiding growth of pseudopodia, see Fig. 3.19bc.

Configuration of excitation and refractoriness exhibited by rule $R(3315)$ reflects the structure of grows which looks like a branching tree or a cluster of diffusion limited aggregation (Fig. 3.19ab); $R(3325), \cdots, R(3328)$ behave similarly. With increasing degree of excitation retention a straight growing structures emerge (Fig. 3.19cd), rules $R(3316), \cdots, R(3318)$.

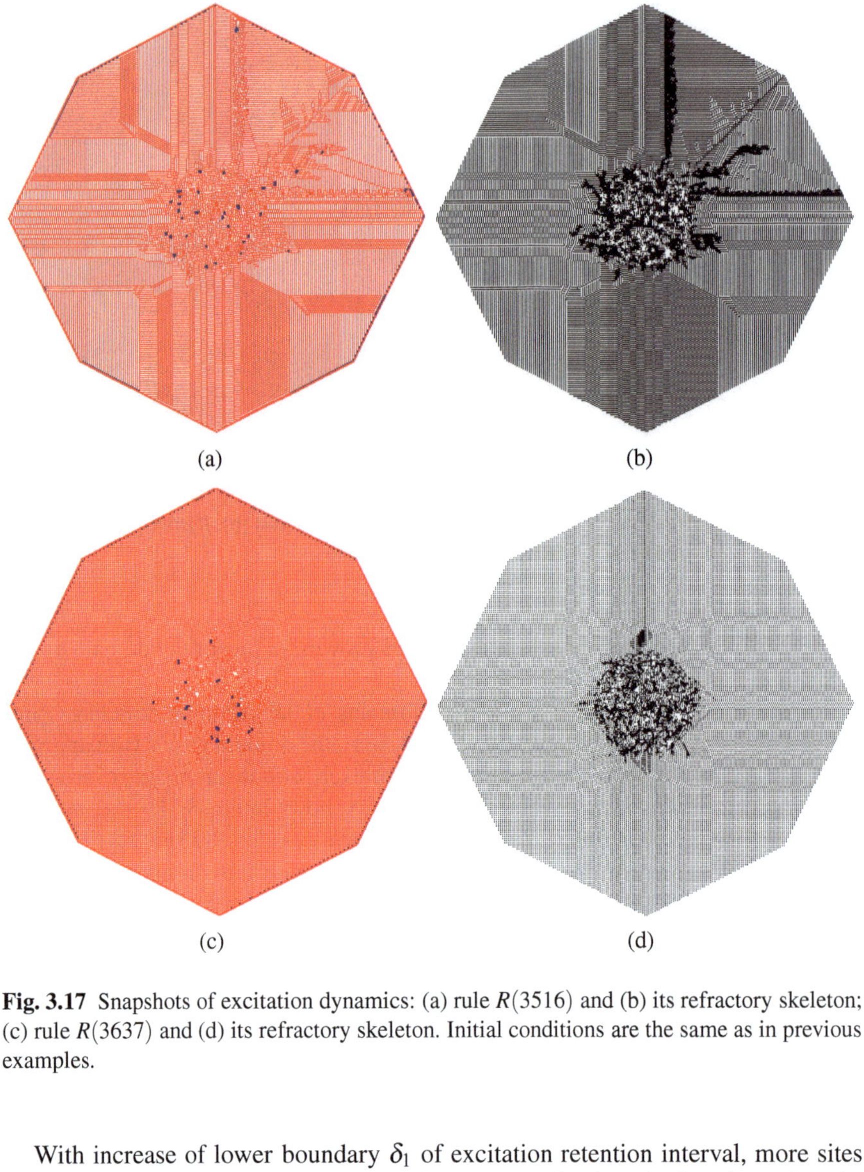

Fig. 3.17 Snapshots of excitation dynamics: (a) rule $R(3516)$ and (b) its refractory skeleton; (c) rule $R(3637)$ and (d) its refractory skeleton. Initial conditions are the same as in previous examples.

With increase of lower boundary δ_1 of excitation retention interval, more sites need to be stimulated initially to give rise to a sustained excitation dynamics. As we can see in example of rule $R(3336)$, Fig. 3.20, the system exhibits some kind of memory, because initial domain of excitation is clearly visible; the same in rules $R(3335),\cdots,R(3338)$.

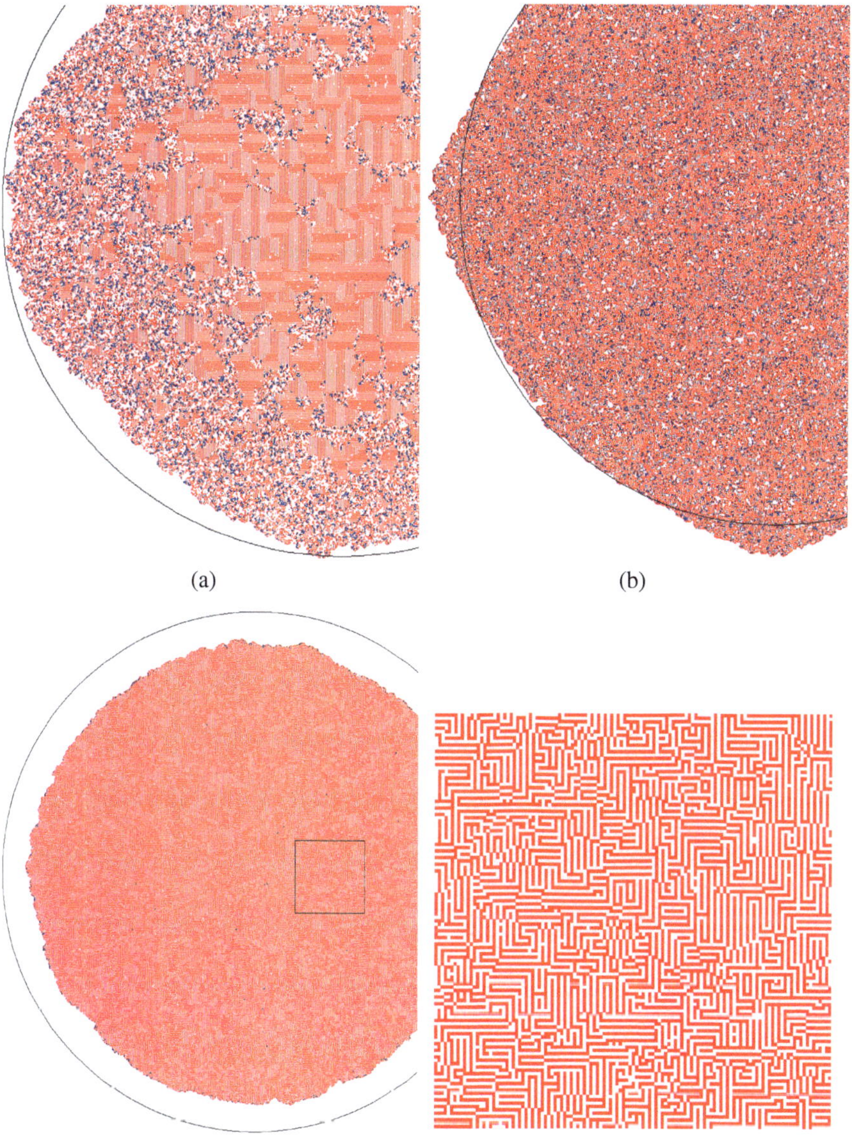

(a) (b)

(c) (d)

Fig. 3.18 Snapshots of patterns growing in cellular automata governed by (a) rule $R(3313)$, snapshot taken at $t = 1512$, (b) rule $R(3435)$, snapshot taken at $t = 891$, (c) rule $R(3314)$, snapshot take at $t = 1030$, (d) rule $R(3314)$, zoomed in fragment of snapshot shown with rectangle in (c). Stimulated domain is a disc with radius 50 cell centred on the lattice, every cell inside the disc initially takes a non-resting state with probability 0.3.

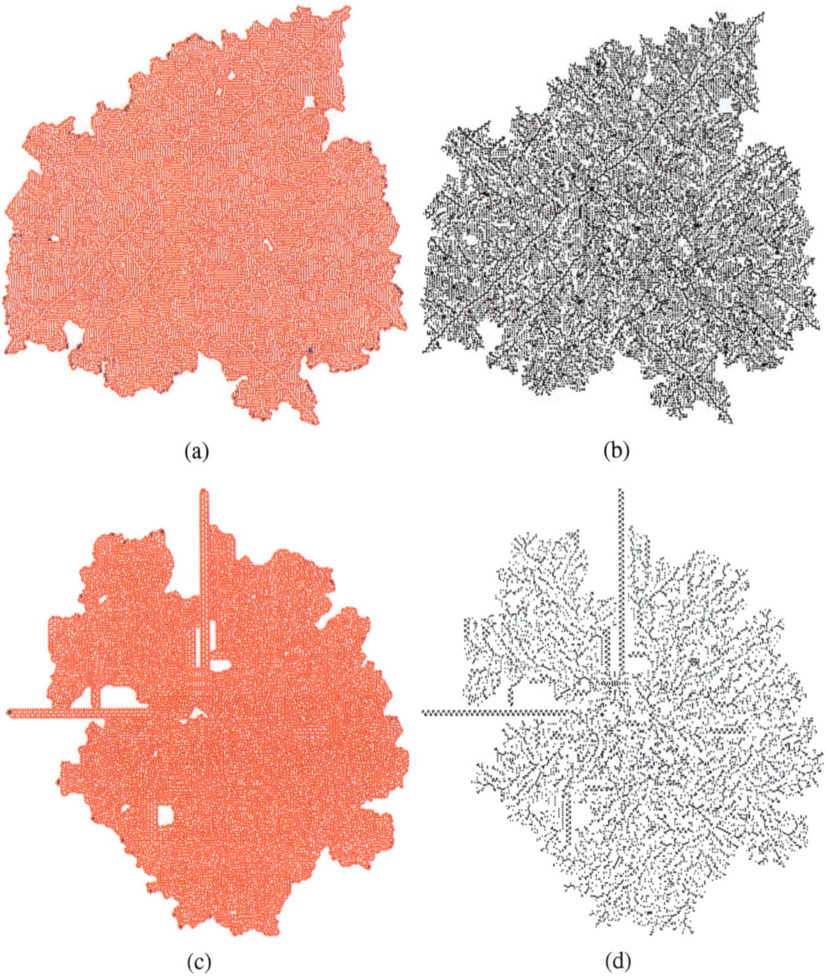

Fig. 3.19 Examples of amoeboid growth on lattice 400×400 cells: (a) snapshot of excitation dynamics of $R(3315)$ made at 272nd step of development, (b) refractory skeleton of $R(3315)$, (c) snapshots of $R(3316)$, (d) refractory skeleton of $R(3316)$. The automata are stimulated by centred rectangle of 100×100 cells, with cell resting with probability 0.9 and takes excited or refractory states with probabilities 0.05 each.

3.6 Not Growing Domains of Excitation

In families of rules discussed below excitation domains do not expand or expand just for few sites beyond boundaries of the initial stimulation domain; at the same time, the automata does not return to entirely resting state.

Thus, in rules (3311), $R(3411)$, $R(3611)$, $R(3711)$, $R(3811)$, $R(4411)$, $R(4511)$, $R(4611)$ tiny stationary localizations are formed, consisting of two excited states,

(a) (b)

Fig. 3.20 Examples of amoeboid growth on lattice 400 × 400 cells: (a) snapshot of excitation dynamics of $R(3336)$ made at 272nd step of development, (b) refractory skeleton of $R(3315)$. The automata are stimulated by centred rectangle of 100 × 100 cells, with cell resting with probability 0.6 and takes excited or refractory states with probabilities 0.2 each.

e.g. ⁛ and ⁚⁚. These localizations are always cushioned by at least one layer of resting cells, therefore they always stay isolated and do not form polymer-like structures.

With increase of retention interval by one unit — rule $R(3312)$ — oscillators (Fig. 3.21) emerge, and they share lattice with stationary localizations.

In rule $R(3323)$ we observe several types of stationary localizations not changing their structure with time (Fig. 3.21c) and at least one type of an oscillating localisation. The oscillatory localisation (Fig. 3.21b) consists of two standing excitations 'emitting' refractory states toward each other.

There are just two types of localizations in rule $R(3334)$, $R(3534)$, $R(3734)$, $R(3834)$: still cluster of 4 adjacent excitation states (as those populating automaton lattice in Fig. 3.21c), and 'rotating' oscillating localisation, example shown in Fig. 3.21d. Rotating localisation consists of still core of four adjacent excitation states and a pair of excitation-refractory states rotating like a typical excitation wave around standing excitation wave. Rules $R(3423)$ and $R(3434)$ exhibit the same spectrum of localizations as $R(3334)$ but 'rotating' oscillators.

In rule $R(4422)$ we can observe only small groups of stationary excitations as shown in Fig. 3.21f.

Only residue oscillating domains can be found in rule $R(3447)$, such domains consist of excited standing localisation and combinations of excited and refractory states traveling along the edges of the standing localisation (Fig. 3.21e); no other types of localizations are found in this rule.

Dynamics in $R(3448)$ is similar to that in $R(3447)$, with more domains. Rule $R(3511)$ exhibits small excited localizations as found in rule $R(3311)$. In rules

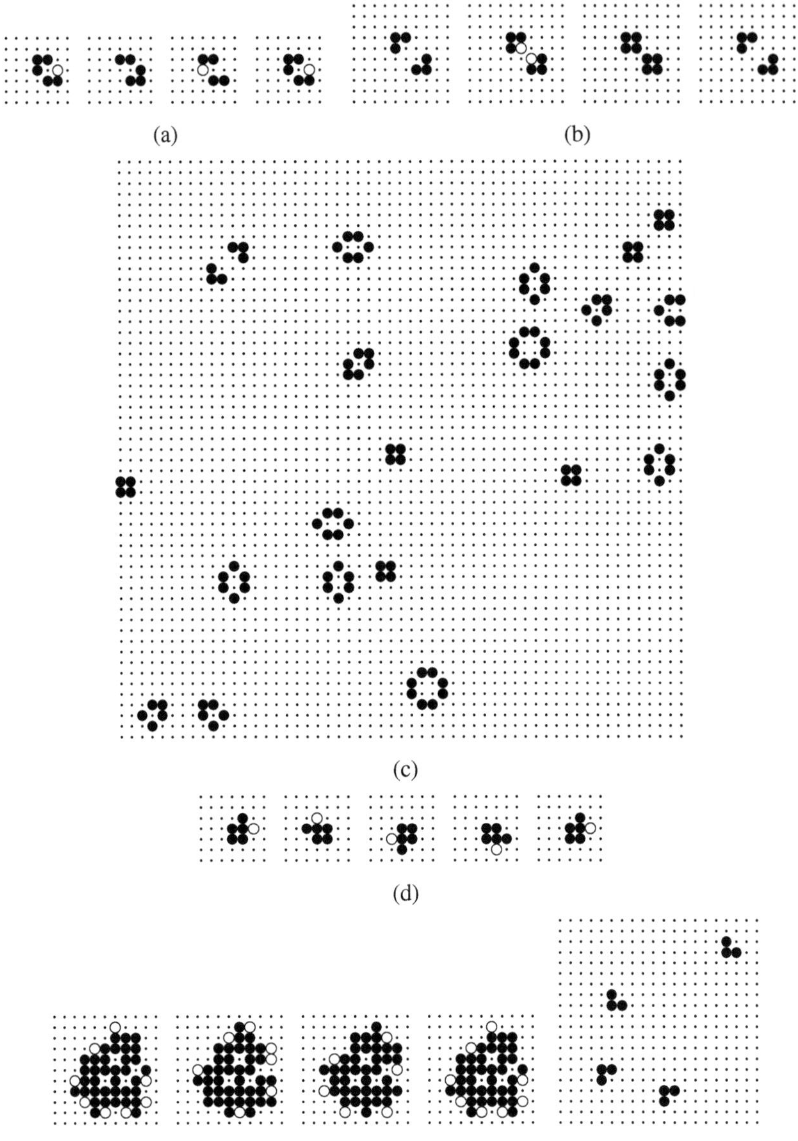

Fig. 3.21 (a) Configurations of oscillator in rule $R(3312)$, time arrowed right, (b) configurations of oscillator in rule $R(3323)$, time arrowed right, (c) variety of still localisations in rule $R(3323)$, (d) anti-clockwise 'rotating' oscillator in rule $R(3334)$, time arrowed right, (e) residue domains in $R(3447)$, time arrowed right, (f) still localizations in $R(4422)$.

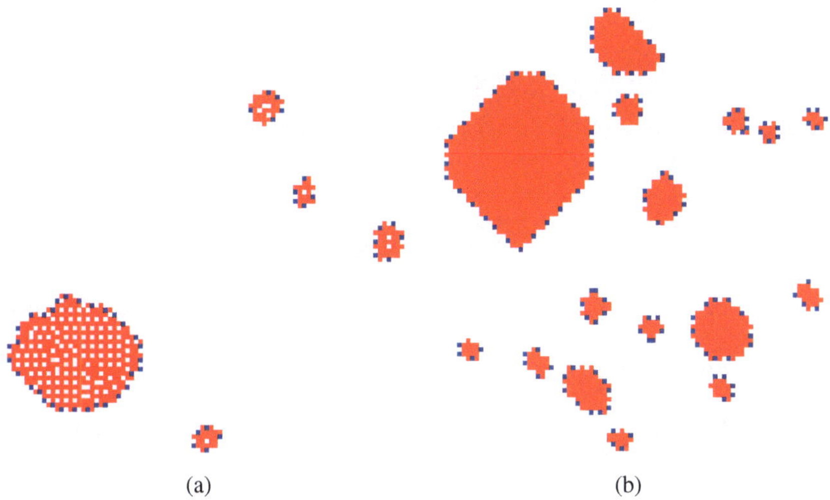

Fig. 3.22 Snapshots of excitation domains found in rules (a) $R(3547)$ and (b) $R(3848)$

$R(3547)$, $R(3647)$ domains of substantial size are formed (Fig. 3.22), they are however never outgrow boundaries of initial stimulation. The domains are becoming much larger in rules $R(3648), \cdots, R(3848)$ (Fig. 3.22b).

Rule $R(4412)$ exhibits tiny stable groups of excited cells, and rule $R(4413)$ also support oscillators as above.

In rule $R(4414)$ labyrinthine-like structure — domains of co-oriented strips of excitation — is formed (Fig. 3.23a). Size of the domains becomes smaller — which causes visual increase of disorder — with an increase of the excitation retention interval, see configuration generated by rule $R(4415)$ (Fig. 3.23b). Further increase of the interval, in rules $R(4416)$ to $R(4418)$ makes configuration even more amorphously-looking, with its extreme disorder in rule $R(4418)$ (Fig. 3.23c).

Number of clusters increases in $R(4423)$ and $R(4424)$ and becomes maximal in rule $R(4425)$ with domains of co-oriented excitation walls (Fig. 3.23d). Similarly in the rule $R(4426)$. In rules $R(4427)$ and $R(4428)$ domains merge and lose their structure, and then become porous pools of excitation as shown in Fig. 3.23e. The same happens in transition $R(4436) \to R(4438)$ (Fig. 3.23f).

In configurations generated by $R(4512)$ more small groups of standing excitation appear, in rule $R(4513)$ they increase in numbers and oscillators emerge.

Rule $R(4514)$ is quite unique — when domain of initial stimulation is a disc, the initial configuration is transformed to the hexagon (Fig. 3.23g) and then stop growing, so hexagon does not grow! However, if domain of initial stimulation is rectangular then its shape remains unchanged! Obviously, it would be extremely interesting to study how other shapes are changed by the rule. The disc to hexagon transformation takes place in rules $R(4515)$ (labyrinthine domains, Fig. 3.23h, there is always excitation at the boundaries between the domains) till $R(4518)$ (almost

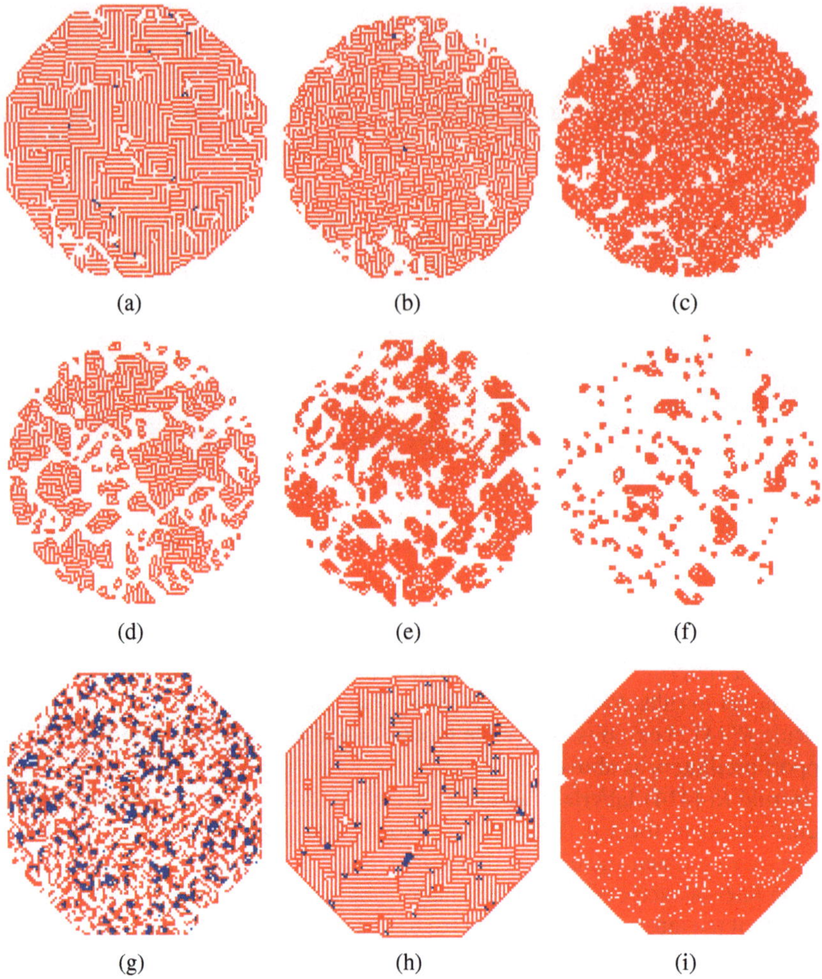

Fig. 3.23 Non-growing configurations in rules (a) $R(4414)$, (b) $R(4415)$, (c) $R(4418)$, (d) $R(4425)$, (e) $R(4428)$, (f) $R(4438)$, (g) $R(4514)$, (h) $R(4515)$, (i) $R(4518)$. Lattice size is 150×150 cells, domain of initial stimulation is a centred disc radius 60 cells; in the domain of initial stimulation each cell takes a non-resting state with probability 0.3.

solid domains, Fig. 3.23i). The same morphological transitions happen in the row of rules $R(4522), \cdots, R(4528)$.

In rule $R(4433)$ there are few blocks, each is a box of four excitation states, number of blocks increases in rule $R(4434)$, and in rule $R(4435)$ blocks merge into large domains.

In rule $R(4525)$ domains of co-oriented excitation stripes are formed (Fig. 3.24) but also closed channels of 'disordered' excitation; in Fig. 3.24 they are visible as vertical and horizontal sub-configurations of excited and refractory states inside the

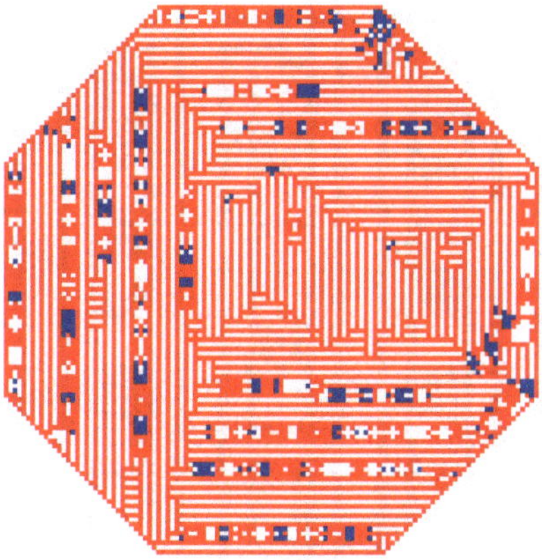

Fig. 3.24 Snapshot of excitation in rule $R(4525)$. Lattice size is 150×150 cells, domain of initial stimulation is a centred disc radius 60 cells; in the domain of initial stimulation each cell takes a non-resting state with probability 0.3.

domains. Detailed dynamics of excitation inside one of the channels is shown in Fig. 3.25.

Then in rule $R(4536)$ dynamics 'return' to normal kind as discussed for the rules above. Transitions from rule $R(4534)$ to $R(4538)$ cause the same morphological transformations of excitation dynamics as transitions from rule $R(4434)$ to $R(4438)$. Essentially, for $R(4abc)$ dynamics is similar to that of $R(45bc)$, $a = 6,7,8$. Further increase of lower boundary of excitation interval, as in rules $R(5,...), \cdots ,R(8,...)$, does not change overall character of dynamics — automata exhibit formation of either sparse configurations of excited sites, or amorphously looking configurations of excitation, sometimes with few oscillators.

3.7 Domains with Small Number of Still Localizations

In this group excitation dies, or just a small number of stationary localizations produced, there are no mobile localizations and no spreading patterns in this group. In rules $R(3a4b)$, $3 \leq a \leq 8$ and $4 \leq b \leq 8$, just one localisation is found: flip-flop pattern shown in Fig. 3.26a. The number of such patterns increases with increase of the upper boundary of excitation retention interval. Also, in rule $R(3336)$ an oscillator shown in Fig. 3.26b appears. In rule $R(3546)$ also extended clusters as in $R(3348)$ could be found.

In rule $R(3347)$ in addition to above described patterns one more stationary oscillator could be observed, see Fig. 3.26c.

Fig. 3.25 Fragments of snapshots of configuration in rule $R(4525)$. Solid discs are excited states, empty discs are refractory states. Time arrowed from (a) to (n).

Rule $R(3348)$ is an extreme case, because localisation are not so rare, and their size increases as well, we can see domains of excited states, where refractory states run along or 'breath' at the boundaries of the standing excitation (Fig. 3.26d).

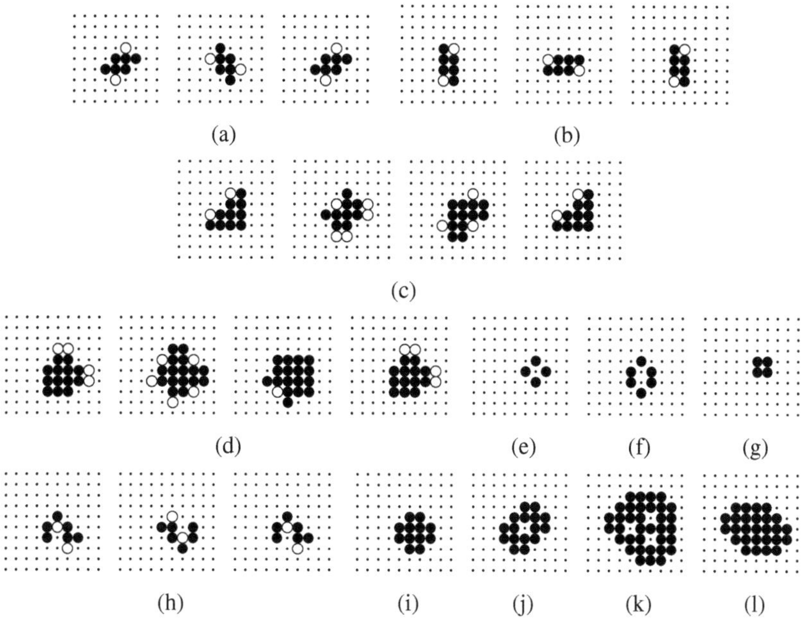

Fig. 3.26 Snapshots of oscillating and still localizations in rule (a) oscillator in $R(3345)$, (b) oscillator $R(3346)$, (c) oscillator in $R(3347)$, (d) oscillator in $R(3348)$, (e) still localisation $R(3322)$, (f) still localisation $R(3322)$, (g) still localisation $R(3333)$, (h) oscillator $R(3422)$, (i) still localisation $R(4447)$, (j) still localisation $R(4447)$, (k) still localisation $R(4747)$, (l) still localisation $R(4748)$. Solid discs are excited states, empty discs are refractory states. In series of snapshots of oscillators time arrowed rightwards.

In family of rules $R(3cab)$, $5 \le a,b \le 8$, $c = 3,\cdots,8$; $R(4cab)$, $4 \le a,b \le 8$, $c = 4,\cdots,8$; $R(5abc), 5 \le a,b,c \le 8$; no localisation are found, automata returns to entirely resting state after a short transient period followed initial stimulation.

In rule $R(3322)$ only rhomboid and hexagonal shaped still clusters of excited states could be found (Fig. 3.26ef), and in the rules $R(3a33)$, $1 \le a \le 8$, $R(3634)$, $R(5533)$, $R(5733)$, only 'boxes' of four excited states (Fig. 3.26g). In rules $R(3a22)$, $1 \le a \le 8$, in addition to above discussed still localisation and oscillators of six excited and two refractory states are discovered (Fig. 3.26h). Rule $R(3446)$ exhibits oscillators as rule $R(3346)$ and extended clusters as rule $R(3348)$.

In rule family $R(4447)$ and $R(4448)$, $R(4457)$, $R(4458)$, $R(5647)$, $R(5648)$ localizations are very rare, but in many trials we snap two most often found types, see Fig. 3.26ij, of 12 and 16 excitation states each, and no refractory states. A bit larger domains emerge in $R(4747)$, Fig. 3.26k, and $R(4748)$, Fig. 3.26l. The same applies to rules $R(5547)$, $R(5747)$, and rule family $R(ab**)$, $a = 5,\cdots,8$, $b = 6,7,8$.

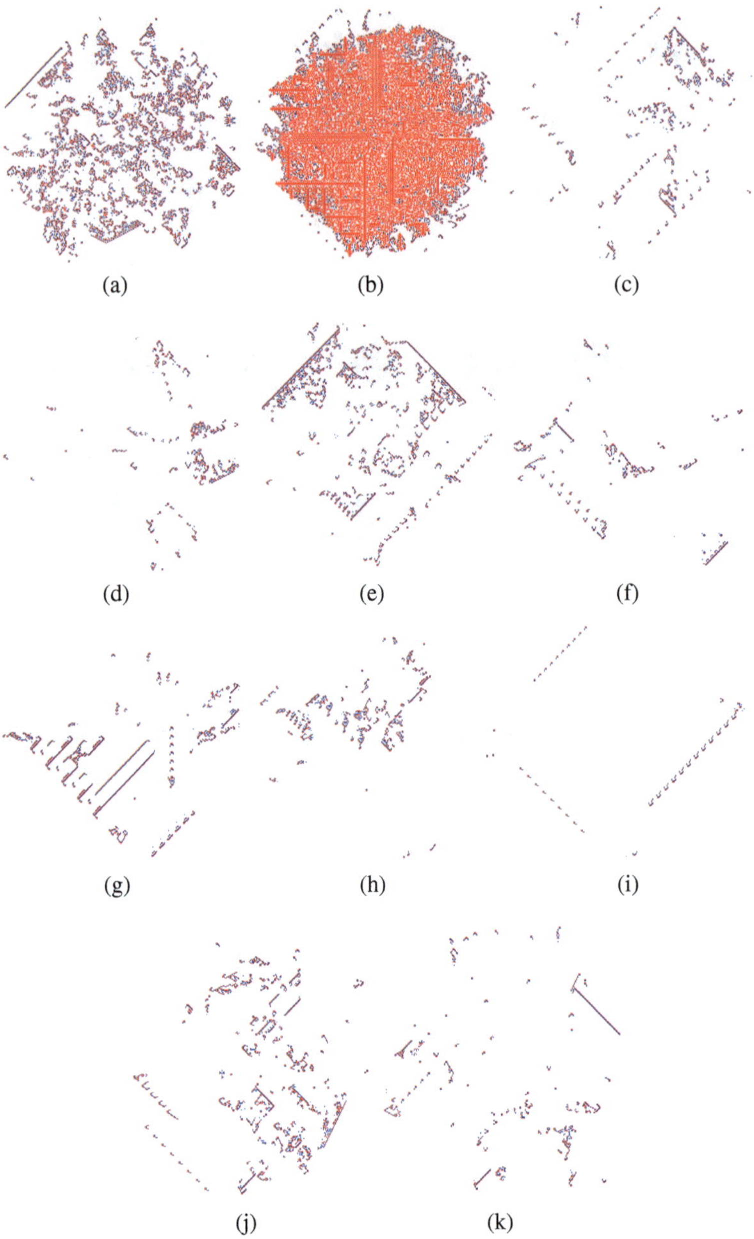

Fig. 3.27 Snapshots of gliders exhibiting configurations. Lattice size 300×300 cells. Domain of initial stimulation is centred disc radius 25 cells. In the domain of initial stimulation each cell takes one of three possible states with probability 0.1 and remains resting otherwise.

3.8 Mobile Localizations

Many of the rules discussed in the chapters support mobile localizations however in many cases the localizations are not easily detectable. This is because when colliding with each other they produce more localizations, and such chain reaction leads cause the lattice to be overpopulated with disordered-looking pattern of dense excitation. In this section we selected rules where mobile localizations, or gliders, are easily noticeable.

These are rules $R(22ab)$, $3 \leq a \leq b \leq 8$. Snapshots of their excitation are presented in Fig. 3.27. Almost all localizations and spreading patterns in these rules have been already described in a context of 2^+-medium, excitable cellular automaton where resting cell becomes excited if it has exactly two excited neighbours [2, 3, 9]. Said that we could not guarantee that novel 'exotic' localizations will not be found during exhaustive search.

Rules $R(3a12)$, $5 \leq a \leq 8$, exhibits the slowest (one cell per five time steps) mobile localisation for the class of rules studied (Fig. 3.28). The localisation looks like a semi-circular wave fragment traveling concave-part forward, with lateral segments protruding toward direction of motion. A relation of between speed and curvature of the wave fragment takes place (in discrete context), concavity is filled with excitation and wave become planar. At this stage however part of the wave collapses, and sites of the medium take refractory states. Remaining fragment — which is convex, like 'classical' wave fragment — continues motion, till it is transformed back to concave fragment. The transformations are cyclical.

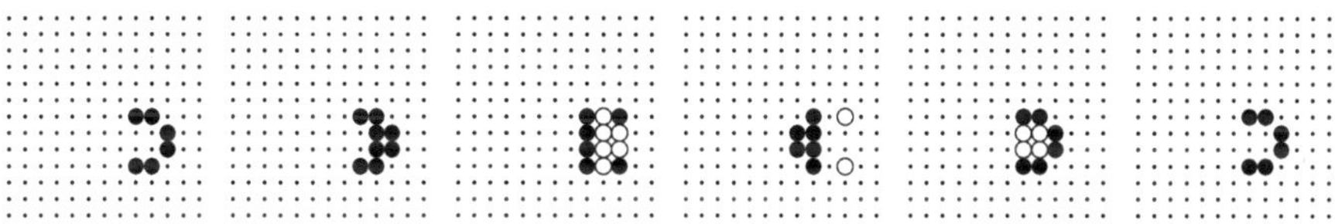

(a)

Fig. 3.28 Glider moving east in rule $R(3512)$, speed $\frac{1}{5}$. Solid discs are excited states, empty discs are refractory states.

3.9 Summary

We can use cell-state transition rule descriptors to roughly predict space-time dynamics of discrete excitable media with retained excitation. If medium is highly excitable, lower boundary of excitation interval equals 1, then any initial shape of stimulation will be transformed to omni-directional growing domain of excitation. These will be domains of almost everywhere excited states if excitation retention interval is wide enough, and domains of mixed states otherwise. If excitability of the medium decreases, lower boundary of excitation interval increases, diamond-shaped growing domains of excitation emerge from patterns of initial stimulation. Further decrease in excitability causes formation of octagonally growing domains of the medium's excitation. Whether all these domains are filled with standing waves of excitation or patterns of intermittent or irregular activities depends on lower and upper boundaries of the excitation retention interval.

In models where excitability is low and lower boundary of excitation retention interval is high, excitation almost does not spread beyond boundaries of initial stimulation domain. Such non-spreading excitation may exhibit wide range of dynamics, including quasi-chaotic, regular, and pattern-forming.

For very narrow intervals of excitation the systems exhibit rare types of growth. Thus amoeboid growth of excitation occurs in cellular automata where every resting cells excites if exactly two neighbours are excited, and the degree of excitation retention is high. Automaton lattices where a resting cell excites when exactly two neighbours are excited exhibit mobile localizations, or gliders, and mobile and stationary generators of mobile localizations, or glider guns. Discovery of stationary localisations is of particular importance in automaton models of excitable medium because these structures can be used in future designs of excitable-medium based unconventional computing devices. With regards to stationary generators of mobile localizations in automata governed by family $R(22ab)$ of rules — their existence fits the findings [98] that traveling waves can bifurcate from standing wave. As to traveling generators of traveling localizations it is also shown experimentally [186] in Belousov-Zhabotinsky system with substrate 1,4-cyclohexanedione that chemical analogs of 'glider guns' do exist. Wave-fragments, or pulses, periodically generate other wave-fragments which travel in the direction opposite to that of their generators. As we demonstrated in rules $R(3a12)$, $5 \leq a \leq 8$, combination of lower excitability and narrow interval of excitation retention may contribute to emergence of mobile localizations, behaving like traveling wave-fragments that change their shapes from concave to planar wave and back.

Chapter 4
Mutualistic Excitation

In a classical Greenberg-Hasting model of excitation a resting cell excites depending on number of its excited neighbours. What if the process of being excited depends also on refractory states? Let every cell x of an automaton imitating mutualistic excitation take three states: resting $\cdot$, excited $\bullet$, refractory $\circ$. and updates its state by the rule

$$x^{t+1} = \begin{cases} \bullet, \text{ if } x^t = \cdot \text{ and } \sigma^t_+(x) \in [\theta_1, \theta_2] \text{ and } \sigma^t_-(x) \in [\delta_1, \delta_2] \\ \circ, \text{ if } x^t = \bullet \\ \cdot, \text{ otherwise} \end{cases} \tag{4.1}$$

A cell x 'calculates' number of excited, $\sigma^t_+(x)$, and refractory, $\sigma^t_-(x)$, neighbours:

$$\sigma^t_a(x) = \sum_{y \in u(x)} \{y : y = a\}, \text{ where } u(x) = \{y : |x - y| <= 1\}$$

Resting cell excites if number of excited neighbours belongs to interval $[\theta_1, \theta_2]$, and number of refractory neighbour to interval $[\delta_1, \delta_2]$. Excited cell become refractory, and refractory cell resting unconditionally, independently on the states of neighbours. We denote the cell-state transition rules by tuples $R(\theta_1 \theta_2 \delta_1 \delta_2)$.

We call the excitation rule (4.1) *mutualistic excitation* because for a cell to be excited not only excited neighbours but also neighbours in refractory states are required. If we interpret the excited and refractory as species, then for each species to persist, or in case of excitation state, to spread to new territories, another species must be present.

The mutualistic excitation rule can be written as the following set of quasi-chemical equations:

$$\begin{aligned} \cdot + a\bullet + b\circ &\rightarrow \bullet \\ \bullet &\rightarrow \circ \\ \circ &\rightarrow \cdot \end{aligned}$$

A. Adamatzky: Reaction-Diffusion Automata, ECC 1, pp. 67–95.
springerlink.com

where $\cdot$ is a substrate (resting cell-state), $\bullet$ is an active form of some reagent (excitation cell-state), and $\circ$ is an inactivated form of the reagent (refractory cell-state). Reaction coefficients a and b could be derived from exact structure of the rule $R(\theta_1 \theta_2 \delta_1 \delta_2)$.

4.1 Phenomenology of Mutualistic Excitation

We have tested behaviour of each rule by 'stimulating' automaton lattice at random. On the initially resting lattice we selected disk of cells centred in the lattice centre, and assigned to each cell of the disc one of three states with probability $\frac{1}{3}$. Amongst over thousand rules studied just one third exhibited excitation dynamics persisting for at least few steps of lattice development. Several classes of rules, based on spatio-temporal dynamics of excitation, were selected.

So-called homogeneous dynamics is most frequently found amongst the studied automata. Excitation in the originally randomly excited disc remains quasi-chaotic without any accentuated features. When excitation pattern spreads excitation at the growing fronts merge forming dense hierarchy of spreading wave fronts (Fig. 4.1a).

Automata which dynamics is very rich with mobile localisations, or gliders, are grouped in the second class. Single gliders are well detectable on the peripheral parts of the original randomly excited domain. However, in the central part of the domain, collisions between gliders produce more glides and therefore the configuration start to look more like a chaotic excitation (Fig. 4.1b).

A domain of 'classical' spiral wave encircled by several fronts of target waves is formed after initial stimulation of the automata belonging to third class of rules. The target waves are in the sites where excited cell are adjacent to refractory cells. Wave-fronts of the target wave merge at the periphery and then propagate as united fronts of target waves (Fig. 4.1c).

Rules of the forth class produce remarkable, very slowly expanding domains of the labyrinth-like configurations (Fig. 4.1d). The patterns even keep their shape, however arbitrary it will be initially for a long time. The wave-front there do not obey curvature-speed principle, and thus looks more like being more close to morphological patterns in living systems than simple excitable systems. Glider are seen on the periphery of the spreading pattern, they move faster then the growing pattern itself.

Fifth class gives us configurations where only mobile localisations (gliders) and generators of mobile localisations (glider guns) exist (Fig. 4.1e).

A slowly growing blob, with quasi-chaotic homogeneous core and labyrinth-like patterned shell, is typical for the rules of sixth class (Fig. 4.1f).

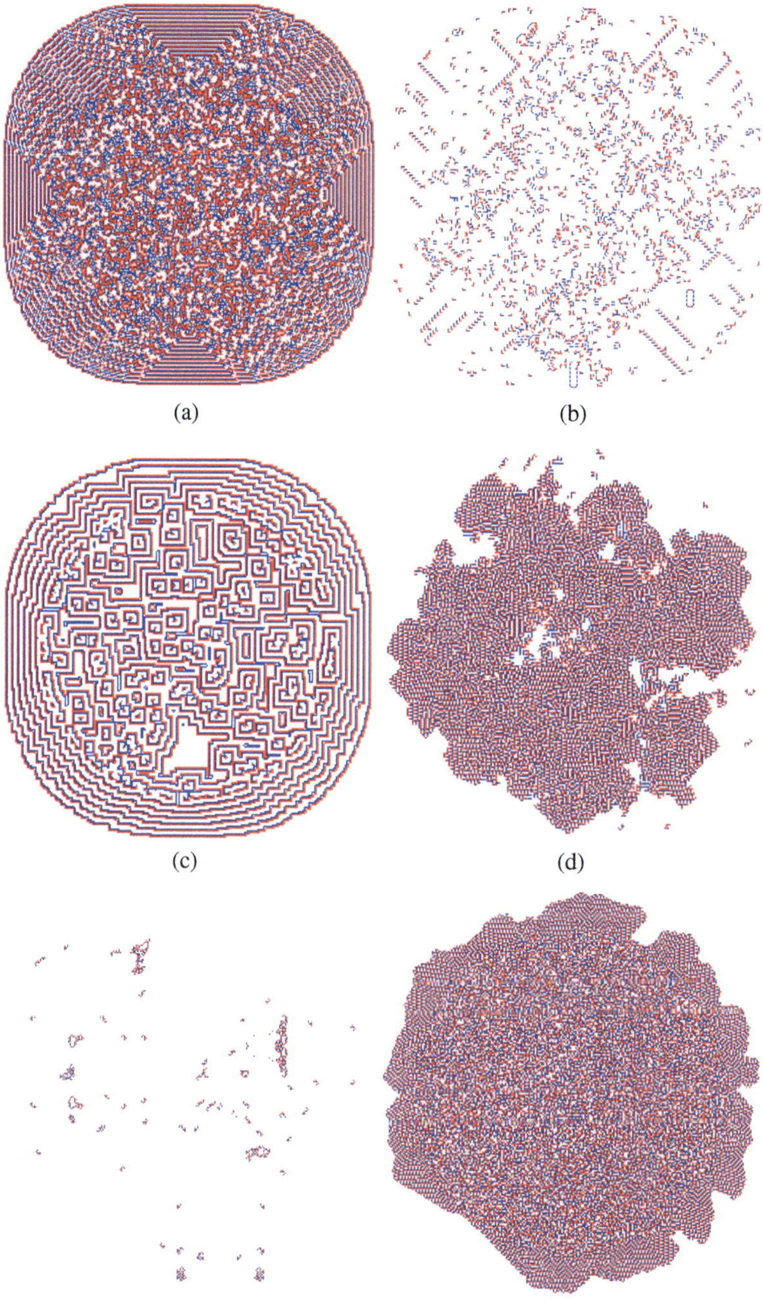

(a) (b) (c) (d) (e) (f)

Fig. 4.1 Snapshots of exemplar configurations generated by functions (a) $R(1306)$, (b) $R(1117)$, (c) $R(1401)$, (d) $R(1428)$, (e) $R(2413)$, (f) $R(2518)$. Cellular automaton 300×300 cells, initially each cell less 100 cells from center of the lattice took one of the state at random.

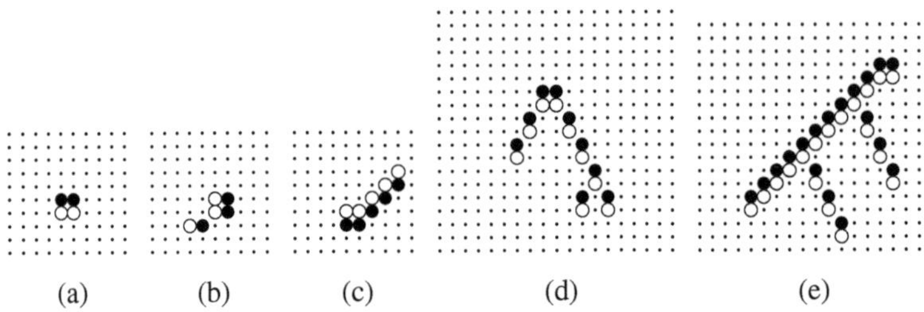

Fig. 4.2 Mobile localisations (a) E_2^2 traveling North, and those derived from E_2^2: (b) shortest extended localisation traveling East, (c) typical extended localisations traveling South, (d) branching extended localisation traveling North, (e) multiply branched traveling North

4.2 Mobile Localisations

When classifying localisations we use term weight. A localisation's weight is a number of cells occupied by the localisation's non-resting states (excitation and refractory states).

Minimal Gliders, Weight Four

The minimal mobile localisation E_2^2 found consists of two excitation states (excitation head) and two refractory states (refractory tail); the localisation travels excited-head forward (Fig. 4.2a). This is the only localisation found in the rule $R(2200)$, it is also present but not unique in $R(2201)$. Localisation E_2^2 and its variation with three excited and three refractory states, are the only gliders in rules $R(2300), R(2400), R(2500), R(2600), R(2700)$, and $R(2800)$. Varieties of the localisation E_2^2 – extended wave-fragments – are shown in Fig. 4.2b–d; they are typical for rules $R(2201)$, $R(2301)$, $R(2401)$, $R(2501)$, $R(2601)$, $R(2701)$ and $R(2801)$. In the rules $R(2202)$, $R(2302)$, $R(2402)$, $R(2502)$, $R(2602)$, $R(2702)$, $R(2802)$ branching localisations are most common, length of lateral branches can vary (Fig. 4.2e).

The glider E_2^2 has weight four states, and speed equivalent to the 'speed of light', i.e. moves along horizontal and vertical rows of the cellular-automaton array. There is another minimal mobile localisation with two excited and two refractory states, the glider $_DE_2^2$, which travels along diagonals of the lattice with speed $\frac{1}{2}$ (Fig. 4.3). The glider is typical for rules $R(1112)$, $R(1113)$, $R(1114)$, $R(1115)$, $R(1116)$, $R(1117)$, and $R(1118)$.

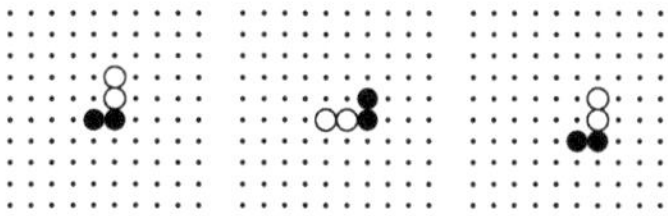

(a) t (b) $t+1$ (c) $t+2$

Fig. 4.3 Glider ${}_{D}E_2^2$ moving South-East

Gliders with Weight Six

Amongst gliders with constant weight six gliders G_{5a}^1 and G_{5b}^1, period two, (Fig. 4.4) have most 'unbalanced' ratio of excited to refractory states. At one step it has five excited and one refractory, and the next step one excited and five refractory states. The gliders change their global state with period two.

These gliders exist in rules $R1215$, $R(1216)$, $R(1217)$, and $R(1218)$.

Glider G_3^3, period four, always has three excited and three refractory states (Fig. 4.4c). This is the only glider ever observed in rules $R(2213)$, $R(2215)$, $R(2216)$, $R(2217)$, and $R(2218)$; and, one of many mobile localisations observed in rules $R(2613)$, $R(2813)$, and $R(2313)$.

Several gliders G_3^3 may couple together to form extended wave-fragments (Fig. 4.5).

Glider E_3^3 is the slowest glider amongst those with weight six: it changes its configuration over period of eight steps (Fig. 4.4d). The glider is very rare but can be found in development of initial random configurations in rules $R(1422)\dots R(1428)$, and rules $R(1522)$, $R(1622)$, $R(1722)$, $R(1822)$, $R(2214)$, and $R(2413)$.

Glider with Weight Changing between Six and Eight

The glider G_{68} alters its configuration in four time steps, during which his weight is varied from six to eight and back (Fig. 4.6). This mobile localisations are typically found in configurations generated by rules $R(1112), R(1113), \dots, R(1118)$.

Glider with Weight Seven or Eight

Cellular automaton governed by rule $R(1323)$ exhibits glider G_{78} (Fig. 4.2), period eight, which updates its weight in the following cyclic pattern $8 \to 8 \to 7 \to 7 \to 8 \to 8 \to 8 \to 7 \to 8$.

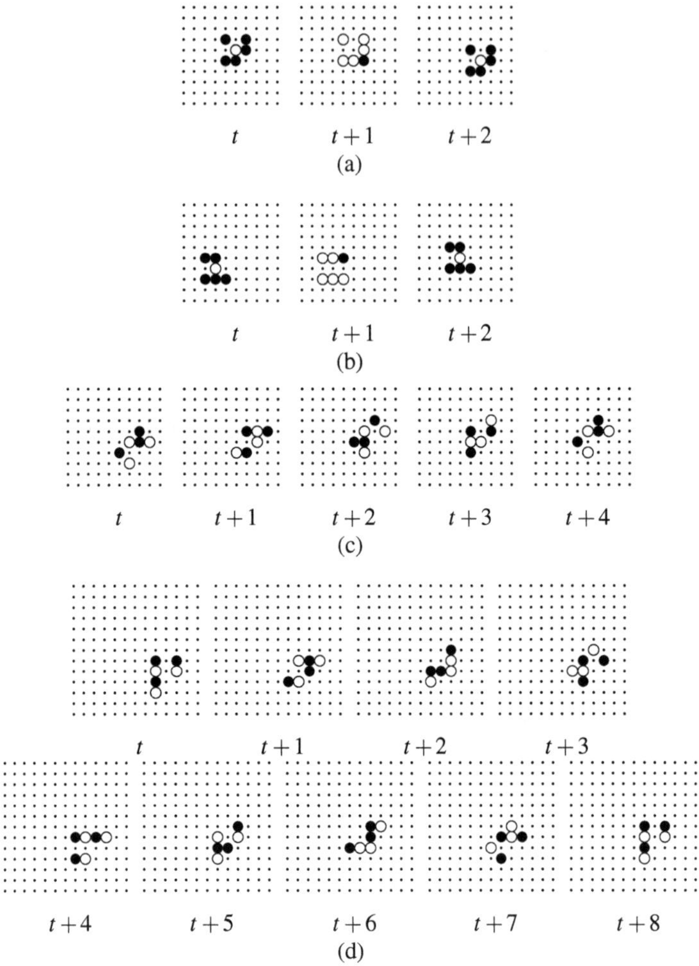

Fig. 4.4 Gliders with constant weight six: (a) glider G^1_{5a} traveling South-East, (b) glider G^1_{5b} traveling North-East, (c) glider G^3_3 traveling South-West, (d)- glider E^3_3 traveling North-West

Gliders with Constant Weight Eight

Rule $R(1214)$ supports glider G_8 weighting eight states, and period two (Fig. 4.8a).

Second glider weighting eight states, E_8, period four, observed in configurations of cellular automaton governed by rule $R(2313)$ (Fig. 4.8b).

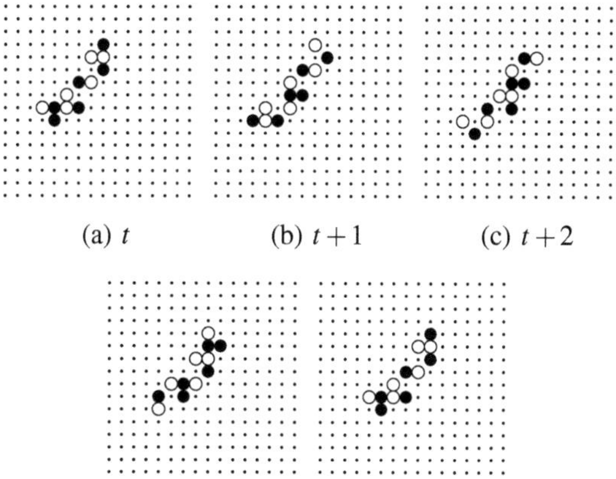

(a) t (b) $t+1$ (c) $t+2$

(d) $t+3$ (e) $t+4$

Fig. 4.5 Two gliders G_3^3 are associated together as a single wave-fragment

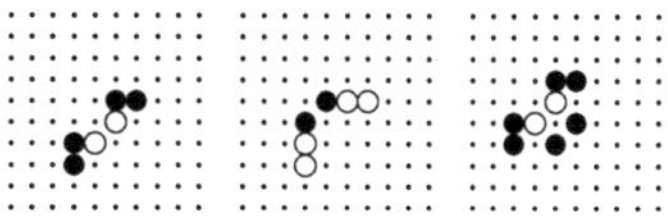

(a) t (b) $t+1$ (c) $t+2$

Fig. 4.6 Glider G_{68} traveling North-West

Gliders with Weight Switching between Eight and Ten

Glider $G_{8/10}$, (Fig. 4.9) period four, changes its weight in the sequence: $10 \to 8 \to 8 \to 10 \to 10$, corresponding variations in number of excited states are $4 \to 4 \to 4 \to 6 \to 4$. The glider $G_{8/10}$ is a typical mobile localisation for rules $R(2313)$, $R(2413)$, and $R(2513)$.

Gliders with Weight Switching between Nine and Ten

Glider $G_{9/10}$ (Fig. 4.10), period eight, changes its weight in the following sequence $10 \to 9 \to 9 \to 10 \to 10 \to 9 \to 9 \to 10 \to 10$. Such localisations are usually found in configurations generated by rules $R(1523)$, $R(1623)$, $R(1723)$.

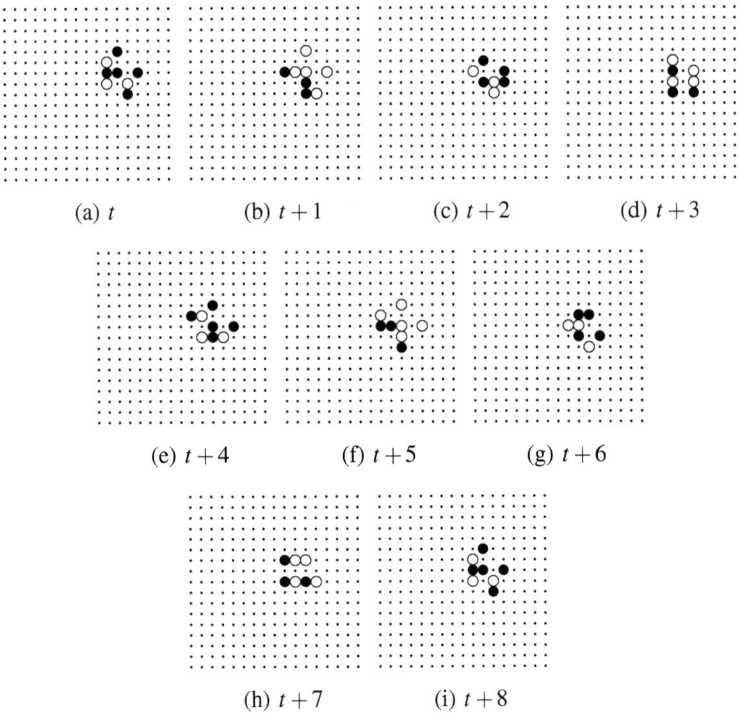

Fig. 4.7 Glider G_{78} traveling South-West

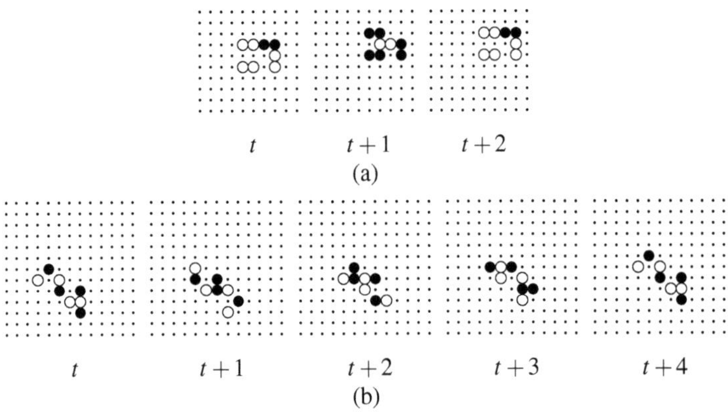

Fig. 4.8 Gliders with weight eight: (a) glider G_8 traveling South-East, (b) glider E_8 traveling North-East

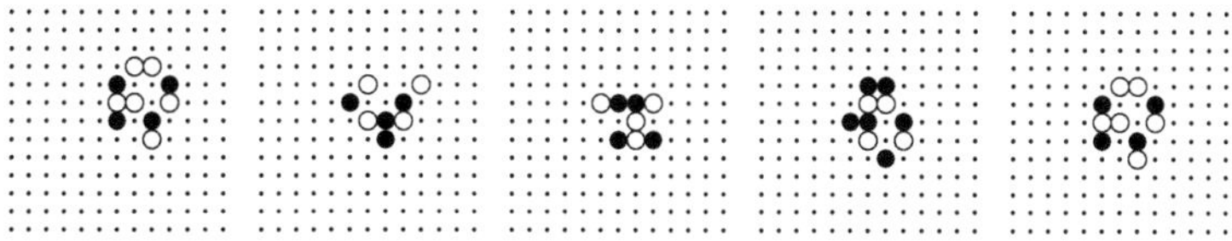

(a) t (b) $t+1$ (c) $t+2$ (d) $t+3$ (e) $t+4$

Fig. 4.9 Glider $G_{8/10}$ traveling South-West

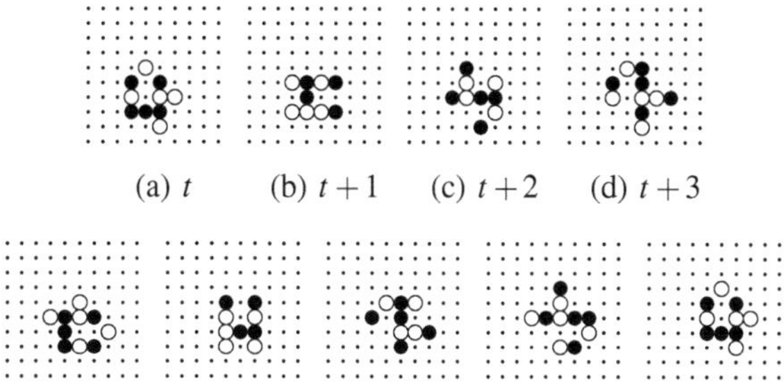

(a) t (b) $t+1$ (c) $t+2$ (d) $t+3$

(e) $t+4$ (f) $t+5$ (g) $t+6$ (h) $t+7$ (i) $t+8$

Fig. 4.10 Glider $G_{9/10}$ traveling North-West

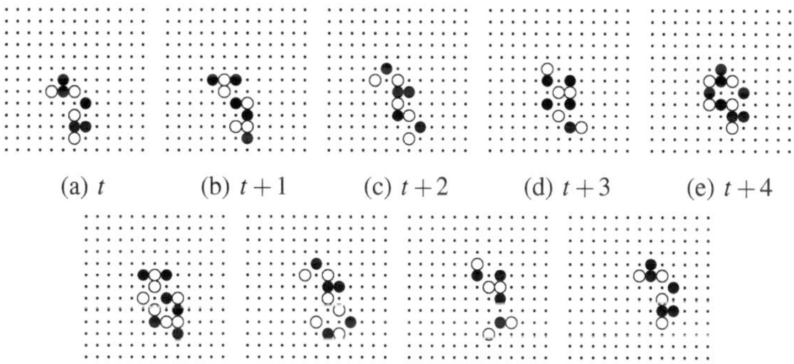

(a) t (b) $t+1$ (c) $t+2$ (d) $t+3$ (e) $t+4$

(f) $t+5$ (g) $t+6$ (h) $t+7$ (i) $t+8$

Fig. 4.11 Glider $G_{9/13}$ traveling North-East

Glider with Weight Switching between Nine and Thirteen

Glider $G_{9/13}$ (Fig. 4.11), period eight, is observed in rules $R(2313)$ and $R(2413)$. It is average weight oscillates around ten and reaches its maximum thirteen just once in the configuration cycle: $9 \rightarrow 10 \rightarrow 10 \rightarrow 10 \rightarrow 12 \rightarrow 13 \rightarrow 11 \rightarrow 9 \rightarrow 9$.

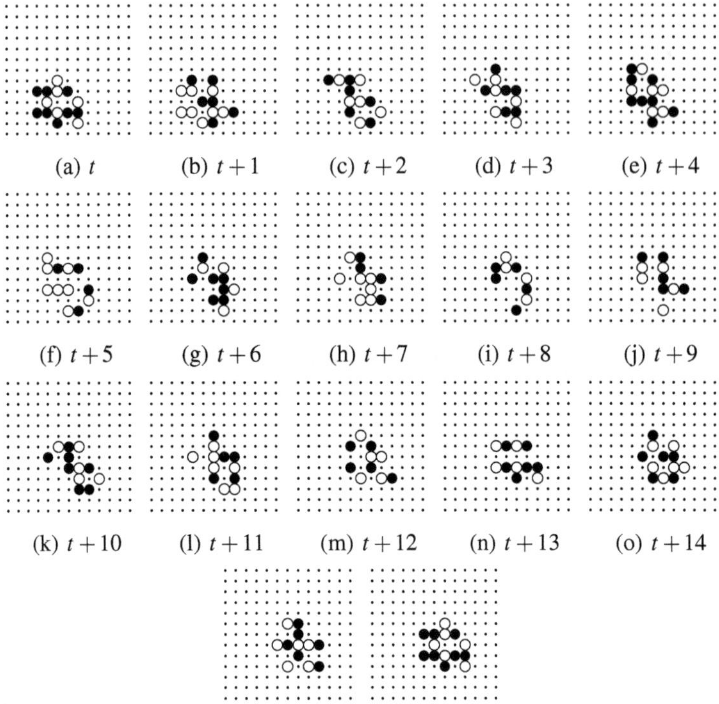

(a) t (b) $t+1$ (c) $t+2$ (d) $t+3$ (e) $t+4$

(f) $t+5$ (g) $t+6$ (h) $t+7$ (i) $t+8$ (j) $t+9$

(k) $t+10$ (l) $t+11$ (m) $t+12$ (n) $t+13$ (o) $t+14$

(p) $t+15$ (q) $t+16$

Fig. 4.12 Glider $G_{(9/14)}$ traveling North-East

Glider with Weight Changing between 9 and 14

The glider $G_{9/14}$ has one of the longest period, it repeats its configuration only once per 17 steps of evolution (Fig. 4.12), observed in rule $R(1423)$. Weight of the glider cycles as follows: $14 \rightarrow 14 \rightarrow 12 \rightarrow 12 \rightarrow 14 \rightarrow 12 \rightarrow 11 \rightarrow 11 \rightarrow 9 \rightarrow 10 \rightarrow 12 \rightarrow 12 \rightarrow 10 \rightarrow 11 \rightarrow 12 \rightarrow 12 \rightarrow 14$.

Glider with Constant Weight 11

The glider G_{11} has period two, and oscillates between configurations with nine excited and two refractory, and two excited and nine refractory states (Fig. 4.2). This glider can be found in the rules $R1315$, $R(1316)$, $R(1317)$.

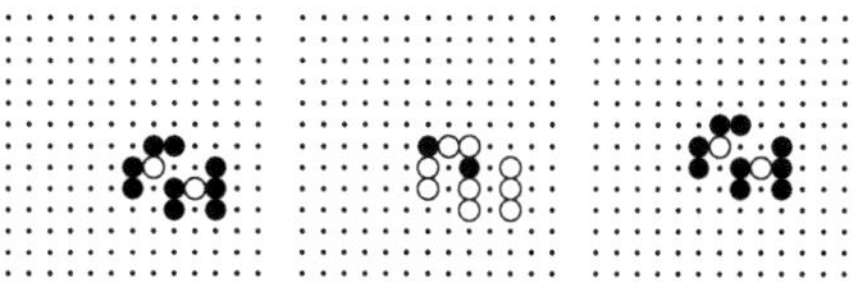

(a) t (b) $t+1$ (c) $t+2$

Fig. 4.13 Glider G_{11} traveling North-West

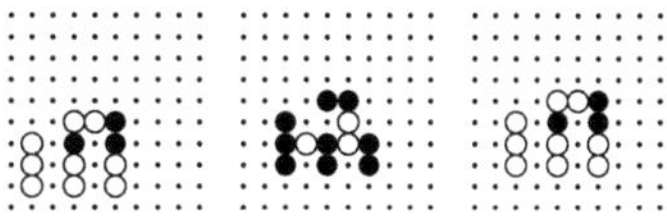

(a) t (b) $t+1$ (c) $t+2$

Fig. 4.14 Glider G_{12} traveling North-East

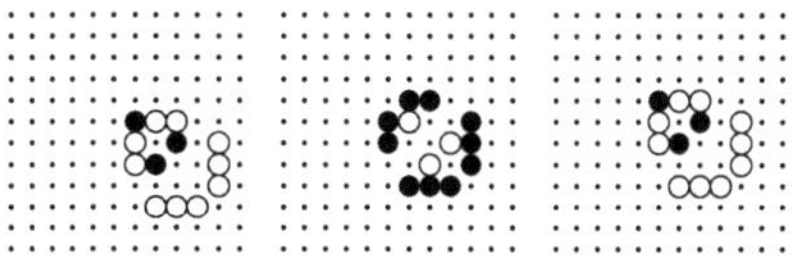

(a) t (b) $t+1$ (c) $t+2$

Fig. 4.15 Glider G_{13} traveling North-West

Glider with Constant Weight 12

The glider G_{12} with constant weight 12 is observed in rules $R(1316)$, $R(1317)$, $R(1318)$. While overall weight of the glider is preserved during evolution, and glider has period two, number of excited states in the glider changes as follows: $3 \rightarrow 9 \rightarrow 3$ (Fig. 4.14).

Glider with Constant Weight 13

Glider G_{13}, as well as G_{19} is a 'relative' of glider G^1_{5a}. The glider has period two, preserves its weight, but number of excited states in the glider configuration switches between three and ten (Fig. 4.15). The glider is typical for rule $R(1214)$.

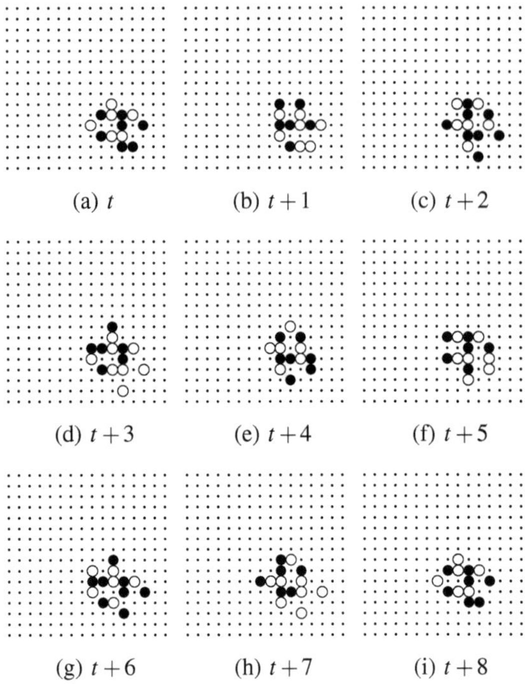

Fig. 4.16 Glider $G_{13/14}$ traveling North-West

Glider with Weight Switching between 13 and 14

Glider $G_{13/14}$ (Fig. 4.16) changes its weight between 13 and 14 during its life cycle of eight steps. The glider is found in rules $R(1523)$, $R(1623)$, and $R(1723)$. The glider's weight varies as follows: $13 \rightarrow 13 \rightarrow 14 \rightarrow 14 \rightarrow 13 \rightarrow 13 \rightarrow 14 \rightarrow 14 \rightarrow 13$.

Glider with Constant Weight 15

Glider G_{15}, weight 15, period two, is one of few gliders traveling along rows or columns of the cellular-automaton lattice (Fig. 4.17). This is a glider typical for rules $R(2203)$, $R(2303)$, $R(2403)$, $R(2603)$, $R(2703)$, $R(2803)$, $R(2201$, and $R(2202)$. The glider is minimal of this type, however it has many heavier variations with larger configurations of tails.

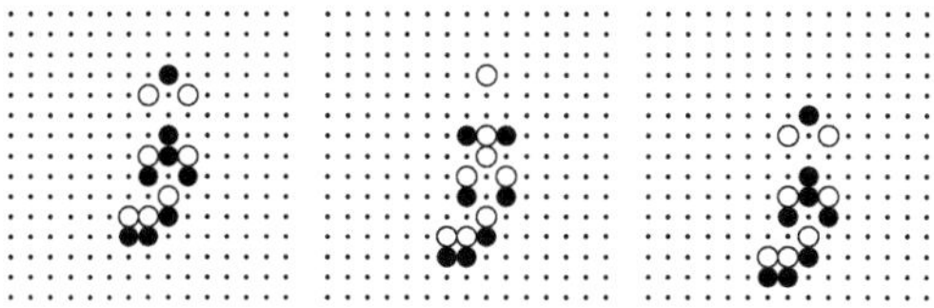

(a) t (b) $t+1$ (c) $t+2$

Fig. 4.17 Glider G_{15} traveling South

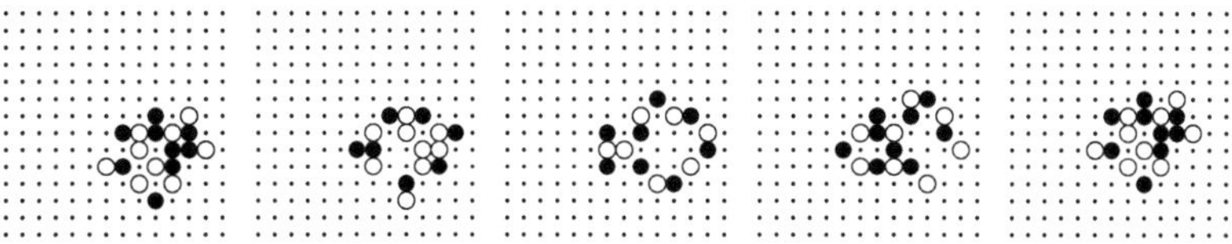

(a) t (b) $t+1$ (c) $t+2$ (d) $t+3$ (e) $t+5$

Fig. 4.18 Glider $G_{15/18}$ traveling North-West

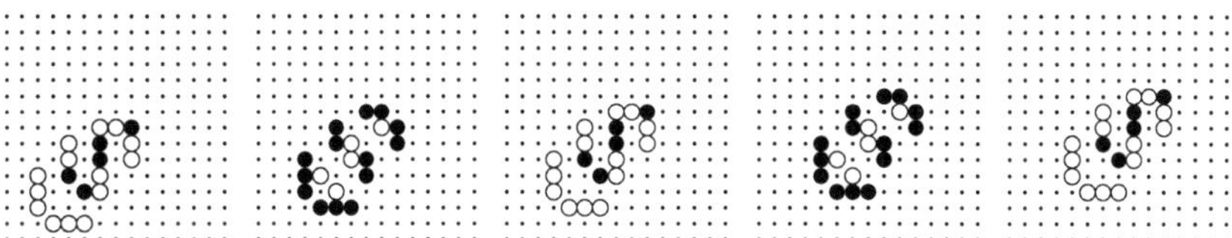

(a) t (b) $t+1$ (c) $t+2$ (d) $t+3$ (e) $t+4$

Fig. 4.19 Glider G_{19} traveling South-East

Glider with Weight Switching between 15 and 18

Glider $G_{15/18}$, found in rules $R(2813)$, $R(2713)$ and $R(2613)$, has period four, and its weight varies as follows: $18 \to 16 \to 15 \to 15 \to 18$ (Fig. 4.18).

Glider with Constant Weight 19

The glider G_{19} relates to G^1_{5a} and G_{13}. It has period four, preserves weight, and switches in number of excited states between 5 and 14 (Fig. 4.2). Cellular automata governed by rules $R(1213)$ and $R(1214)$ exhibit such glider in their development.

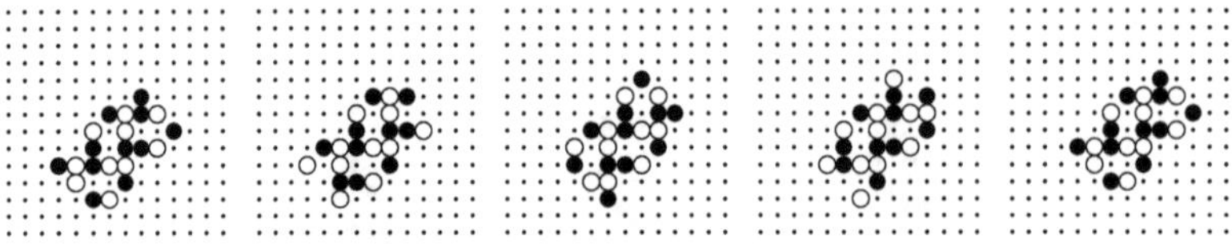

(a) t (b) $t+1$ (c) $t+2$ (d) $t+3$ (e) $t+4$

Fig. 4.20 Glider $G_{20/22}$ traveling North-East

Glider Changing Its Weight from 21 to 22

Glider $G_{20/22}$ (Fig. 4.20), period four, observed in rules $R(2314)\dots R(2318)$, alters its weight gradually as $21 \to 21 \to 22 \to 21 \to 21$. Its excitation component switches between 10 and 11 every two steps of automaton evolution as follows: $11 \to 10 \to 11 \to 10 \to 11$.

4.3 Stationary Localisations

Localisations with Weight Seven to Eight

The smallest stationary localisation L_4^3 (Fig. 4.21abcd) updates its structure with period three, and consists of three excitation states and four refractory states (Fig. 4.21b), or three refractory states and four excited states (Fig. 4.21c). It has four refractory and four excited states in its largest form (Fig. 4.21ad). The localisation is a typical one in rules $R(1423)$ and $R(1422)$.

The localisation L_7^a (Fig. 4.21i–m), typical for rule $R(1622)$ oscillates with period four. The localisation preserves its weight, however number of excited states changes as follows: $3 \to 4 \to 3 \to 4 \to 3$.

Localisation with Weight Eight

The second minimal stationary localisation L_4^4 (Fig. 4.21efgh), consists of four excited and four refractory states, and 'oscillates' with period three. This is the only localisations found in rule $R(2323)$, $R(2324)$, $R(2325)$, $R(2326)$, $R(2327)$, $R(2328)$, $R(2423)$, $R(2523)$, $R(2723)$, $R(2823)$, and L_4^4 is one of two observable localisations in rules $R(2424)$, $R(2524)$, $R(2425)$, $R(2426)$, $R(2427)$, $R(2525)$, $R(2526)$, $R(2527)$, $R(2528)$, $R(2622)$, $R(2623)$, $R(2626)$, $R(2627)$, $R(2628)$, $R(2724)$, $R(2725)$, $R(2726)$, $R(2727)$, $R(2824)$, $R(2825)$, $R(2826)$, $R(2827)$, $R(2828)$, and one of several localisations in rules $R(2214)$, $R(2215)$, $R(2216)$.

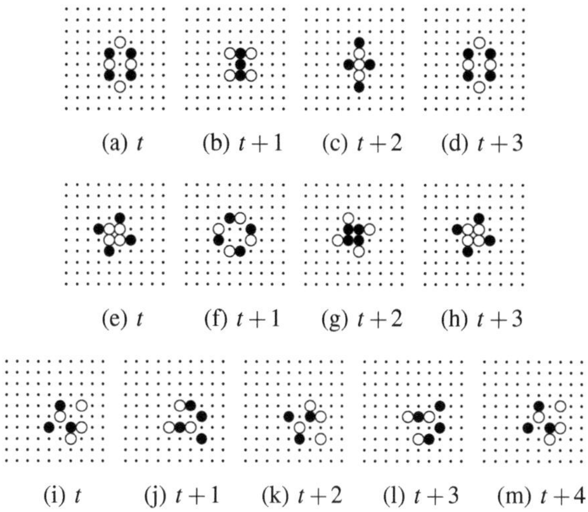

Fig. 4.21 (a)–(d) Stationary localisation L_4^3. (e)–(h) Stationary localisation L_4^4, (i)–(m) stationary localisation L_7^a.

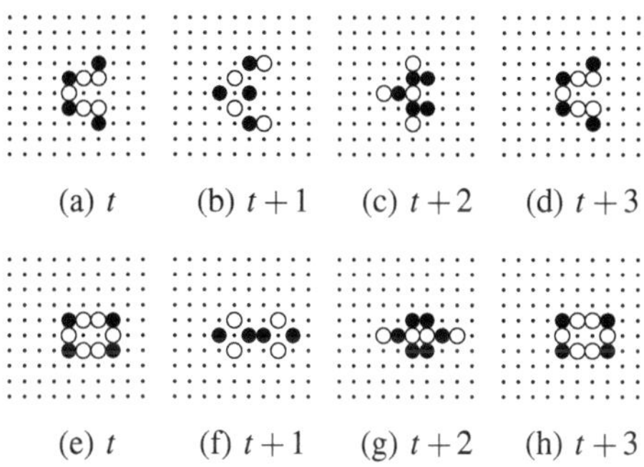

Fig. 4.22 localisations weighting between eight and ten. (a)–(d) localisation $L_{8/9}$; (e)–(h) localisation $L_{8/10}$.

Localisation with Weight Switching between Eight and Ten

Rules $R(2215)\ldots R(2218)$ support two stationary localisations weighting between eight and ten. $L_{8/9}$, which weight oscillates, has period three, and updates its weight as follows $9 \to 8 \to 9 \to 9$ (Fig. 4.22a–d). localisation $L_{8/10}$ changes its weight in the following cycle: $10 \to 8 \to 10 \to 8$ (Fig. 4.22e–h).

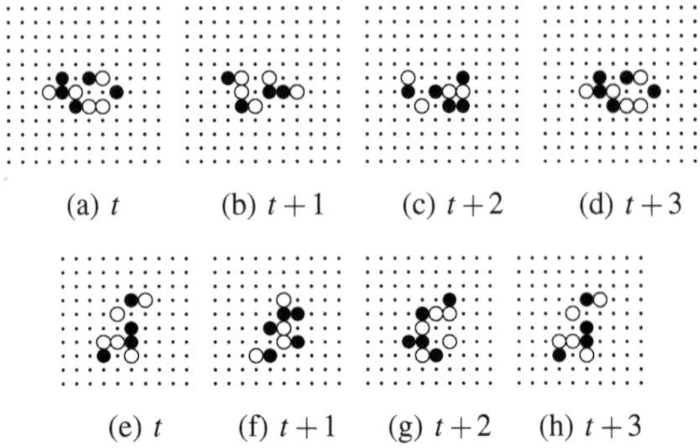

Fig. 4.23 Localisations with weight varying between nine and ten: (a)–(d) $L^a_{9/10}$, (e)–(h) $L^b_{9/10}$

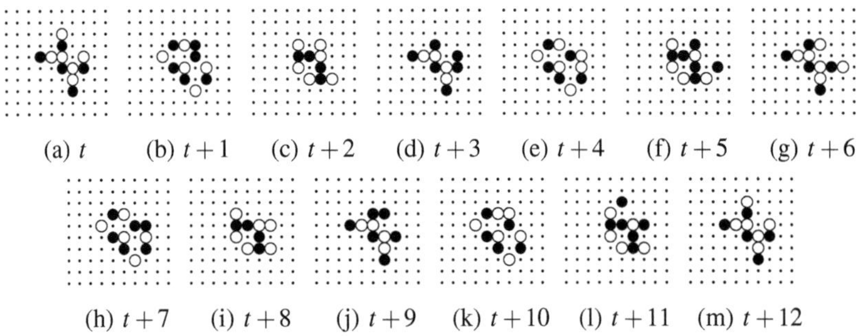

Fig. 4.24 Stationary localisation $L_{10/11}$

Localisations with Weight Switching between Nine and Ten

Localisations weighting between nine and ten are quite common amongst the rules studied. These are localisations $L^a_{9/10}$, rules $R(2214)$ and $R(2413)$, (Fig. 4.23a–d), and $L^b_{9/10}$, rules $R(2215) \dots R(2218)$, (Fig. 4.23e–h). Both localisations oscillate with period four, and their weights are changed as follows: $10 \rightarrow 9 \rightarrow 9 \rightarrow 10$ ($L^a_{9/10}$) and $9 \rightarrow 9 \rightarrow 10 \rightarrow 9$ ($L^b_{9/10}$).

Stationary Localisations with Weight Switching between 10 and 11

Localisation $L_{10/11}$ is observed in rule $R(1423)$. It oscillates with period 12, and changes its weight as $11 \rightarrow 11 \rightarrow 10 \rightarrow 10 \rightarrow 11 \rightarrow 11 \rightarrow 11 \rightarrow 11 \rightarrow 10 \rightarrow 10 \rightarrow 11 \rightarrow 11 \rightarrow 11$ (Fig. 4.24).

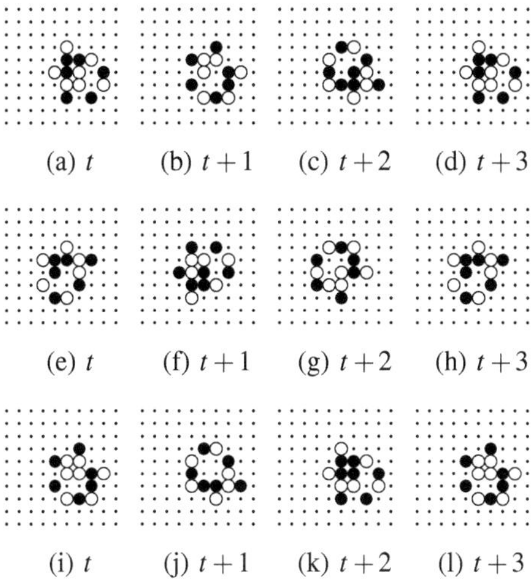

Fig. 4.25 Localisation weighting between 12 and 13: (a)–(d) $L^a_{12/13}$, (e)–(h) $L^b_{12/13}$ (i)–(l) $L^c_{12/13}$

Localisations with Weight Switching between 12 and 13

Rule $R(2624)$ exhibits localisation $L^a_{12/13}$ oscillating with period three. The pattern changes its weight in the following sequence: $13 \rightarrow 12 \rightarrow 13 \rightarrow 13$. Despite oscillations of weight number of excited states stays constant six (Fig. 4.25a–d). In localisation $L^b_{12/13}$ number of excited states is always seven, and weight oscillates with period three as follows: $12 \rightarrow 13 \rightarrow 13 \rightarrow 12$ (Fig. 4.25e–h).

Pulsating localisation $L^c_{12/13}$ is observed in rule R(2525 and $R(2725)$, the localisation loses its central core becoming ring-line, then restores its centre in cycling configuration change of four steps (Fig. 4.25i–l.)

Localisations $L^d_{12/13}$, observed in rule $R(2827)$, (Fig. 4.26a–d), and $L^e_{12/13}$, rule $R(2828)$, (Fig. 4.26e–h), very similar to previously described, hence differ in few sites of their configurations.

Also, localisations $L^f_{12/13}$ (rule $R(2427)$), $L^g_{13/13}$ (rules $R(2424)$, $R(2426)$, $R(2427)$) are shown in Fig. 4.27.

Localisation $L^g_{12/13}$ (Fig. 4.27i–), observed in rule $R(2824)$, resembles in its behaviour non-expanding spiral wave in excitable systems, rotating clock-wise. Another localisation, $L^h_{12/13}$, populating lattice of cellular automata governed by rule $R(2826)$ is shown in Fig. 4.28.

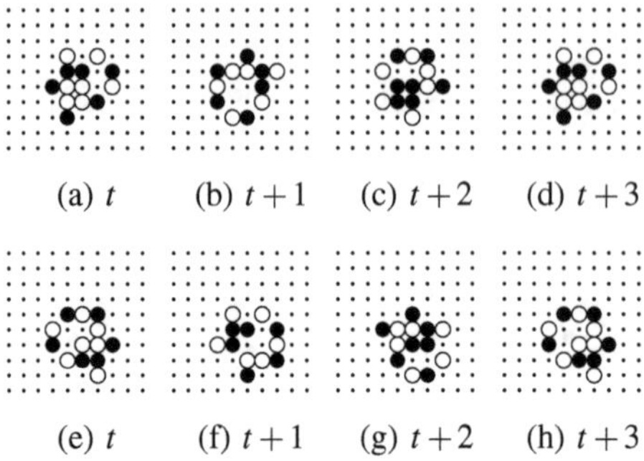

Fig. 4.26 (a)–(d) Localisation $L^d_{12/13}$, (e)–(h) localisation $L^e_{12/13}$

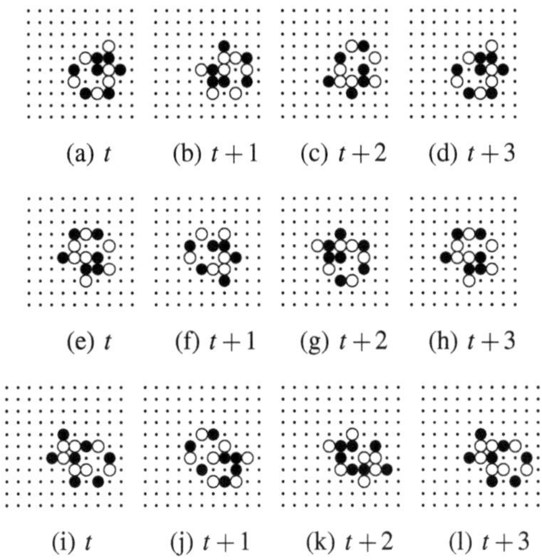

Fig. 4.27 (a)–(d) Localisation $L^f_{12/13}$, (e)–(h) localisation $L^f_{12/13}$, (i)–(l) localisation $L^g_{12/13}$

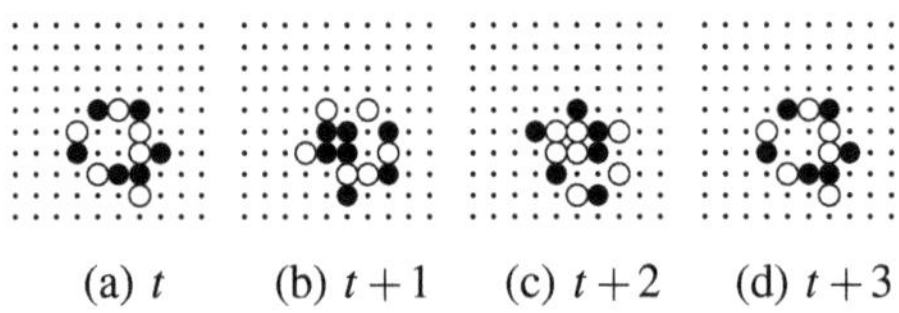

Fig. 4.28 Localisation $L^h_{12/13}$

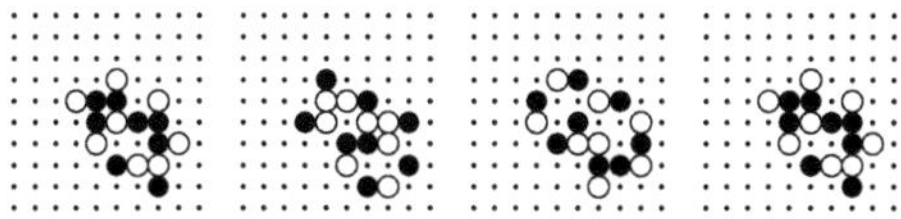

(a) t (b) $t+1$ (c) $t+2$ (d) $t+3$

Fig. 4.29 Localisation L_{16}

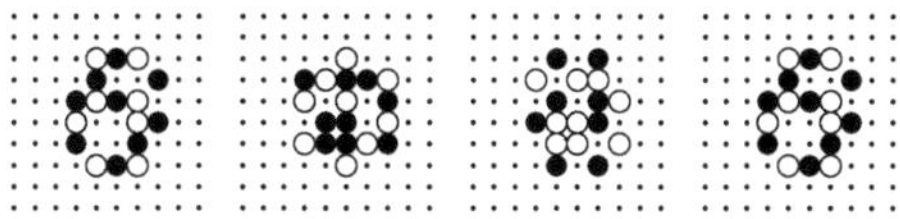

(a) t (b) $t+1$ (c) $t+2$ (d) $t+3$

Fig. 4.30 Localisation $L_{17/18}$

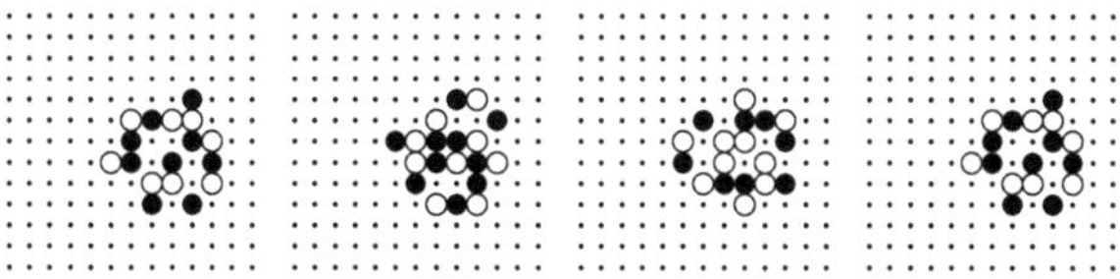

(a) t (b) $t+1$ (c) $t+2$ (d) $t+3$

Fig. 4.31 Localisation $L_{17/19}$

Localisation with Constant Weight 16

Amongst few stationary localisations with conserved weight, localisation L_{16} is discovered in development of cellular automaton governed by rule $R(2526)$ (Fig. 4.29). Not only the localisation keeps its weight constant but the number of excited and refractory states in the localisation's configuration is preserved as well.

Localisation with Weight Switching between 17 and 18

Rule $R(2725)$ supports localisation $L_{17/18}$, which alters its weight between 17 to 18 and back to 17 (Fig. 4.30).

Localisation with Weight 17 to 19

localisation with weight oscillating between 17 and 19 is observed in rule $R(2428)$ (Fig. 4.31). During four steps of the localisation life-cycle its weight changes as $17 \rightarrow 19 \rightarrow 18 \rightarrow 17$.

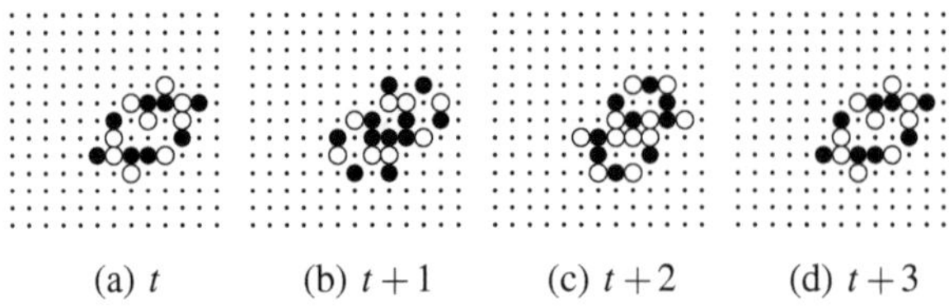

Fig. 4.32 Localisation $L_{17/20}$

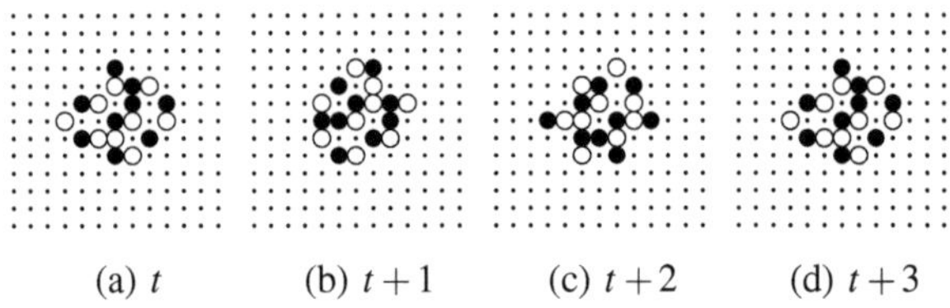

Fig. 4.33 Localisation L_{18}

Localisation with Weight between 17 and 20

localisation $L_{17/20}$ keeps its weight changing as $17 \to 19 \to 20 \to 17$ (Fig. 4.32); however, its excited component oscillates as $8 \to 11 \to 9 \to 8$. This localisation can be found in rule $R(2525)$.

Localisation with Weight 18

Localisation L_{18} populates lattices of cellular automata governed by rule $R(1823)$ (Fig. 4.33). It oscillates with period three and preserves its weight. Ratio of excited and refractory states in the localisation's configuration is constant as well.

Localisation with Weight between 18 and 19

Rule $R(2625)$ supports stationary localisation $L_{18/19}$, which weight increases to 19 and then drops back to 18 during the localisation's life-cycle: $18 \to 19 \to 19 \to 18$ (Fig. 4.34). The corresponding changes in excited states are $9 \to 10 \to 9 \to 9$.

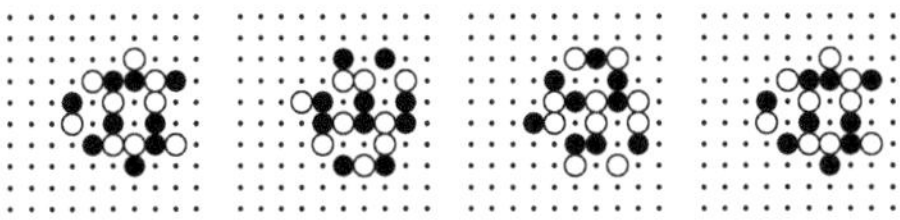

(a) t (b) $t+1$ (c) $t+2$ (d) $t+3$

Fig. 4.34 Localisation $L_{18/19}$

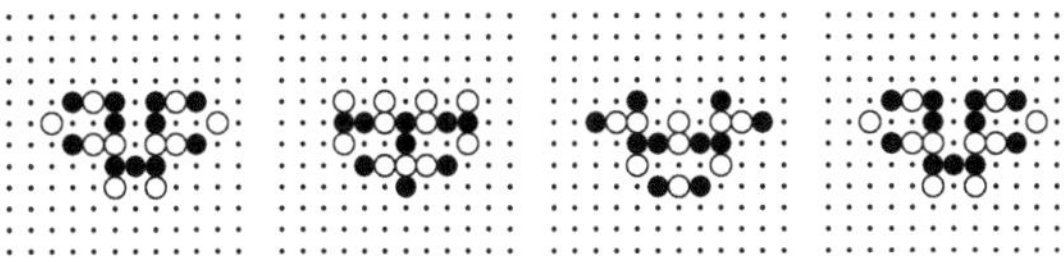

(a) t (b) $t+1$ (c) $t+2$ (d) $t+3$

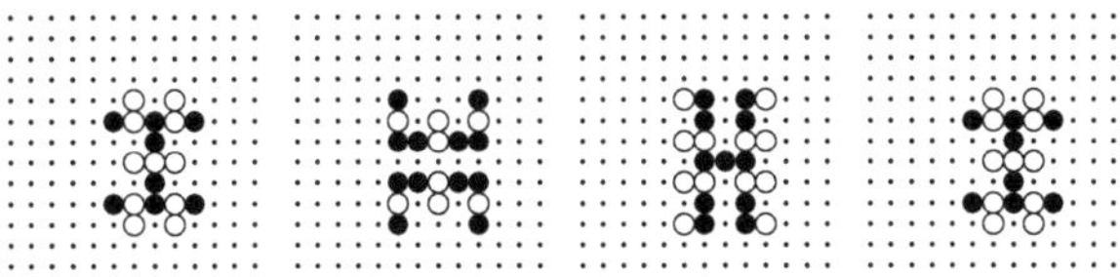

(e) t (f) $t+1$ (g) $t+2$ (h) $t+3$

Fig. 4.35 (a)–(d) Localisation $L_{19/21}$, (e)–(h) localisation $L_{19/23}$

Localisation with Weight Changing between 19 and 23

Cellular automata governed by rules $R(1523)$, $R(1623)$, $R(1723)$ produce localisation $L_{19/23}$ and $L_{19/21}$, period three. Weight of localisation $L_{19/23}$ is changed in the localisation life-cycle as $19 \to 20 \to 23 \to 19$ (Fig. 4.35e–h), while localisation $L_{19/21}$ updates its weight as $21 \to 20 \to 19 \to 21$. These are amongst few symmetrical localisations.

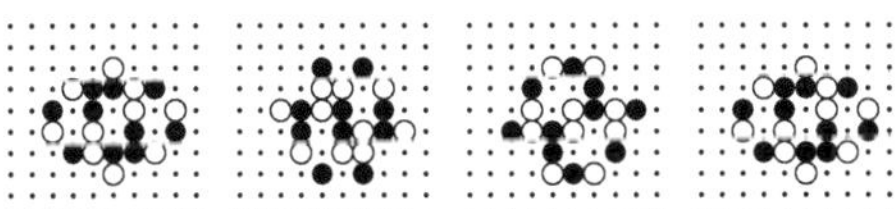

(a) t (b) $t+1$ (c) $t+2$ (d) $t+3$

Fig. 4.36 Localisation L_{20}

Localisation with Constant Weight 20

Localisation L_{20}, observed in rule $R(2724)$, does not change its weight in its three-step life-cycle (Fig. 4.36), total number of excited and refractory states also remains constant. So, we can call this localisation 'combinatorial'.

4.4 Huge Localisations

There is a variety of non-minimal localisations observed in cellular automata models of mutualistic excitation mode. We consider just most common, or in some cases most remarkable examples.

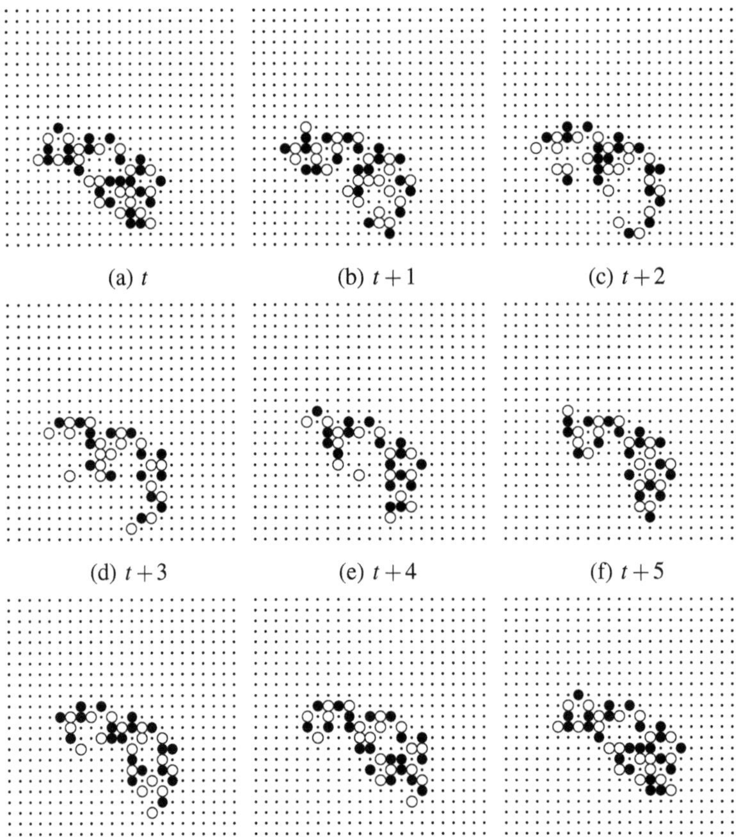

(a) t (b) $t+1$ (c) $t+2$ (d) $t+3$ (e) $t+4$ (f) $t+5$ (g) $t+6$ (h) $t+7$ (i) $t+8$

Fig. 4.37 Large mobile localisation observed in rule $R(2713)$, traveling North-East

Huge Mobile Localisations

Huge mobile localisations, with weight over 30, are not uncommon in the rules investigated. They are however very sensitive to perturbations. If not taken care of immediately after birth, the localisations are usually destroyed in collision with smaller localisations and gliders. An example of very large mobile localisation weight of which exceeds 43 is show in Fig. 4.37. It has period eight and moves North-East.

The large traveling localisation shown in Fig. 4.38 resembles propagating wave-fragments in sub-excitable Belousov-Zhabotinsky system [14, 219]. A wave-fragment growth from the centre (Fig. 4.38kl), excitation spreads towards ends of the fragment (Fig. 4.38lm), where it dies (Fig. 4.38n).

Even more complicated wave-fragment is shown in Fig. 4.39. Head of the pattern has a hollow structure, with excitation developing on the contour of the head, and thus causing forward motion to the whole pattern.

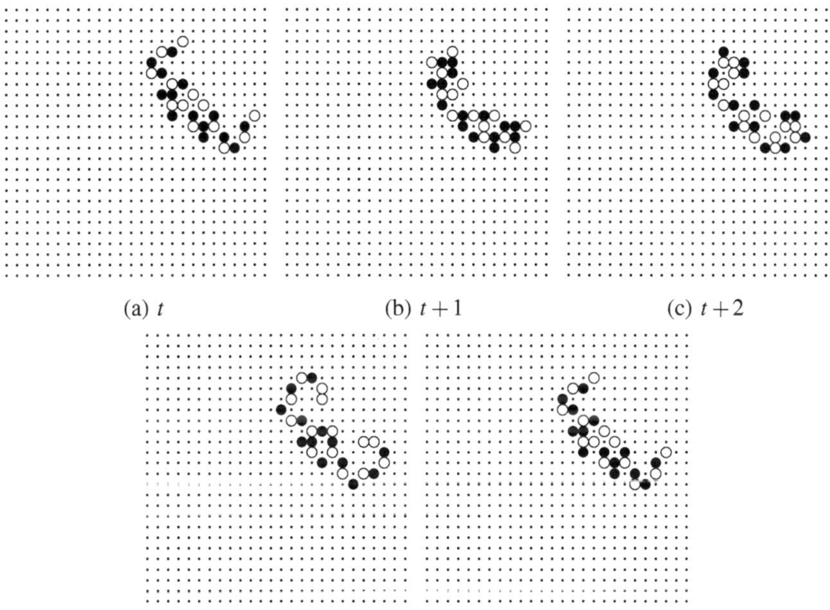

(a) t (b) $t+1$ (c) $t+2$ (d) $t+3$ (e) $t+4$

Fig. 4.38 Large mobile localisation discovered in rule $R(2813)$, it also exists in rules $R(2613)$ and $R(2713)$

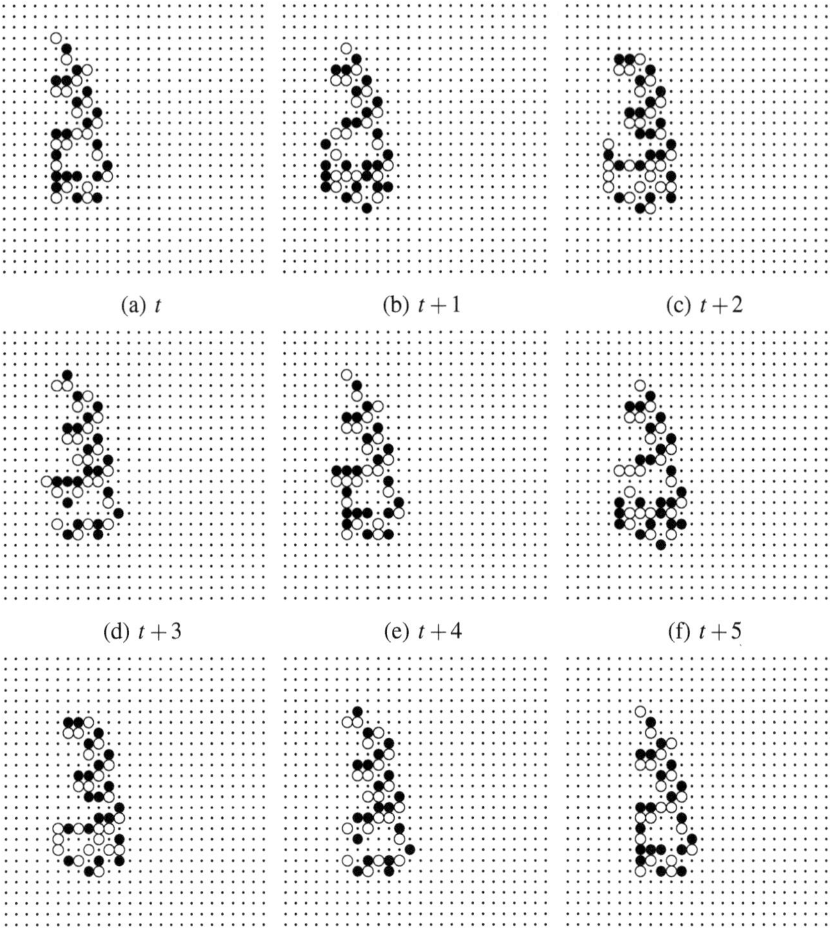

(a) t (b) $t+1$ (c) $t+2$

(d) $t+3$ (e) $t+4$ (f) $t+5$

(g) $t+6$ (h) $t+7$ (i) $t+8$

Fig. 4.39 Complex traveling pattern observed in rule $R(2813)$, also can be found in rules $R(2814)$ and $R(2818)$

Huge Stationary Localisations

In addition to small stationary localisations described above, rules $R(2827)$ and $R(2425)$ exhibits variety of large localisation, for example the localisation shown in Fig. 4.40. Both localisations have period four.

Snapshots of exemplar stationary localisations of different shapes observed in rules $R(2528)$, $R(2428)$, $R(2727)$, $R(2626)$ are shown in Fig. 4.41, period of most huge localisation is four. This indicates that a length of life-cycle does not correlated with size of the configuration. Shapes of the localisations do not change, and

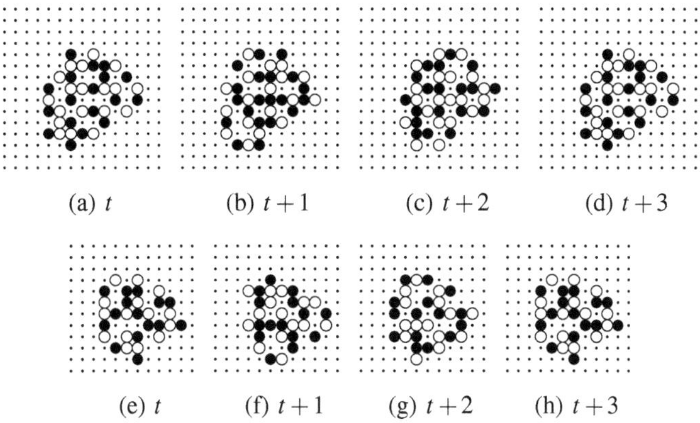

Fig. 4.40 Huge stationary localisations: (a)–(d) Stationary localisation with weight exceeding 35, exhibited in rule $R(2827)$, (e)–(h) Stationary localisation with weight exceeding 29, exhibited in rule $R(2425)$

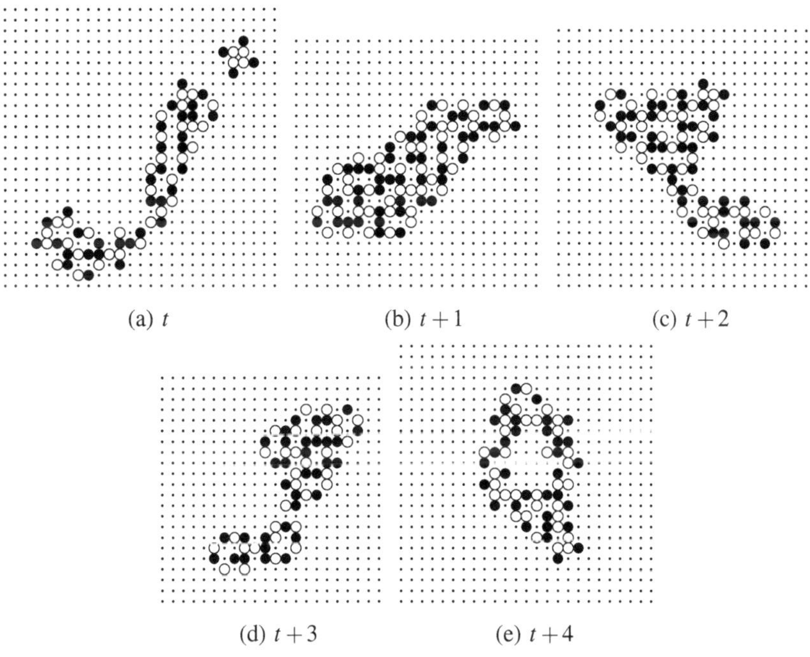

Fig. 4.41 Different shapes of huge stationary localisations: (a)–(b) two types of huge localisations observed in rule $R(2528)$, (c) rule $R(2428)$, (d) rule $R(2727)$, (e) rule $R(2626)$

oscillations in configurations are mainly to traveling excitation patters along the localisation contour. We can see in Fig. 4.41e that empty space inside the localisation is never filled.

Glider	Weight	Imbalance	Density	Period
E_2^2	4	0	1	1
$_DE_2^2$	4	0	2/3	2
G_{5a}^1	6	2/3	2/3	2
G_{5b}^1	6	2/3	1/2	4
G_3^3	6	0	3/8	3
E_3^3	6	0	3/8–1/2	4
G_{68}	6–8	1/3–1/2	3/8–1/2	3
G_{78}	7–8	1/7–1/4	7/16–2/5	8
G_8	8	1/2	2/3	2
E_8	8	0	3/10–8/25	4
$G_{8/10}$	8–10	0–1/5	1/3–1/2	4
$G_{9/10}$	9–10	0–1/9	1/2–3/4	8
$G_{9/13}$	9–13	0–1/9	13/24–3/8	8
$G_{9/14}$	9–14	0–1/7	2/5–14/25	16
G_{11}	11	7/11	11/20	2
G_{12}	12	3/4	3/5	2
G_{13}	13	10/13	13/25	2
$G_{13/14}$	13–14	0–1/7	7/18–13/30	8
G_{15}	15	1/15	3/8–5/12	2
$G_{15/18}$	15–18	0–1/15	5/14–3/7	4
G_{19}	19	14/19	19/49	4
$G_{20/22}$	20–22	0–1/21	3/8–11/28	4

Fig. 4.42 Characteristics of mobile localisations, gliders

4.5 Characterising Localisations

The following characteristics were used to analyse sets of localisations discovered. Weight is a number of non-resting cells in localisation. Imbalance is an absolute difference between excited and refractory states in a localisation divided by weight of the localisation. Density is localisation's weight divided by size of a rectangle occupied by the localisation. Period is a number of steps localisation needs to restore its state.

From the table in Fig. 4.42 we see that period of gliders is — in overall — increasing with increase of glider's weight till the weight reaches ten non-resting cells. Then period starts to decrease. Localisation with maximum period, longest time to restore their states, have weight around ten. Density, in general, decreases with the increase of glider's weight. The more balanced is glider the larger its period of oscillation. Highly imbalanced gliders usually change their state in small number of time steps.

Period of a stationary localisation does not depend on the localisation's weight at all, see Fig. 4.43. Imbalance and density of stationary localisations non-linearly depend on the localisations weight. Namely, both imbalance and density increases with increasing weight, till reach their maximum at localisations weighting around

localisation	Weight	Imbalance	Density	Period
L_4^3	7-8	0–1/7	7/15–7/9	3
L_4^4	8	0	1/2	3
L_7^a	7	0	7/16	4
$L_{8/9}$	8–9	0–1/9	2/5–9/15	3
$L_{8/10}$	8–10	0–1/5	1/2–10/12	3
$L_{9/10}^a$	9–10	0–1/9	5/9–9/15	3
$L_{9/10}^b$	9–10	0–1/9	9/20–1/2	3
$L_{10/11}$	10–11	1/11–1/5	11/30–5/8	12
$L_{12/13}$	12–13	0–1/13	12/25 –13/25	3
L_{16}	16	0	4/9	3
$L_{17/18}$	17–18	1/17	17/30	3
$L_{17/19}$	17–19	1/19–1/17	17/36–19/36	3
$L_{17/20}$	17–20	1/17	17/42–10/21	3
L_{18}	18	0	9/21	3
$L_{18/19}$	18–19	0–1/19	1/2–19/30	3
$L_{19/21}$	19–21	0–1/10	21/50–4/7	3
$L_{19/23}$	19–23	1/23–1/5	19/35–23/35	3
L_{20}	20	0	10/21	3

Fig. 4.43 Characteristics of stationary localisations

10. Then average values of these two characteristics decrease and arrive at their minimum for the localisations weighting 16 non-rest cells. After that both imbalance and density start to increase again (Fig. 4.43).

4.6 Excitation Rules Rich with Localisations

What are the richest — in terms of number of different species of localisations — excitation rules? The answer is in matrices of localisations' frequencies shown in Fig. 4.44. Each matrix $\mathbf{M} = (M_{ij})_{0 \leq i,j \leq 8}$ is constructed as follows. Initially, all entries of the matrix are nils. Then for each localisation, either it is mobile or stationary one, found in configurations generated by rule $R(abcd)$:
$M_{ij} \leftarrow M_{ij} + 1$ if $i \in [\theta_1, \theta_2]$ and $j \in [\delta_1, \delta_2]$. When all entries are updated, each entry M_{ij} shows a number of species of mobile or stationary localisations supported by rule: resting cell excites if it has i excited neighbours and j refractory neighbours.

From the matrices (Fig. 4.44) we see that rule $R(2222)$ is the most rich in localisations, it supports 11 species of gliders, and 17 species of stationary localisations. The automata governed by the excitation intervals which include rule $R(2222)$ are extended versions of the 2^+-medium, [2, 3, 9].

(a)

$[\theta_1,\theta_2]$ \ $[\delta_1,\delta_2]$	0	1	2	3	4	5	6	7	8
0	0	0	0	0	0	0	0	0	0
1	0	3	7	8	4	4	3	3	2
2	2	7	11	11	5	5	4	4	3
3	2	6	10	10	3	3	2	2	1
4	2	4	6	6	0	0	0	0	0
5	2	3	4	4	0	0	0	0	0
6	2	2	4	4	0	0	0	0	0
7	2	2	4	4	0	0	0	0	0
8	2	2	2	2	0	0	0	0	0

(b)

$[\theta_1,\theta_2]$ \ $[\delta_1,\delta_2]$	0	1	2	3	4	5	6	7	8
0	0	0	0	0	0	0	0	0	0
1	0	4	2	0	0	0	0	0	0
2	4	17	15	11	0	10	5	5	4
3	1	14	13	9	7	3	3	2	0
4	1	14	13	8	7	3	3	2	0
5	0	12	10	8	6	2	2	1	0
6	0	11	9	7	5	2	2	1	0
7	0	5	5	3	2	1	1	0	0
8	0	2	2	1	1	1	1	0	0

Fig. 4.44 Localisations frequency matrices for (a) mobile localisations, or gliders, and (b) stationary localisations

For mobile localisations 'richness' has its maximum in $R(2323)$ (10 and 11 species), and then gradually decrease, as if forming diffusive profile (Fig. 4.44a), a bit biased toward direction of increasing indexes of columns and rows of the matrix. Frequency profile of stationary localisations (Fig. 4.44b) has its maximum at $M_{22} = 17$ and then values are decreasing in the direction of simultaneous increase of indexes of the matrix.

4.7 Summary

We explored a simple modification of cell-state transition function of an excitable automaton: excitation of a cell depends not only on a number of excited neighbours, as in classical models of excitation, but also on a number of refractory neighbours. Every resting cell excites if there is between θ_1 and θ_2 excited neighbours and between δ_1 and δ_2 refractory neighbours, $0 \leq \theta_1 \leq \theta_2 \leq 8$, $0 \leq \delta_1 \leq \delta_2 \leq 8$, and $\theta_1 + \theta_2 + \delta_1 + \delta_2 \leq 8$. In terms of biological ecosystems such rule can be interpreted as a mutualism. Both excitation and refractory species derive benefits of their interaction, or at least co-existence in the same local domain (cell neighbourhood). Refractory species 'benefit' because every excited species becomes refractory. Excited species 'benefit' because chances of a resting cell to excited a boosted by refractory neighbours. Cell-state transition rules supporting rich localisation dynamics have correspond to an obligatory mutualism because $\delta_1 > 0$.

We could not establish an immediate direct correspondence between our abstract model and any real systems. One could say that Belousov-Zhabotinsky reaction would be a closest match in terms of phenomenology of excitation: in sub-excitable media traveling wave-fragments, or quasi-particles, are induced by random configuration of illumination, light noise [255]. Collision dynamics of quasi-particles generated in Belousov-Zhabotinsky systems is remarkably similar to that observed in cellular automaton models, and bear huge potential for implementation of architectureless non-classical computing devices [12, 14]. BZ-AOT modification [249] of Belousov-Zhabotinsky reaction is very promising because it allows for co-existence of classical spot, stripe and labyrinthine patterns. A simple surface oxidations systems exhibiting mobile localisations [217] are also showing excellent possibilities for achievement of experimental implementation of the studied systems with mutualistic excitation mode.

Chapter 5
Dynamical Excitation Intervals: Diversity and Localisations

Back in 1998 [7], we introduced an excitable cellular automaton, where a resting cell is excited if a number of its excited neighbours belong to a fixed interval $[\theta_1, \theta_2]$. The interval $[\theta_1, \theta_2]$ is called an excitation interval. For two-dimensional cellular automaton with eight-cell neighbourhood $1 \leq \theta_1 \leq \theta_2 \leq 8$. We found that by tuning θ_1 and θ_2 we can persuade the automaton to imitate almost all kinds of excitation dynamics, from classical target and spiral waves observed in physical and chemical excitable media to wave-fragments inhabiting sub-excitable media [7]. How does excitation dynamics change if we allow boundaries of the excitation interval to change during the automaton development? We partially answer the question in present chapter by making the interval $[\theta_1^t(x), \theta_2^t(x)]$ of every cell x to be dynamically updatable at every step t depending on state of the cell x and numbers of excited and refractory neighbours in the cell x's neighbourhood. All cells update their intervals independently on each other but by the same rule and simultaneously.

Let x^t and x^{t+1} be states of a cell x at time steps t and $t+1$, and $\sigma_+^t(x)$ and be $\sigma_-^t(x)$ be sums of excited and refractory neighbours in cell x's neighbourhood $u(x) = \{y : |x-y|_{L_\infty} = 1\}$. Cell x updates its state by the following rule:

$$x^{t+1} = \begin{cases} +, \text{ if } x^t = \cdot \text{ and } \sigma_+^t(x) + \in [0_1^t(x), 0_2^t(x)] \\ -, \text{ if } x^t = + \\ \cdot, \text{ otherwise} \end{cases}$$

A resting cell is excited if a number of its neighbours belongs to excitation interval $[\theta_1^t(x), \theta_2^t(x)]$, where $1 \leq \theta_1^t(x), \theta_2^t(x) \leq 8$. The boundaries $\theta_1^t(x)$ and $\theta_2^t(x)$ are dynamically updated depending on cell x's state and numbers of x's excited $\sigma_+^t(x)$ and refractory $\sigma_-^t(x)$ neighbours. A natural way to update boundaries is by increasing or decreasing their values as follows:

$$\theta_1^{t+1}(x) = \xi(\theta_1^t(x) + \Delta_1 \phi(\sigma_+^t(x) - \sigma_-^t(x)))$$

$$\theta_2^{t+1}(x) = \xi(\theta_2^t(x) + \Delta_2 \phi(\sigma_+^t(x) - \sigma_-^t(x)))$$

A. Adamatzky: Reaction-Diffusion Automata, ECC 1, pp. 97–114.
springerlink.com

where

$$\Delta_1 = \begin{cases} T_1, \text{ if } x = + \\ T_3, \text{ if } x = - \\ 0, \text{ if } x = 0 \end{cases} \quad \Delta_2 = \begin{cases} T_2, \text{ if } x = + \\ T_4, \text{ if } x = - \\ 0, \text{ if } x = 0 \end{cases}$$

and $\phi(a-b) = 1$ if $a > b$, 0 if $a = b$ and -1 if $a < b$, and $\xi(a) = 1$ if $a < 1$ and 8 if $a > 8$. Excitation interval update rules are determined by values of $T_1, \cdots T_4$. We therefore address these functions as tuples $E(T_1T_2T_3T_4)$ which range from $E(-1-1-1-1)$ to $E(1111)$.

The experiments are conducted on a cellular array of $n \times n$ cells with periodic boundary conditions. In a typical experiment we perturb resting cellular array with a localised domain of excitation, wait till transient period is over (1000 iterations is enough) and most excitation patterns collide and disappear or persist indefinitely, and then analyse three configurations: configuration of excitation represented by an array of cells states x^t, and configurations of lower $\theta_1^t(x)$ and upper $\theta_2^t(x)$ boundaries of excitation intervals.

Initially $\theta_2^0(x) = 8$ for any x. In experiments we considered initial conditions $\theta_1^0(x) = 1$ and $\theta_1^0(x) = 2$. Space-dymamics of the automata were investigated in the following scenarios of initial perturbations of resting cellular arrays:

- (++)-start, $\theta_1 = 2$: all cells are resting but two neighbouring cells are assigned excited state,
- R1-start: let D be a disc radius $n/4$ centred in the array L of $n \times n$ cells, all cells are resting but cells lying in D are assigned excited states with probability 0.2 and $\theta_1^0(x) = 1$ for any x,
- R2-start: all cells are resting but cells lying in D get excited states with probability 0.2 and $\theta_1^0(x) = 2$ for any x,
- D1-start: all cells are resting but cells lying in D get excited states with probability 0.1 or refractory states with probability 0.1 and $\theta_1^0(x) = 1$ for any x,
- D2-start: all cells are resting but cells lying in D get excited states with probability 0.1 or refractory states with probability 0.1 and $\theta_1^0(x) = 2$ for any x,
- (−+)-start: all cells are resting but one cell is excited and its western neighbour is refractory,
- (− + +)-start all cells are resting but one cell is excited, its first order western neighbour is excited and its second order western neighbour is refractory.

Cell states are represented by colours and grey levels as follows: excited state + is red (c. 76 grey), resting state is white and refractory state − is blue (c. 28 grey). Colour values of excitation interval boundaries θ_1 and θ_2 are following: 1 is white, 2 is green or 150 grey, 3 is yellow or 226 grey, 4 is blue or 28 grey, 5 is magenta or 104 grey, 6 is cyan or 178 grey, 7 is red or 76 grey, and 8 is black. See examples of colours and grey levels in Tab. 1.2.

Table 5.1 Diversity of excitation configuration is measured by selecting seven functions with highest values $\mathscr{E}$ and S (columns $\mathscr{E}$-S) and diversity of excitation interval configuration by selecting seven functions with highest values $\mathscr{E}_1$ and $\mathscr{E}_2$ (columns $\mathscr{E}_1$-$\mathscr{E}_2$) , i.e. Shannon entropy calculated on configurations of θ_1 and θ_2. Functions exhibiting highest morphological diversity in their groups are underlined.

(++)-start, $\theta_2 = 2$		R1-start		R2-start		D1-start		D2-start		(−+)-start		(− + +)-start	
S-$\mathcal{E}$	$\mathcal{E}_1$-$\mathcal{E}_2$	S-$\mathcal{E}$	$\mathcal{E}_1$-$\mathcal{E}_2$	S-$\mathcal{E}$	$\mathcal{E}_1$-$\mathcal{E}_2$	S-$\mathcal{E}$	$\mathcal{E}_1$-$\mathcal{E}_2$	S-$\mathcal{E}$	$\mathcal{E}_1$-$\mathcal{E}_2$	S-$\mathcal{E}$	$\mathcal{E}_1$-$\mathcal{E}_2$	S-$\mathcal{E}$	$\mathcal{E}_1$-$\mathcal{E}_2$
000-1 0100 1010 11-10 1100 1101 1111	-10-1-1 -11-1-1 00-1-1 01-1-1 01-10 11-1-1 11-10		-1-1-11 -11-11 1-1-1-1 1-1-11 1-11-1 1-111 11-11	-100-1 0100 0101 1-10-1 1-100 1000 1101	-1-1-1-1 -1-11-1 -1-110 -10-1-1 -1011 -1111	0-100 0-101 0001 0100 0101 1-100 1000	-1-1-1-1 -1-1-10 -1-101 -10-1-1 -11-1-1 1-1-11 11-1-1	0100 1-10-1 1-100 1-101 100-1 1000 1101	-1-11-1 -1-110 -1-111 -10-1-1 -1011 -11-1-1 -1111	-1-100 0-1-1-1 0-100 0-101 0001 1-101 1001	-1-1-1-1 -1-1-10 -1-1-11 -1-101 -10-1-1 -10-11 -1001	000-1 0100 1-101 11-10 1100 1101 1111	1-101 100-1 1001 11-11 110-1 111-1 1110

5.1 Morphological Diversity

We evaluated morphological diversity of configurations of excitation and interval boundaries using Shannon entropy and Simpson's index. Let $W = \{\circ, +, -\}$ be a set of all possible configurations of a nine-cell neighbourhood $w(x) = u(x) \cup x$, $x \in \mathbf{L}$. Let c be a configuration of automaton, we calculate a number of non-resting neighbourhood configurations as $\eta = \sum_{x \in \mathbf{L}} \varepsilon(x)$, where $\varepsilon(x) = 0$ if for every resting x all its neighbours are resting, and $\varepsilon(x) = 1$ otherwise. The Shannon entropy is calculated as $\mathscr{E} = -\sum_{w \in W}(\nu(w)/\eta \cdot ln(\nu(w)/\eta))$, where $\nu(w)$ is a number of times the neighbourhood configuration w is found in automaton configuration c. Simpson's index is calculated as $S = 1 - \sum_{w \in W}(\nu(w)/\eta)^2$. The measures $\mathscr{E}$ and S were calculated on configurations of cell-states and interval boundaries after long transient period ($t = 1000$), sufficient enough (for the lattice size 200×200 cells) for any perturbation to span the array and settle down in one of the global states.

The diversity of excitation patterns is evaluated using S-$\mathscr{E}$ plots. See examples of S-$\mathscr{E}$ plots for D1-start, $\theta_1^0(x) = 1$, in Fig. 5.1 and (++)-start, $\theta_1^0(x) = 2$, in Fig. 5.2. Distributions of functions by their values of Shannon entropies for θ_1 and θ_2 are illustrated in Figs. 5.3 and 5.4

Top seven functions showing highest values of diversity indices, e.g. those encircled in examples Fig. 5.1–5.4, are grouped in Tab. 5.1 for various scenarios of initial start.

Finding 6. *There is no match between functions which produce configurations of excitation with highest morphological diversity and configurations of interval boundaries with highest morphological diversity..*

Amongst functions listed in Tab. 5.1 only function $E(1-101)$ gets into top seven functions with highest diversity of both excitation and interval boundaries for scenario $(-++)$-start. Exemplar configurations of excitation and interval boundaries generated by automaton governed by $E(1-101)$ are shown in Fig. 5.10b. Function

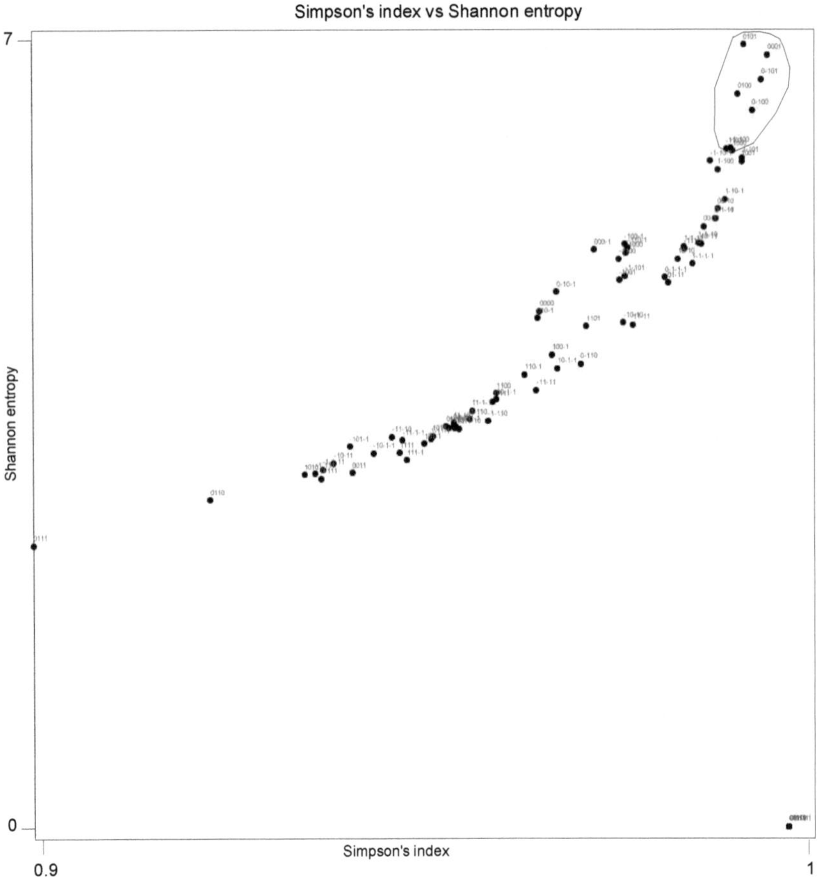

Fig. 5.1 Morphological diversity of functions for D1-start, $\theta_1^0(x) = 1$ for all x: Simpson's index S (horizontal axis) vs Shannon entropy $\mathscr{E}$ (vertical axis) for configuration of excitable array of 200×200 cells with periodic boundary condition, recorded at $t = 1000$. Encircled data points are seven functions with highest morphological diversity specified in column D1-start, S-$\mathscr{E}$ in Tab. 5.1.

$E(1-101)$ is also amongst functions with highest diversity for $D2$- and $(-+)$-start. The function governs the following update of the excitation interval boundaries.

Low boundary $\theta_1(x)$ of excitation of cell x is updated only if cell x is excited. The boundary $\theta_1(x)$ increases if cell x has more excited neighbours than refractory neighbours, $\theta_1(x)$ decreases if number of refractory neighbours of x exceeds number of excited neighbours. The boundary $\theta_1(x)$ does not change if cell x has the same number of excited neighbours as refractory neighbours. Upper boundary $\theta_2(x)$ increases if cell x's dissents with excitation-refractoriness ratio in its neighbourhood: $x^t = +$ and $\sigma_+^t(x) < \sigma_-^t(x)$ or $x^t = -$ and $\sigma_+^t(x) > \sigma_-^t(x)$. The boundary $\theta_2(x)$ decreases if cell x conforms to excitation-refractoriness ratio in its

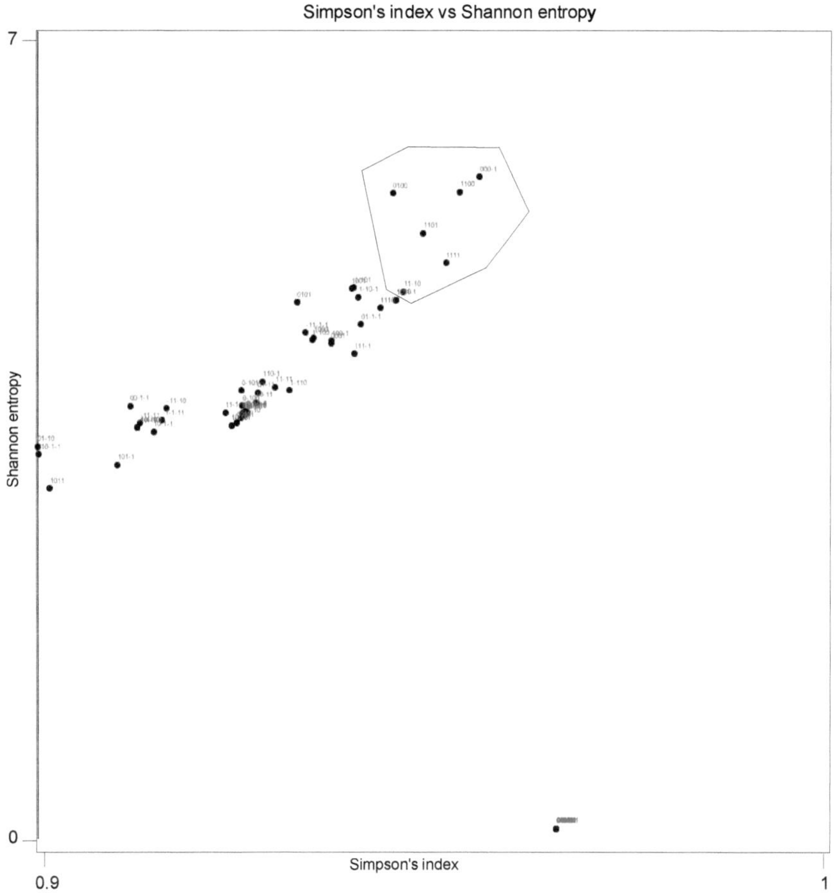

Fig. 5.2 Morphological diversity of functions for (++)-start, $\theta_1^0(x) = 2$ for all x: Simpson's index S (horizontal axis) vs Shannon entropy $\mathscr{E}$ (vertical axis) for configuration of excitable array of 200×200 cells with periodic boundary condition, recorded at $t = 1000$. Encircled data points are seven functions with highest morphological diversity specified in column (++)-start, S-$\mathscr{E}$ in Tab. 5.1.

neighbourhood: $x^t = +$ and $\sigma_+^t(x) > \sigma_-^t(x)$ or $x^t = -$ and $\sigma_+^t(x) < \sigma_-^t(x)$. Increase of θ_1 and decrease of θ_2 lead to decrease cell's excitability. Thus we can characterise function $E(1-101)$ as follows: excitability of a cell decreases if the cell dissents with its neighbourhood and increases otherwise.

Finding 7. *Functions $E(-1-101)$ and $E(-1011)$ generate configurations of excitation interval boundaries with highest morphological diversity for three and two types of initial stimulation, respectively, and functions $E(000-1)$ and $E(0100)$ generate configurations of excitation with highest morphological diversity for two types of initial stimulation.*

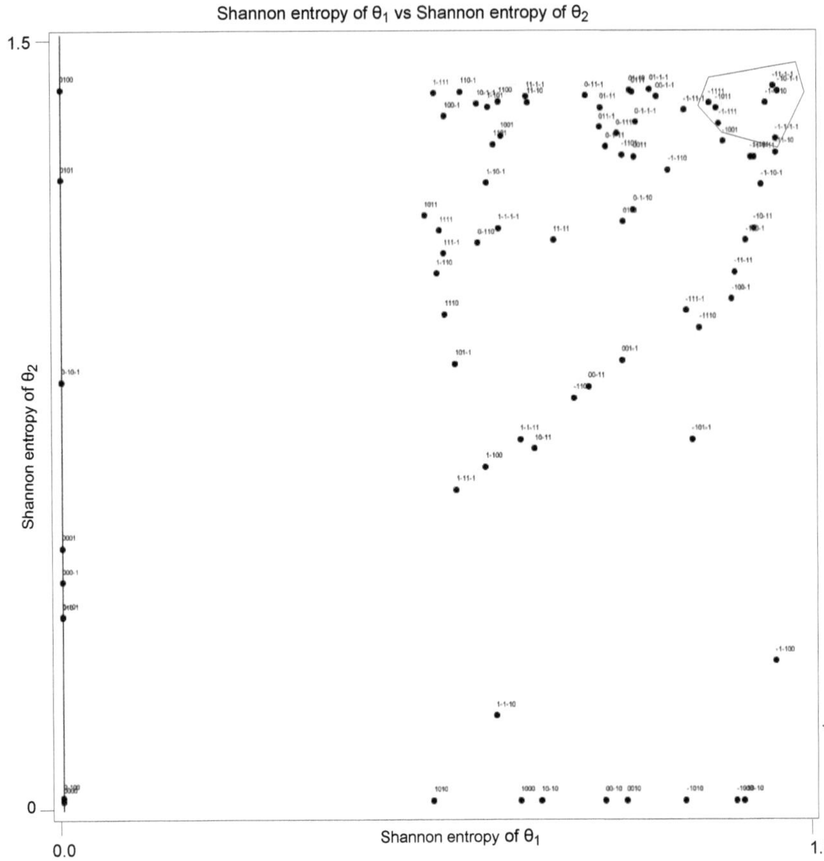

Fig. 5.3 Morphological diversity of functions for D1-start, $\theta_1^0(x) = 1$ for all x: Shannon entropy $\mathscr{E}_1$ for configuration of θ_1 (horizontal axis) vs Shannon entropy $\mathscr{E}_2$ for configuration of θ_2 (vertical axis), recorded at $t = 1000$. Encircled data points are seven functions with highest morphological diversity specified in column D1-start, $\mathscr{E}_1$-$\mathscr{E}_2$ in Tab. 5.1.

The function $E(-1-101)$ generates configurations with highest morphological diversity of θ_1 and θ_2 configurations for R2-, D1- and $(-+)$-starts, see examples in Fig. 5.7b and 5.9b, and $E(-1011)$ generates highest morphological diversity configurations of θ_1 and θ_2 for R2- and D2-starts (Figs. 5.6b and 5.8b). The functions $E(000-1)$ and $E(0100)$ produce highest morphological diversity configurations of excitation for $(++)$- and $(--+)$-starts ($E(000-1)$) and R2- and D2-starts ($E(0100)$). See examples for $E(000-1)$ in Figs. 5.5a and 5.10a and $E(0100)$ in Figs. 5.6a and 5.8a.

Finding 8. *Function* $E(0100)$ *generates most morphologically diverse excitation patterns in larger, comparing to other functions, number of initial conditions.*

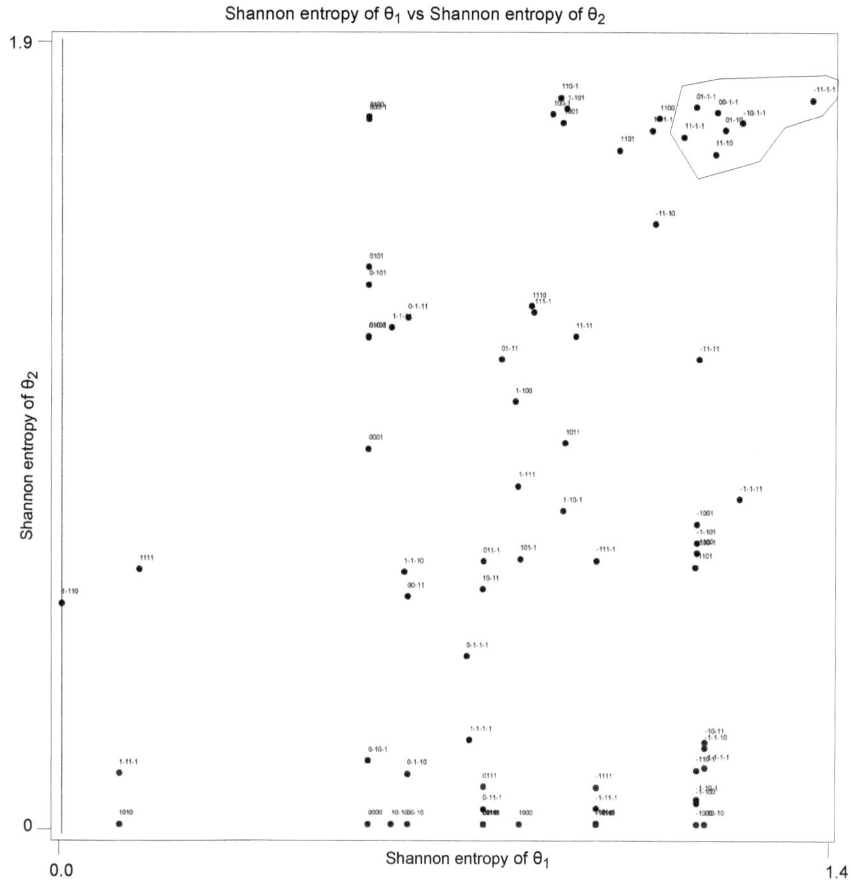

Fig. 5.4 Morphological diversity of functions for (++)-start, $\theta_1^0(x) = 2$ for all x: Shannon entropy $\mathscr{E}_1$ for configuration of θ_1 (horizontal axis) vs Shannon entropy $\mathscr{E}_2$ for configuration of θ_2 (vertical axis), recorded at $t = 1000$. Encircled data points are seven functions with highest morphological diversity specified in column (++)-start, $\mathscr{E}_1$-$\mathscr{E}_2$ in Tab. 5.1.

Function $E(0100)$ is amongst top seven functions with highest morphological diversity of excitation in (++)-, R2-, D1-, D2- and $(-++)$-starts (Tab. 5.1). The function generates most morphologically diverse excitations in D2-start. Examples of configurations generated by $E(0100)$ are shown in Figs. 5.6a and 5.8a. The function $E(0100)$ gives us a proof that space-time dynamics of excitation can be tuned by changing only upper boundary of the excitation interval while lower boundary fixed. Value $\theta_1(x)$ is not updated. Value $\theta_2(x)$ increases if cell x is excited and it has more excited than refractory neighbours, the value $\theta_2(x)$ decreases if cell x is excited and has more refractory than excited neighbours. Excitability of a cell decreases if the cell dissents with excitation ratio in its neighbourhood, and increases otherwise.

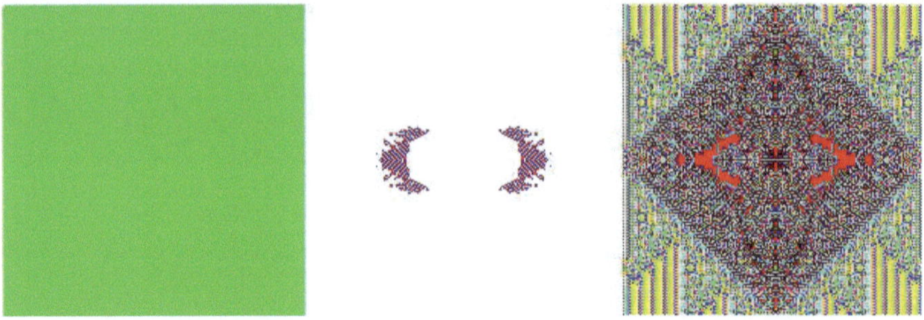

(a) ++-start, $E(000-1)$

(b) ++-start, $E(-11-1-1)$

(c) R1-start, $E(1-111)$

Fig. 5.5 Examples of most morphologically diverse configurations generated in (++)-start (ab) and R1-start (c). (a) Configurations with highest morphological diversity of excitation generated by function $E(000-1)$. (b) Configurations with highest morphological diversity of interval boundaries θ_1 and θ_2 generated by function $E(-11-1-1)$. (c) Configurations with highest morphological diversity of interval boundaries θ_1 and θ_2 generated by function $E(1-111)$. Configurations of θ_1 (left), excitation (middle) and θ_2 (right) are taken in 200×200 cell array, at $t = 1000$.

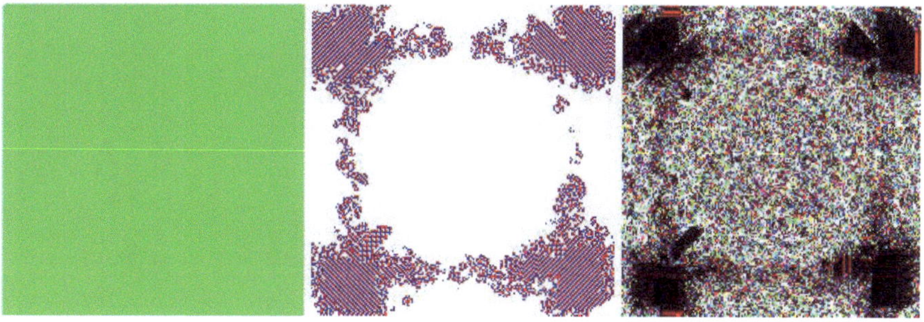

(a) R2-start, $E(0100)$

(b) R2-start, $E(-1011)$

Fig. 5.6 Examples of most morphologically diverse configurations generated in R2-start. (a) Configurations with highest morphological diversity of excitation generated by function $E(0100)$. (b) Configurations with highest morphological diversity of interval boundaries θ_1 and θ_2 generated by function $E(-1011)$. Configurations of θ_1 (left), excitation (middle) and θ_2 (right) are taken in 200×200 cell array, at $t = 1000$.

Finding 9. *Function $E(-10-1-1)$ generates most morphologically diverse patterns of interval boundaries in larger, comparing to other functions, number of initial conditions.*

Function $E(-10-1-1)$ gets in top seven functions with highest morphological diversity of θ_1 and θ_2 patterns for $(++)$-, R2-, D1-, D2-, and $(-+)$-starts (Tab. 5.1). In automata governed by this function $\theta_2(x)$ is not updated if cell x is excited. Otherwise, $\theta_1(x)$ and $\theta_2(x)$ increase if refractory neighbours outnumber excited neigbours in the cell x's neighbourhood and decrease if excited neighbours outnumber refractory neighbours.

Finding 10. *Upper boundary of excitation interval more significantly affects morphological diversity of configurations generated than lower boundary of the interval does.*

There are only two functions, $E(1010)$ and $E(1000)$, where only lower boundary θ_1 is updated in Tab. 5.1. Function $E(1010)$ represents a situation when $\theta_1(x)$ is

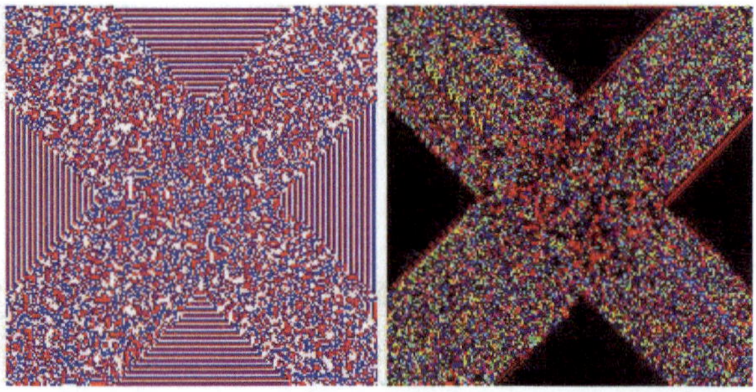

(a) D1-start, $E(0101)$

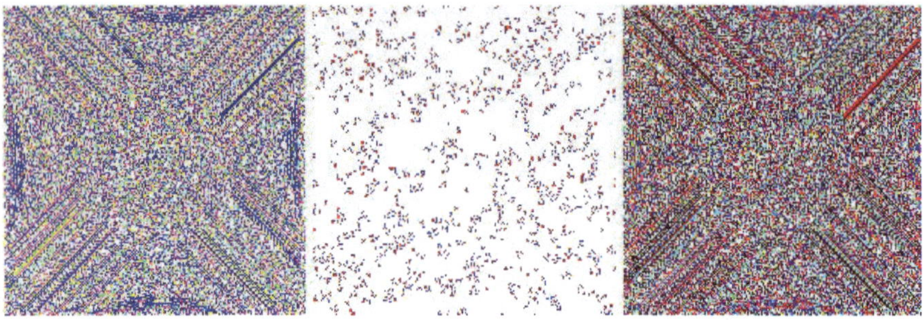

(b) D1-start, $E(-1-101)$

(c) D1-start, $E(11-1-1)$

Fig. 5.7 Examples of most morphologically diverse configurations generated in D1-start. (a) Configurations with highest morphological diversity of excitation generated by function $E(0101)$. (b) Configurations with highest morphological diversity of interval boundaries θ_1 and θ_2 generated by function $E(-1-101)$. (c) Configurations with highest morphological diversity of interval boundaries θ_1 and θ_2 generated by function $E(11-1-1)$. Configurations of θ_1 (left), excitation (middle) and θ_2 (right) are taken in 200×200 cell array, at $t = 1000$.

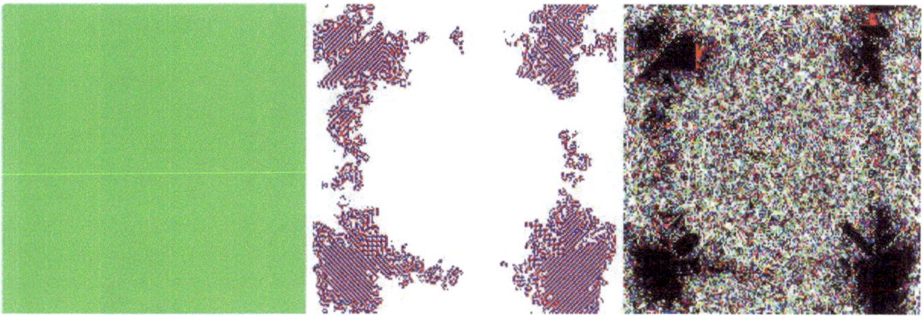

(a) D2-start, $E(0100)$

(b) D2-start, $E(-1011)$

Fig. 5.8 Examples of most morphologically diverse configurations generated in D2-start. (a) Configurations with highest morphological diversity of excitation generated by function $E(0100)$. (b) Configurations with highest morphological diversity of interval boundaries θ_1 and θ_2 generated by function $E(-1011)$. Configurations of θ_1 (left), excitation (middle) and θ_2 (right) are taken in 200×200 cell array, at $t = 1000$.

updated independently of a state of cell x: $\theta_1(x)$ increases if number of excited neighbours exceeds number of refractory neighbours, and $\theta_1(x)$ decreases if refractory neighbours outnumber excited neighbours. In automata governed by function $E(1000)$ value of $\theta_1(x)$ is updated as above but only if cell x is excited.

There are several functions with highest morphological diversity which represent fixed lower boundary and dynamical upper boundary, e.g. $E(000-1)$ (Figs. 5.5a and 5.10a), $E(0100)$ (Figs. 5.6a and 5.8a), $E(0101)$ (Fig. 5.7a), $E(0-100)$, $E(0-101)$, and $E(0001)$ (Tab. 5.1).

5.2 Generative Diversity and Localisations

Generative diversity characterises how many different configurations are generated during space-time development of automaton starting with entirely resting configuration but one to three cells in a non-resting state. We consider two starting conditions: $(++)$-start and $\theta_1^0(x) = 2$ for any x; $(-+)$-start and $\theta_1^0(x) = 1$; and,

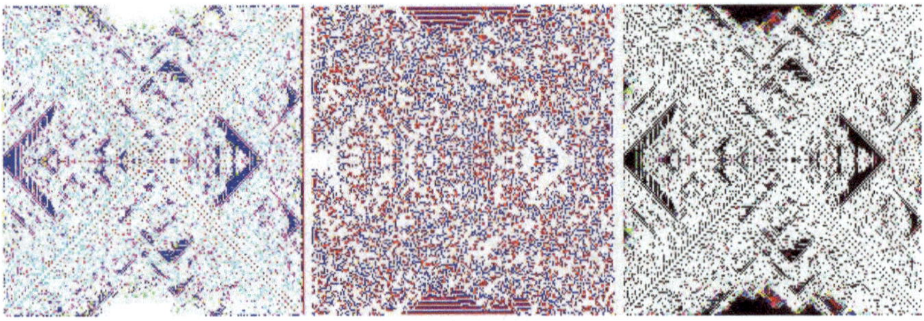

(a) $(-+)$-start, $E(-1-100)$

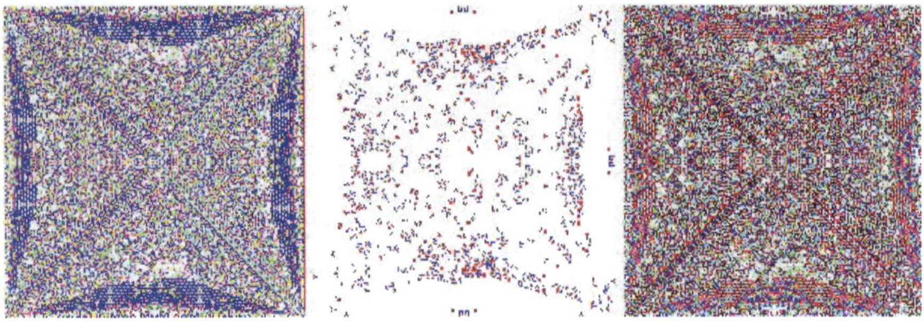

(b) $(-+)$-start, $E(-1-101)$

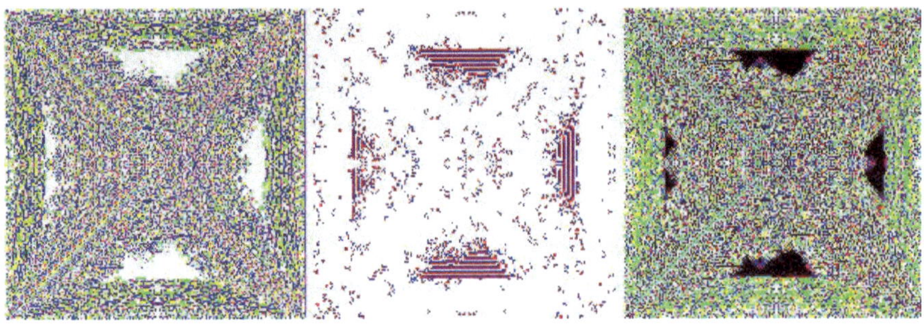

(c) $(-+)$-start, $E(-1001)$

Fig. 5.9 Examples of most morphologically diverse configurations generated in $(-+)$-start. (a) Configurations with highest morphological diversity of excitation generated by function $E(-1-100)$. (b) Configurations with highest morphological diversity of interval boundaries θ_1 and θ_2 generated by function $E(-1-101)$. (c) Configurations with highest morphological diversity of interval boundaries θ_1 and θ_2 generated by function $E(-1001)$. Configurations of θ_1 (left), excitation (middle) and θ_2 (right) are taken in 200×200 cell array, at $t = 1000$.

$(-++)$-start and $\theta_1^0(x) = 2$. Generative diversity is evaluated using Shannon entropy and space-filling (a ratio of cells in a non-resting state). Functions generating configurations with maximum Shannon entropy and minimum space-filling are assumed to have higher generative diversity, see example in Fig. 5.12.

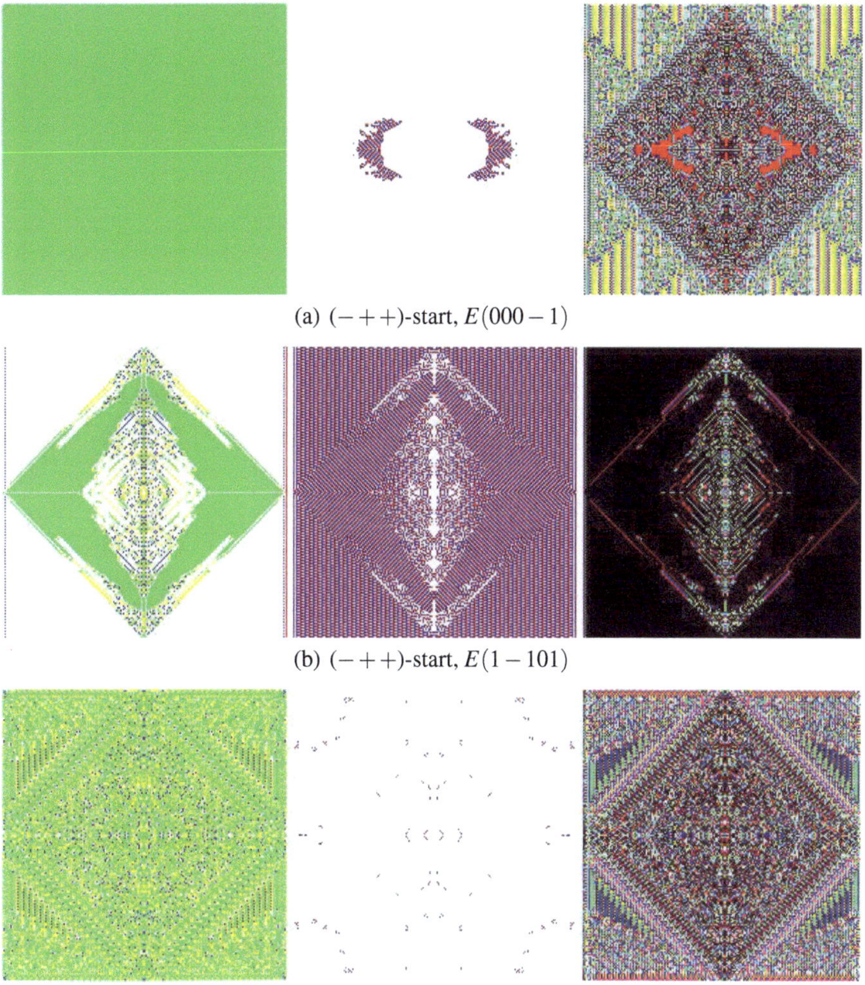

(a) $(-++)$-start, $E(000-1)$

(b) $(-++)$-start, $E(1-101)$

(c) $(-++)$-start, $E(110-1)$

Fig. 5.10 Examples of most morphologically diverse configurations generated in $(-++)$-start. (a) Configurations with highest morphological diversity of excitation generated by function $E(000-1)$. (b) Configurations with highest morphological diversity of interval boundaries θ_1 and θ_2 generated by function $E(1-101)$. (c) Configurations with highest morphological diversity of interval boundaries θ_1 and θ_2 generated by function $E(110-1)$. Configurations of θ_1 (left), excitation (middle) and θ_2 (right) are taken in 200×200 cell array, at $t = 1000$.

Seven functions with highest generative diversity are listed in Tab. 5.2. Configurations of excitation and interval boundaries for $E(-11-1-1)$, $++$-start, shown in Fig. 5.5b, $E(-1-101)$, $(-+)$-start, in Fig. 5.9b and configurations generated by functions $E(1111)$, $E(11-10)$ and $E(0-1-1-1)$ in Fig. 5.12.

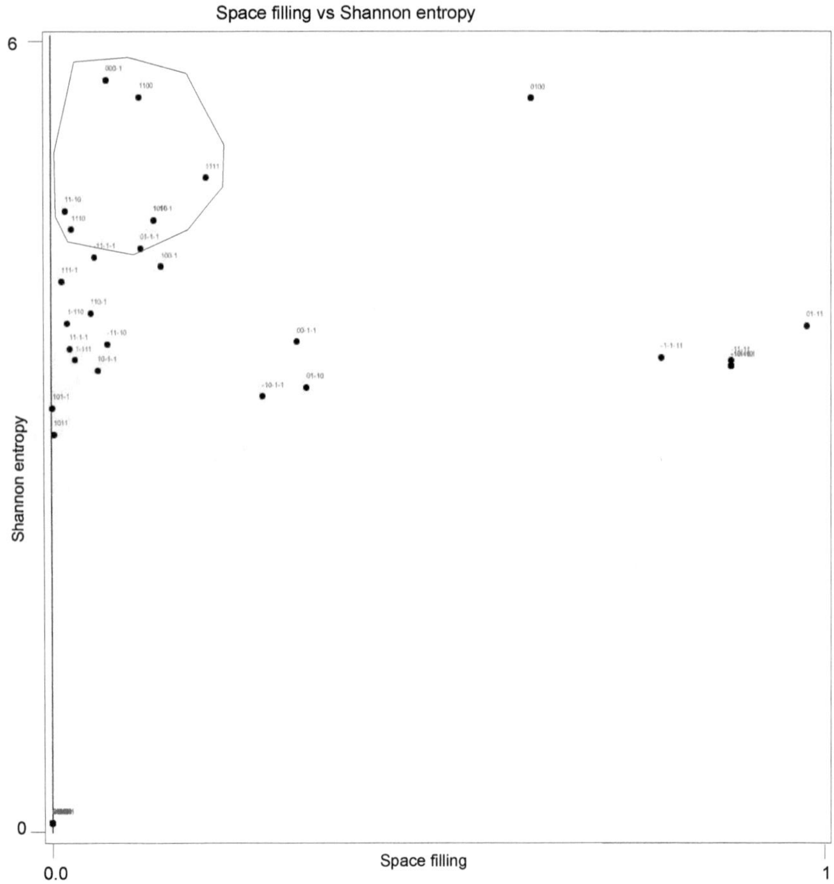

Fig. 5.11 Generative diversity of functions for $(++)$-start, $\theta_1^0(x) = 2$ for all x: Space filling (horizontal axis) vs Shannon entropy (vertical axis) for configuration of excitable array of 200×200 cells with periodic boundary condition, recorded at $t = 1000$. Encircled data points are seven functions with highest generative diversity specified in column $(++)$-start, in Tab. 5.2.

Table 5.2 Seven functions with highest generative diversity for $(++)$- and $(-++)$-start, $\theta_1 = 2$ (first column) and $(-+)$-start, $\theta_1 = 1$ (second column)

$(++)$- and $(-++)$-start, $\theta_1^0 = 2$	$(-+)$-start, $\theta_1^0 = 1$
$-11-1-1$	$-1-101$
$01-1-1$	$0-1-1-1$
$100-1$	$00-1-1$
1010	$01-1-1$
$11-10$	$01-11$
1110	$10-1-1$
1111	$11-11$

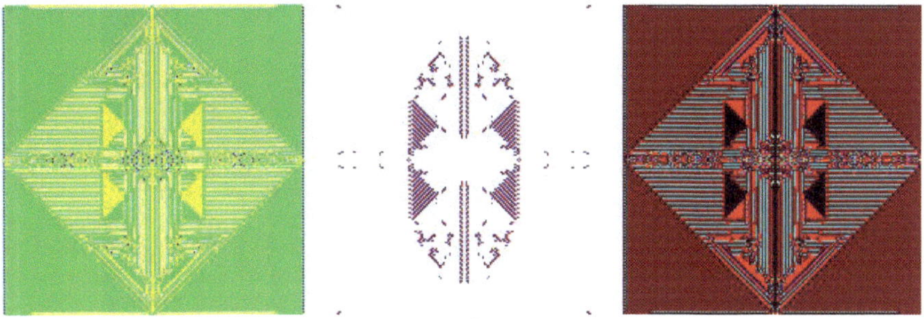

(a) (++)-start, $\theta_1^0 = 2$, $E(1111)$

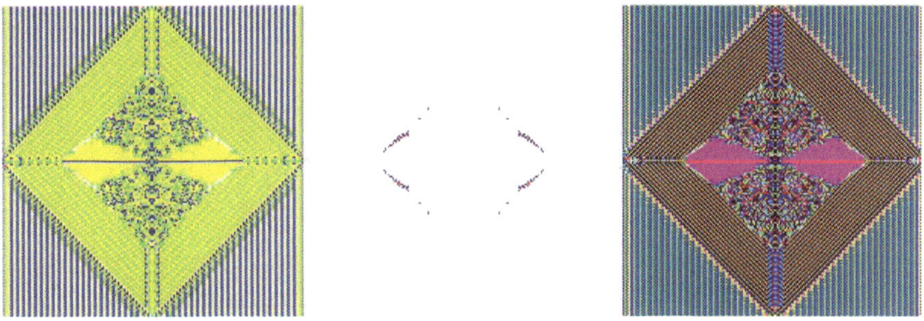

(b) (++)-start, $\theta_1^0 = 2$, $E(11-10)$

(c) (−+)-start, $\theta_1^0 = 1$, $E(0-1-1-1)$

Fig. 5.12 Examples of configurations generated by functions with highest generative diversity. (a) $E(1111)$, (b) $E(11-10)$, (c) $E(0-1-1-1)$. Automaton array has 200×200 cells, configurations of θ_1 (left), excitation (middle) and θ_2 (right) at $t = 1000$.

Finding 11. *Most localizations generated in (++)- and (− + +)-starts are stationary.*

Around half of the functions generate configurations with localizations in case of R2-start (43 functions), D1-start (41 functions), and D2-start (43 functions). We focus on functions which produce localizations from few-cell perturbations. There are nine functions for (++)-start and eight function for (−+)-start , $\theta_1^0 = 2$ (Tab. 5.3).

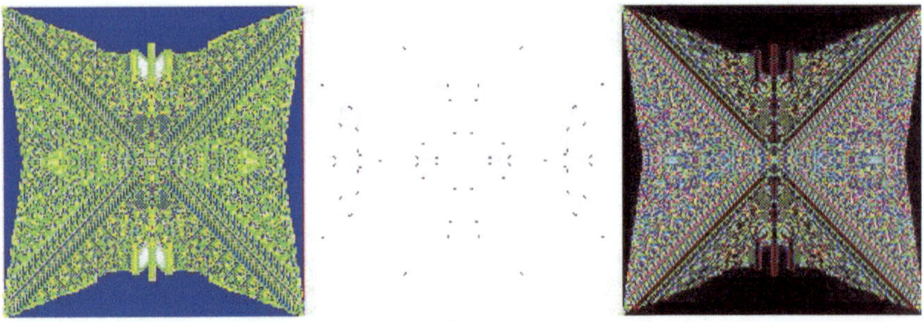

(a) (++)-start, $\theta_1^0 = 2$, $E(00-1-1)$

(b) (++)-start, $\theta_1^0 = 2$, $E(101-1)$

(c) (++)-start, $\theta_1^0 = 2$, $E(1110)$

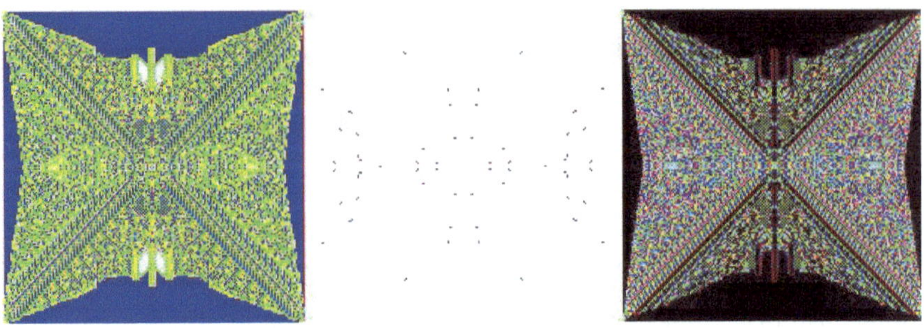

(d) (−+)-start, $\theta_1^0 = 1$, $E(00-1-1)$

Fig. 5.13 Examples of configurations with localised excitations developed in (++)-start (abc) and (−+)-start (d) scenarios. Size of cellular array is 200×200 cells, configurations of θ_1 (left), excitation (middle) and θ_2 (right) at $t = 1000$.

Table 5.3 Functions supporting localizations in (++)- and $(-++)$-start (left column) and $(-+)$-start (right column)

(++)-start	$(-+)$-start
$10-1-1$	$-10-1-1$
$11-1-1$	$-11-10$
$11-10$	$00-1-1$
$100-1$	$01-1-1$
$101-1$	$01-10$
$110-1$	$10-1-1$
$111-1$	$11-1-1$
1011	$11-10$
1110	

Examples of configurations generated by functions $E(00-1-1)$, $E(101-1)$, $E(1110)$ and $E(00-1-1)$ are shown in Fig. 5.13.

Finding 12. *Functions $E(11-10)$ and $100-1$ are amongst top seven functions with highest generative diversity supporting localised excitation dynamics in scenario of (++)-start. Functions $E(00-1-1)$, $E(01-1-1)$, $E(10-1-1)$ and $E(11-1-1)$ are amongst top seven functions with highest generative diversity supporting localised excitation dynamics in scenario of $(-+)$-start*

See Tabs. 5.2 and 5.3. Configurations generated by function $E(11-10)$ are exemplified in Fig. 5.12b and function $E(00-1-1)$ in Fig. 5.13a. Functions $E(100-1)$ and $E(00-1-1)$ are the functions with minimal updates of excitation interval boundaries. In automata governed by function $E(00-1-1)$ interval boundaries $\theta_1(x)$ and $\theta_2(x)$ are updated only if the cell x is in refractory state: both boundaries decrease if excited neighbours of x outnumber refractory neighbours, and they increase if refractory neighbours dominate. In automata governed by function $E(100-1)$ the boundary $\theta_1(x)$ increases if the excited cell x has more excited neighbours than refractory ones, and $\theta_1(x)$ decreases if the cell x has more refractory neighbours. The boundary $\theta_2(x)$ is updated only if cell x is refractory: $\theta_2(x)$ decreases if excited neighbours outnumber refractory neighbours, and it increases otherwise.

5.3 Summary

Excitable cellular automata with dynamical excitation interval exhibit a wide range of space-time dynamics based on an interplay between propagating excitation patterns which modify excitability of the automaton cells. Such interactions lead to formation of standing domains of excitation, stationary waves and localised excitations. We analysed morphological and generative diversities of the functions studied and characterised the functions with highest values of the diversities. Amongst other intriguing discoveries we found that upper boundary of excitation interval more significantly affects morphological diversity of configurations generated than lower boundary of the interval does and there is no match between functions which produce configurations of excitation with highest morphological diversity and configurations of interval boundaries with highest morphological diversity. Potential directions of futures studies of excitable media with dynamically changing excitability may focus on relations of the automaton model with living excitable media, e.g. neural tissue and muscles, novel materials with memristive properties, and networks of conductive polymers.

Chapter 6
Excitable Delaunay Triangulations

Given a planar finite set $\mathbf{V}$ the Delaunay triangulation [92] $\mathscr{D}(\mathbf{V}) = \langle \mathbf{V}, \mathbf{E} \rangle$ is a graph subdividing the space onto triangles with vertices in $\mathbf{V}$ and edges in $\mathbf{E}$ where the circumcircle of any triangle contains no points of $\mathbf{V}$ other than its vertices. Neighbours of a node $v \in \mathbf{V}$ are nodes from $\mathbf{V}$ connected with v by edges from $\mathbf{E}$.

The set $\mathbf{V}$ is constructed as follows. We take a disc-container of radius 480 and fill it with up to 15,000 disc-nodes. We assume that each disc-node has radius 2.5, thus a minimal distance between any two nodes is 5. The Voronoi diagram, and its dual triangulation, are appropriate representations of such identical sphere packing on two-dimensional surface, where planar points of $\mathbf{V}$ represent centres of the spheres.

We define density ϕ as a ratio of areas occupied by discs to area of the disc container. Examples of triangulations for densities $\phi = 0.0027, 0.027, 0.136$ and 0.407, corresponding to number of nodes packed 100, 1000, 5000, and 15000, are shown in Fig. 6.1.

The Delaunay automaton is defined in the following way. A node $v \in \mathbf{V}$ is a finite state machine. Every node updates its state in discrete time depending on states of its neighbours. All nodes update their states simultaneously. Nodes can have different number of neighbours therefore we better use totalistic node-state update function, where a node updates its state depending on just the numbers of different node-states in its neighbourhood.

We assign three states — resting ($\circ$), excited ($+$) and refractory ($-$) — to nodes of $\mathbf{V}$. A resting node excites depending on a number of excited neighbours. If a node is excited at time t the node takes refractory state at time step $t+1$, independently on states of its neighbours. Transition from refractory to resting state is also unconditional.

Let $\nu(v) = \{u \in \mathbf{V} : (vu) \in \mathbf{E}\}$ be node v's neighbourhood, v^t a state of node v at time step t, $\sigma^t(v)$ a number of excited neighbours of v at step t, and $d(v)$ be a degree, or a number of neighbours $|\nu(v)|$, of node v. Then the node-state transition functions can be defined as follows.

A. Adamatzky: Reaction-Diffusion Automata, ECC 1, pp. 115–133.
springerlink.com

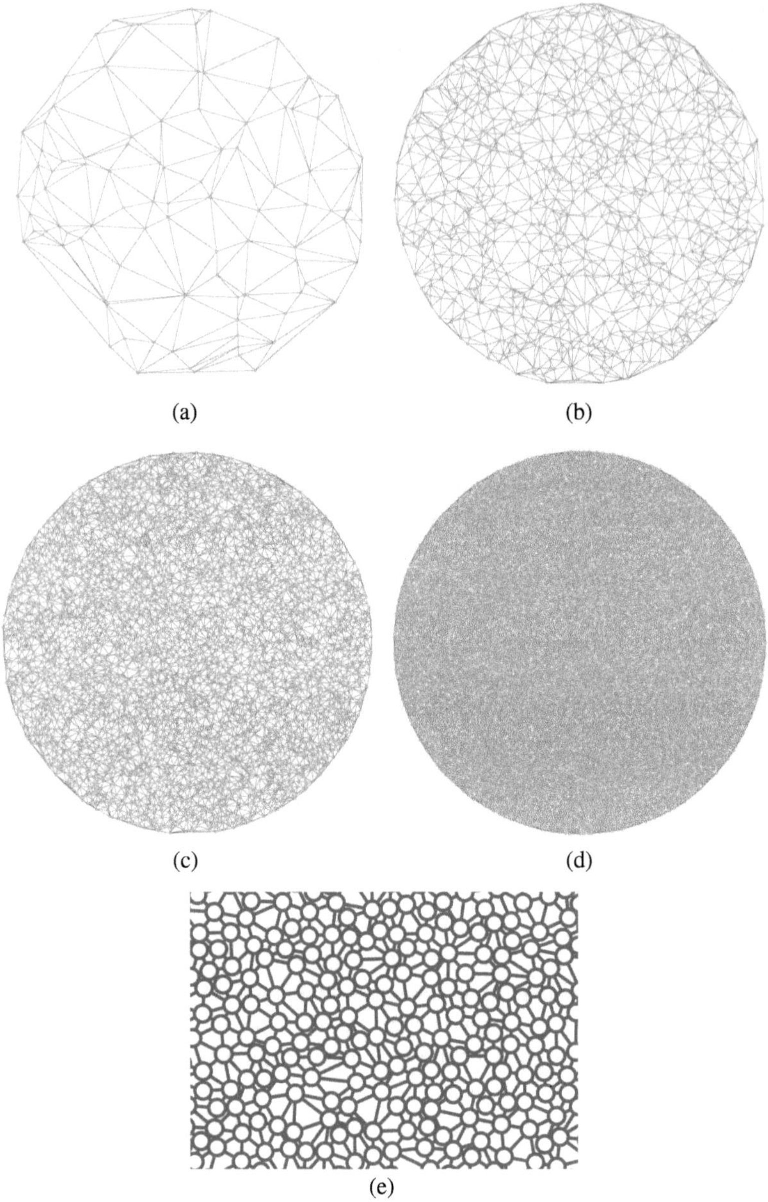

Fig. 6.1 Examples of triangulations for densities (a) $\phi = 0.0027$, (b) $\phi = 0.027$, (c) $\phi = 0.136$, (d) $\phi = 0.407$, (e) fragment of triangulation (a) is zoomed on

- *Absolute excitability*:

$$v^{t+1} = \begin{cases} +, \text{ if } \sigma^t(v) \in \mathbf{S} \\ -, \text{ if } v^t = + \\ \circ, \text{ if } v^t = - \end{cases} \quad (6.1)$$

  Node v excites if $\sigma^t(v) \in \mathbf{S}$, where $\mathbf{S}$ is a set of natural numbers. For example, $\mathbf{S} = \{2,4\}$ means a resting node excites if it has two or four excited neighbours. The rule includes threshold excitation $\sigma^t(v) \geq \theta$, θ is a natural number. For example, $\theta = 2$ means a resting node excites if it has at least 2 excited neighbours, i.e. $\mathbf{S} = \{2,3,4,\cdots\}$.
- *Relative excitability*:

$$v^{t+1} = \begin{cases} +, \text{ if } \frac{\sigma^t(v)}{d(v)} > \varepsilon \\ -, \text{ if } v^t = + \\ \circ, \text{ if } v^t = - \end{cases} \quad (6.2)$$

  Node v excites if $\rho^t(v) = \frac{\sigma^t(v)}{d(v)} > \varepsilon$, where $0 \leq \varepsilon \leq 1$. This condition bring more 'fairness' in excitation process. Some nodes can have less neighbours than other nodes, thus measuring excitation of neighbourhood just by number of excited neighbours would not be 'fair'. It is feasible to calculate a ratio $\rho^t(v)$ of excited neighbours $\sigma^t(v)$ to a total number of neighbours $d(v)$.

6.1 Structural Properties of Delaunay Automata

Propagation of excitation in Delaunay automata depends on structural properties of the triangulation. We found that in case of sparse distribution of disc-nodes (Fig. 6.1a), $\phi = 0.0027$, majority of nodes have five then six and four neighbours each (Fig. 6.2a). With increase of the density (Fig. 6.1b–d) maximum of degree distribution is shifted to six neighbours per node (with probability $p(6) = 0.45$), five nodes ($p(5) = 0.27$) and seven nodes ($p(7) = 0.22$) (Fig. 6.2a). The deviation of the distribution significantly decreases with increase of the density (Fig. 6.2b). The degree distributions found experimentally conform to classical results on packing of equal spheres which is between 5.5 for sparse packing and 6.4 for a close yet random packing [51], and mean degree of 6 for randomly packed sphere with density 0.64 [128].

Let every node of a triangulation takes just two states: '0' (unoccupied) and '1' (occupied). A resting node becomes occupied if it has at least η neighbours in state '1'. Initially just one (for $\eta = 1$) or two (for $\eta = 2$) nodes are assigned state '1'. Dynamics of occupation, measured in a ratio α of nodes occupied by time step t, is shown in Fig. 6.3.

A speed of propagation of state '1' measured in a ratio of nodes occupied in step t increases with decrease of occupation threshold η. If we measure speed in discrete time steps, we will see that it also decreases with increase of the density ϕ.

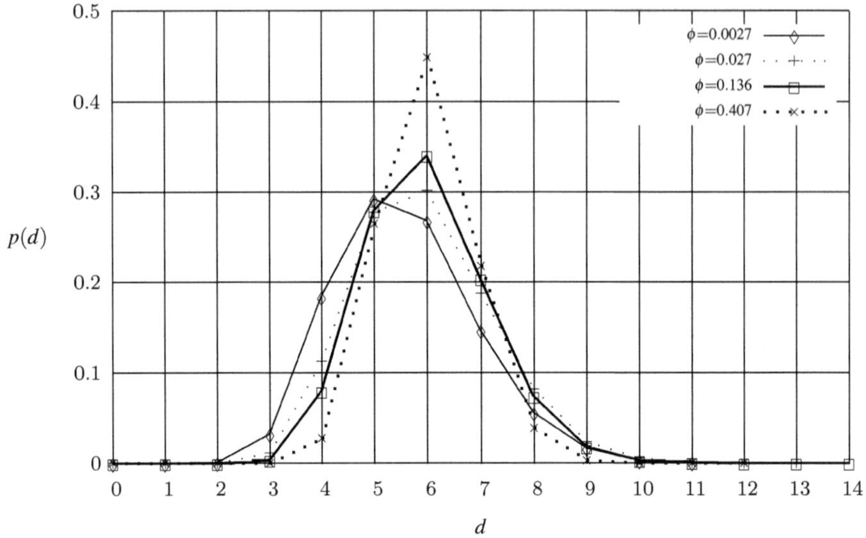

(a)

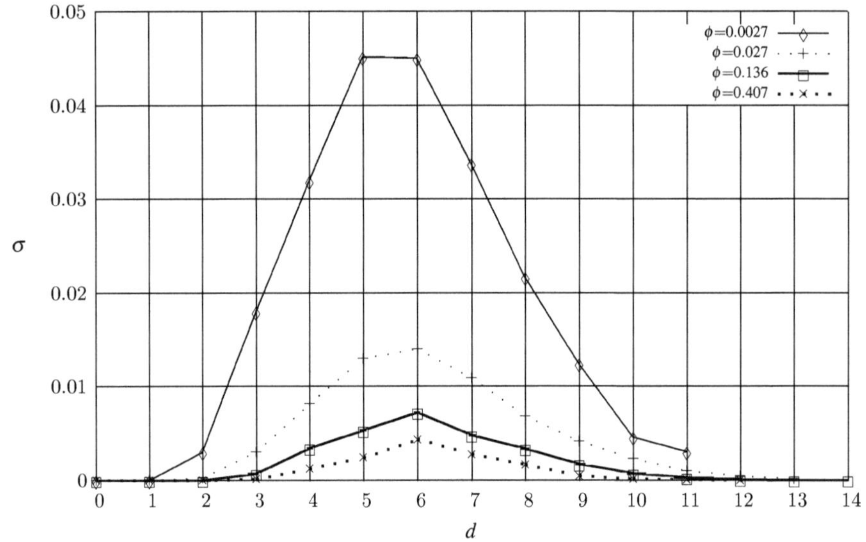

(b)

Fig. 6.2 Distribution of degree d probabilities $p(d)$ (a) and standard deviation σ of distribution $p(d)$ (b) for several densities ϕ of disc-nodes

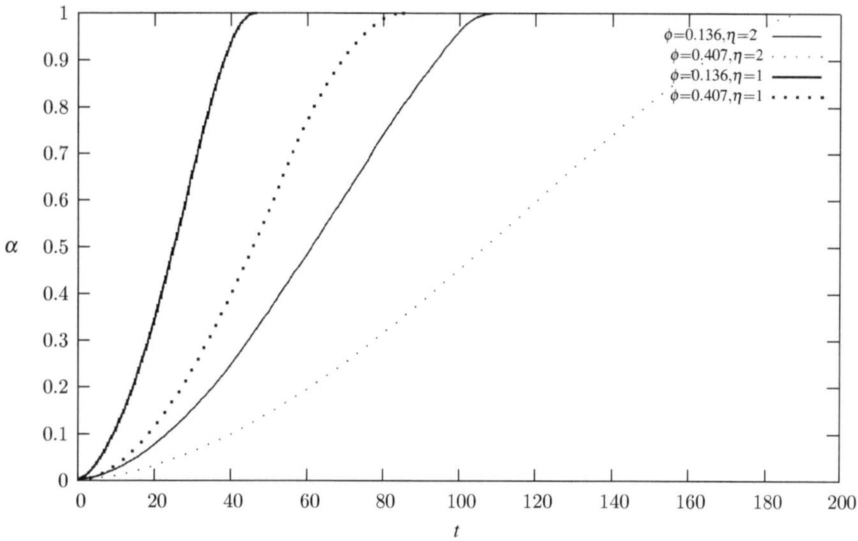

Fig. 6.3 Dynamics of occupation α of triangulations from single-node perturbation; α is a ratio of occupied nodes at time step t to a total number of nodes in the triangulation. Graph presents dynamics of occupation for densities $\phi = 0.136$ and $\phi = 0.407$, and occupation thresholds $\eta = 1$ and $\eta = 2$.

However, the distance covered in any period of time remains comparable between triangulations with different number of nodes.

Finding 13. *Threshold η of node occupation determines a shape of propagating occupation fronts.*

For $\eta = 2$ occupation wave-front is convex, shaping into planar by the end of propagation (Fig. 6.4b). In case of lower threshold of occupation, $\eta = 1$, propagation wave near the edges of triangulation moves quicker then inside the triangulation core (Fig. 6.4a). Thus wave front becomes concave. The part of wave front travelling along edges of triangulation reaches the side of triangulation opposite to the initial perturbation side quicker then the front propagating inside the triangulation. Thus wave front becomes closed (Fig. 6.4a). The domain surrounded by a nested group of target waves, in the western part of the triangulation in Fig. 6.4a, is the place where the occupation waves traveling from east collapse.

Finding 14. *Perturbation propagates faster along edges of triangulation when threshold of node perturbation is low.*

An edge of triangulation is a set of nodes lying on segments of the convex hull of **V**. Spatial structures of node neighbourhoods at the edge of triangulation may be responsible for the particulars of propagations described above.

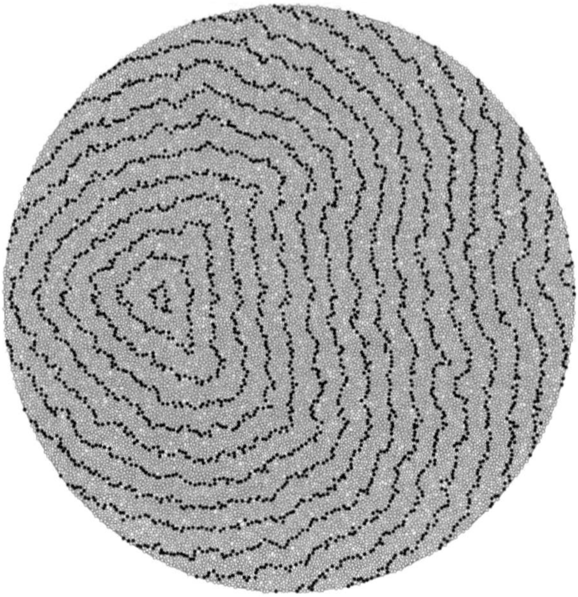

(a)

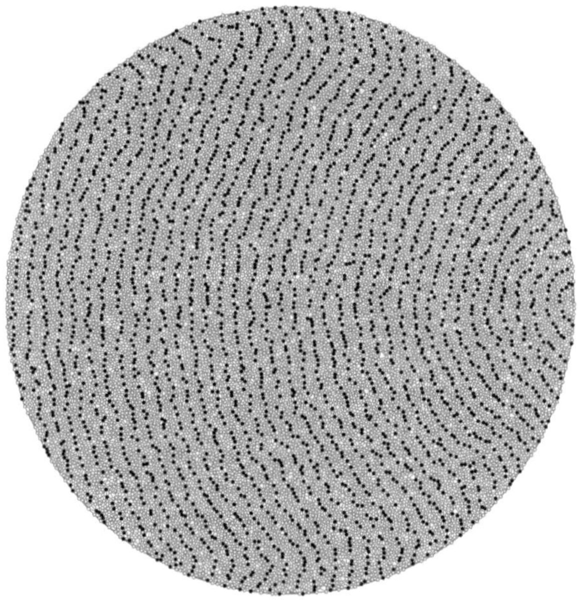

(b)

Fig. 6.4 Time lapsed contours of activation in triangular lattices with density $\phi = 0.407$ and activation threshold (a) $\eta = 1$ and (b) $\eta = 2$. Propagating wave front is converted to a contour every 10th step of simulation. Initially eastmost nodes are activated.

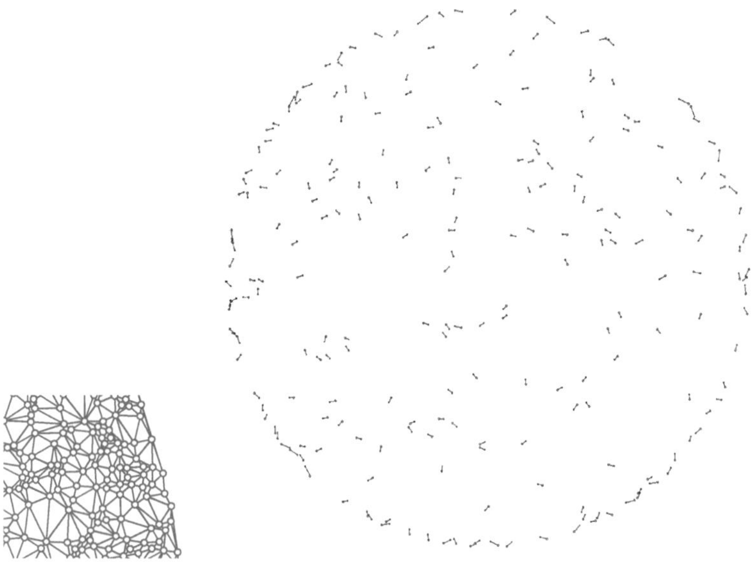

(a) (b)

Fig. 6.5 Geometry of neighbourhoods: (a) a fragment of triangulation, density $\phi = 0.136$, with visible neighbourhoods of edge nodes, edge of triangulation is on the right, (b) vectors from nodes to their geometric centres, only vectors which length exceeding eight units are shown. Density is $\phi = 0.136$.

Typically neighbours of each node are distributed more or less equally around the node while nodes belonging to the edge of triangulation have their neighbours located towards inside of the triangulation or on the edge (Fig. 6.5a). We can integrate spatial distribution of neighbours of node v as a vector $p = \overline{vg}$ from the node v to geometric centre g of the node v's neighbourhood. Edge nodes of triangulation have usually longer vectors p, see Fig. 6.5b. This is why a perturbation propagates faster on or near edges of triangulation for a low threshold of occupation (Fig. 6.4a). Increase of occupation threshold is disadvantageous for edge nodes which is reflected in changed shape of the growing pattern (Fig. 6.4b).

6.2 Absolute Excitability

If a resting node v excites when it has at least one excited neighbour, rule (6.1) $\sigma^t(v) \geq 1$, 'classical' excitation waves are observed (Fig. 6.6ab). By 'classical' we mean that excitation waves annihilate when they reach edges of triangulation, two waves merge when they collide one with another, a single-site perturbation initiates a circularly propagating excitation wave, and refractory state is a necessary component in seeds of target-wave generators.

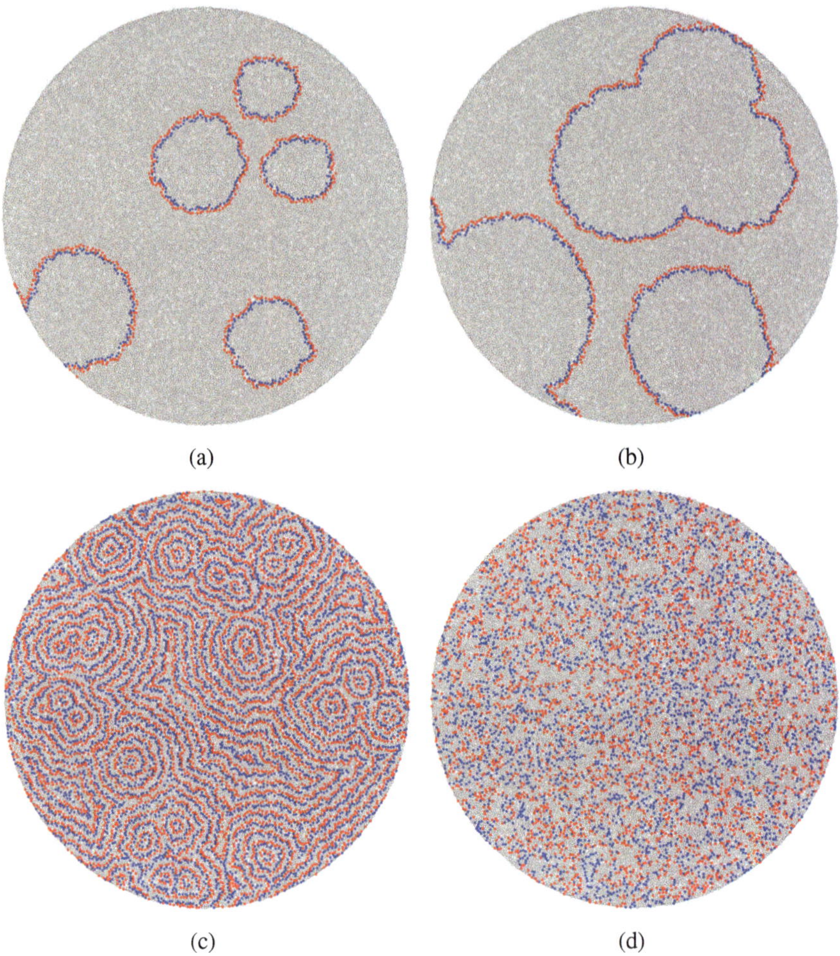

Fig. 6.6 Examples of excitation dynamics for rule (6.1) $\sigma^t(v) \geq 1$ (a)–(c) and rule (6.1) $\sigma^t(v) = 1$ (d): (ab) waves generated by singular excitations merge when collide, (b) random initial configuration (at the beginning of simulation a node got excited state with probability 0.1, refractory state with probability 0.1, and resting state with probability 0.8) develops into a configuration of several target-wave generators; (c) configuration of excitation developed in 300 time steps from initial configuration where all but one nodes are resting. Density of disc-nodes in triangulation is $\phi = 0.407$.

The excitable triangulation behaves less conventionally when nodes are selective in their excitability. Consider the rule (6.1) $\mathbf{S} = \{1\}$: a resting node excites if it has exactly one excited neighbour. Single site excitation leads to formation of circular waves. The waves do merge when collide with each other and also they may form generators of target waves at the sites of their collision (Fig. 6.6c).

Finding 15. *Let a resting node of Delaunay triangulation excite if exactly one neighbour is excited, then a generator of target-waves can be produced by a single-site excitation.*

For $\sigma^t(v) \geq 2$ no excitation persists. However when resting node excites if exactly two or three of its neighbours are excited we obtain a quasi-chaotic excitation dynamics (Fig. 6.6d). This happens when nodes follow state update rules (6.1) for $\mathbf{S} = 2$ and $\mathbf{S} = \{2,3\}$.

A delay in recovery from refractory states modifies space-time dynamics of excitation. Assume that if a node took refractory state then the node stays in the refractory state for δ time steps. In automata governed by rule (6.1) with excitability $\sigma^t(v) \geq 1$ initial random disturbances — nuclei of excited and refractory states — will not lead to formation of wave generators (as in rule $\sigma^t(v) \geq 1$ without node-state recovery delay). However a phenomenon of excitation backfiring is still observed. Travelling excitation wave can backfire with localised excitations. These localised excitations pass through the wave's refractory tail and initiate generation of new wave fronts (Fig. 6.7).

Finding 16. *Delaunay excitable automata governed by rules of absolute excitability exhibit the following phenomena:*

- *threshold activation rules cause formation of classical excitation wave fronts,*
- *rules relying on exact number of excited neighbours show formation of target wave generators during collision between ordinary circular waves,*
- *when recovery of a node from refractory state to resting state is delayed propagating wave fronts backfire with localised excitations, which cause formation of target-wave generators.*

6.3 Relative Excitation

We found that we can initiate a persistent excitation patterns for $0 \leq \varepsilon < 0.25$. For $\varepsilon \geq 0.25$ any initially invoked excitation quickly ceases. Automata excited by rule (6.2) with $\varepsilon = 0$ exhibit persistent global oscillations because every node autonomously follows the cycle $\circ \rightarrow + \rightarrow - \rightarrow \cdots$. For excitability range $0 < \varepsilon < 0.09$ Delaunay automata exhibit classical excitation waves. Single excitation generates single circular wave, clusters of excited and refractory states may become generators of target wave. When a propagating wave front hits edge of triangulation the wave disappears. Two colliding wave-fronts merge or annihilate.

When threshold of excitation increases to $\varepsilon = 0.09$ edges of the triangulation may become reflective for excitation waves. If an excitation wave front hits an edge of the triangulation the front reverses it velocity vector and again propagates inside the

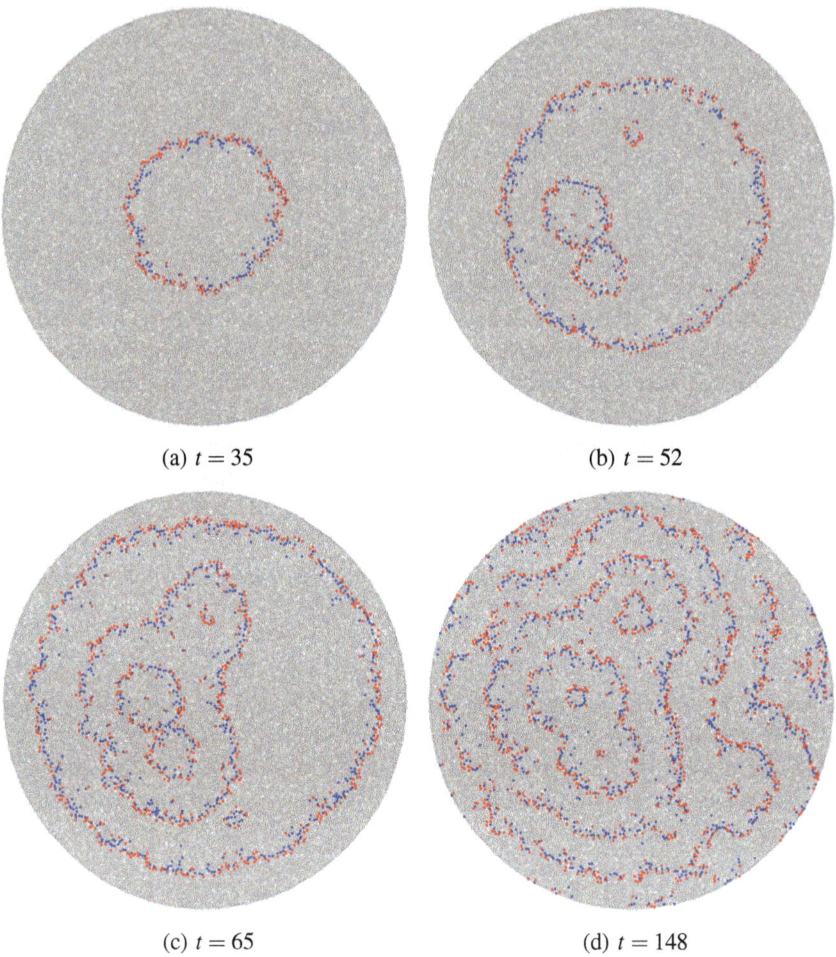

Fig. 6.7 Example of wave generators formed due to backfiring of excitation from a single propagating wave front. Rule $\sigma^t(v) \geq 1$ with delayed (by 10 time steps) recovery from refractory state. Density of disc-nodes in triangulation is $\phi = 0.407$.

triangulation. An example is shown in Fig. 6.8. Initially graph is in resting state. We excited one node. A circular wave of excitation propagates outwards the initial perturbation (Fig. 6.8a). When the wave front reaches edge of the triangulation it starts to disappear and almost annihilate. However in some part of triangulation edge domains of centrifugal wave-fragments evoke centripetal wave-fragments (Fig. 6.8bc). Fragments of the centripetal wave closer to edge of triangulation propagate quicker then those inside the triangulation. Therefore the wave encircles the triangulation (Fig. 6.8de) and collapses inside the triangulation (Fig. 6.8f).

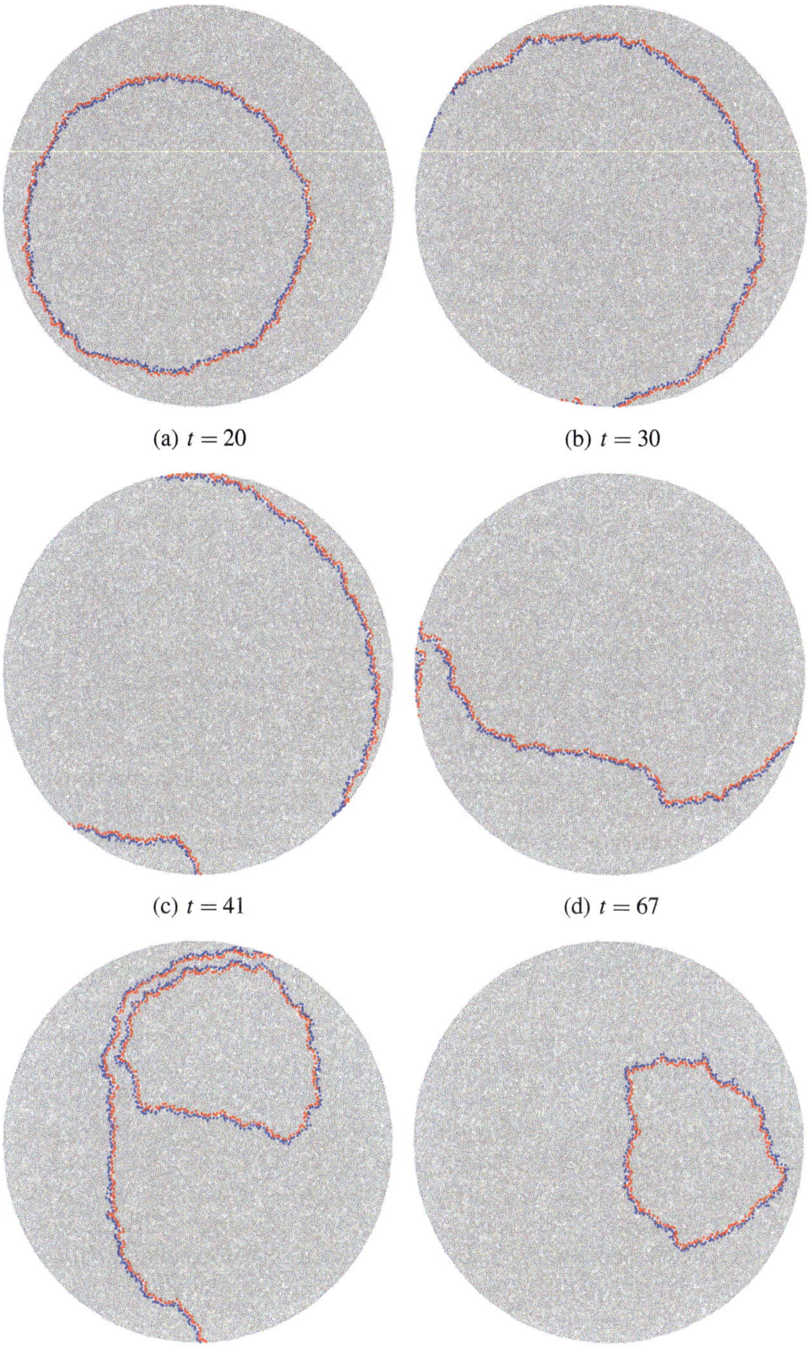

(a) $t = 20$ (b) $t = 30$

(c) $t = 41$ (d) $t = 67$

(e) $t = 100$ (f) $t = 132$

Fig. 6.8 Dynamics of excitable triangulation for rule (6.2) $\varepsilon = 0.091$. Initially just a single node is excited. Density of disc-nodes in triangulation is $\phi = 0.407$.

The reflection of waves can be observed for values $0.09 < \varepsilon < 0.11$. Further increase of excitation threshold brings up the phenomenon of excitation backfiring again.

Triangulations, which nodes are excited if ratio ε is at least 0.11, exhibit backfiring — generators of target waves are formed behind the wave front initiated by a single excitation (Fig. 6.9). Due to inhomogeneous structures of node neighbourhoods the wave front (Fig. 6.9a) breaks up (Fig. 6.9b). A gap is formed and some part of the front folds backward. A spiral wave or waves are formed. They give rise to a succession of target waves (Fig. 6.9cde). Eventually original wave front disappears at the edge of triangulation but the triangulation remains filled with sources of target waves (Fig. 6.9f).

With ε increasing from 0.11 to 0.17 a number of wave generators forming behind propagating wave front increases considerably. Thus, in networks excited by rule (6.2) with $\varepsilon > 0.14$ a generator is formed almost immediately after single-excitation wave starts its propagation. The earlier in time a generator is born the more likely the generator will dominate the triangulation. Therefore in many cases we can observe just few generators close to the site of original excitation (Fig. 6.10).

With ε exceeding 0.17 the automaton approaches regime of sub-excitability. Single site excitation no longer leads to formation of a circular wave. However we observe travelling and stationary localized excitations, distant analogous to wave-fragments in sub-excitable Belousov-Zhabotinsky medium [89]. Initial local perturbations lead to propagating localised excitations. The excitations later can form a localised, and not changing it is outer shape, domain of excitation, see example in Fig. 6.11, or a slowly growing domain, which may eventually occupy the whole triangulation (Fig. 6.12). Usually, two or three wave-fragments are formed near the site of initial activation of resting triangulation. These wave-fragments travel outward the perturbation loci. Some time after their initiation the wave-fragment backfires few excited micro-localizations which may form generators of wave-fragments. Thus the domain of excitation is born. Sometimes generators of wave-fragments emerge due to folding of the 'wings' of a wave-segment backwards. Folded parts of wave-fragment interact with each other and thus they produce the generator.

The growing domains guided by mobile localisations are typical for excitability values $0.17 < \varepsilon < 0.2$. When ε exceed 0.2 the domains seize propagating. The excitation can still persist but in a minuscule quantity. A random initial configuration, where every node gets one of three states equiprobably, is either transformed to a totally resting state or to a resting configuration with one or two tiny oscillators. Examples of most common oscillators are shown in Fig. 6.13.

What are degrees of nodes occupied by non-resting states of the oscillators? We stimulated triangulations with random configurations of excited and refractory states (each state is assigned with probability $\frac{1}{3}$), waited till all excitation but minimal oscillator ceases and recorded degrees of nodes occupied by the oscillator states. We found that nodes occupied by excited or refractory states in the minimal oscillator (Fig. 6.13c) have 4, 6, 7 or 8 neighbours. In larger oscillators nodes have degrees 4 to 8 or 4 to 9.

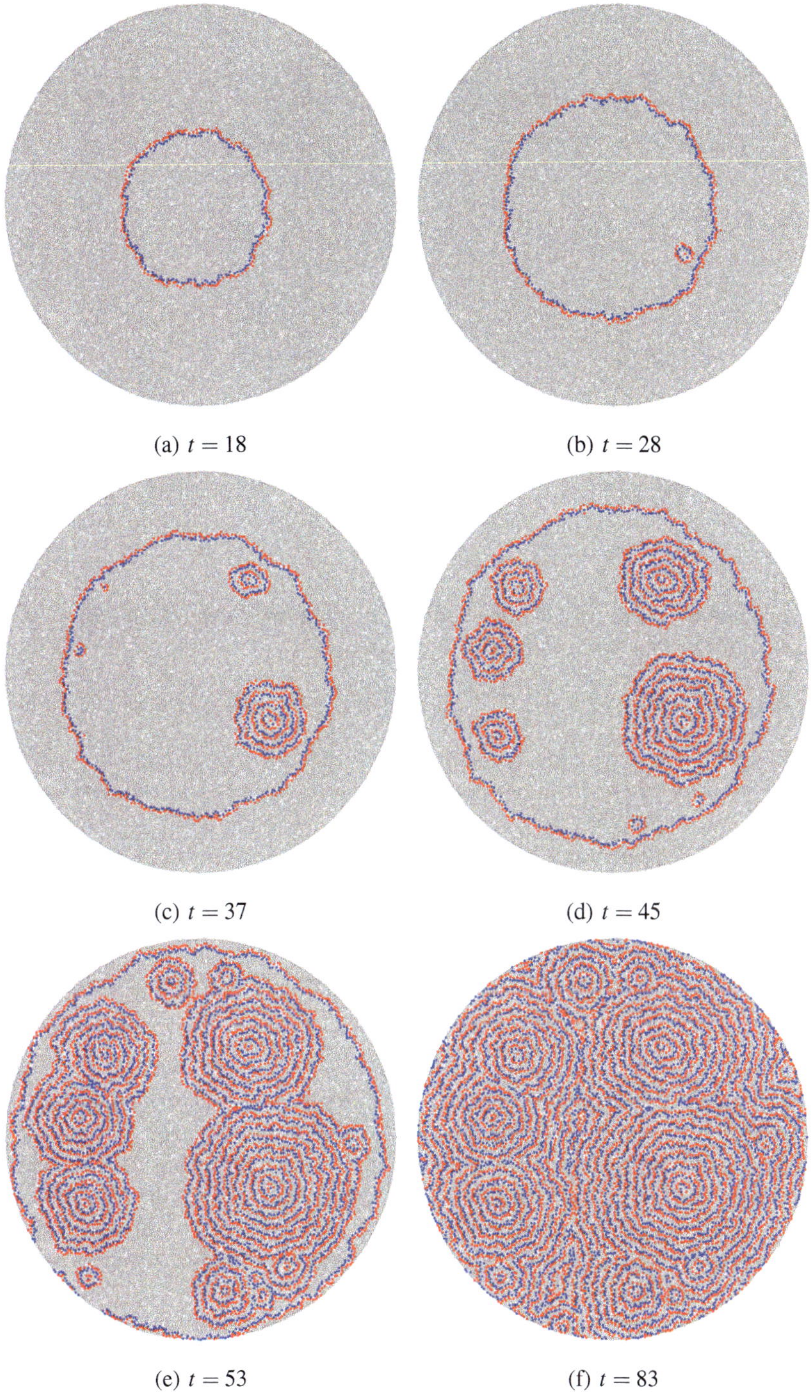

Fig. 6.9 Example of backfiring of excitation wave fronts which leads to formation of generators of target waves in excitable triangulation, density $\phi = 0.407$, for rule (6.2) with $\varepsilon = 0.11$

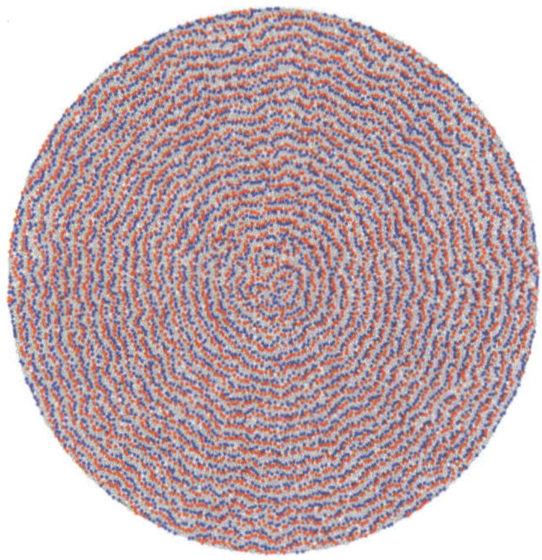

Fig. 6.10 Example of generators formed close to a site of original perturbation. Automaton is excited by rule (6.2) with $\varepsilon = 0.150$. Density of disc-nodes in triangulation is $\phi = 0.407$.

Proposition 6.1. *Diversity of node degrees is a necessary requirement for existence of an oscillator in a sub-excitable triangulation.*

The oscillators survive for $0.2 \leq \varepsilon < 0.25$. For $\varepsilon \geq 0.25$ no excitation persist at all.

Finding 17. *Delaunay excitable automata governed by rules of relative excitability with dimensionless threshold of excitation ε exhibit the following phenomena:*

ε	*phenomenon*
0	*global oscillations*
$0 < \varepsilon < 0.09$	*classical excitation waves*
$0.09 \leq \varepsilon < 0.11$	*waves are reflected from edges*
$0.11 \leq \varepsilon < 0.17$	*waves backfire with excitation*
$0.17 \leq \varepsilon < 0.2$	*localised wave-fragments lead growing tips of branching domains*
$0.2 \leq \varepsilon < 0.25$	*only tiny oscillating domains are present*
$0.25 \leq \varepsilon$	*no excitation persists*

The finding is illustrated in Fig. 6.14. Is there any structural meaning of ε-boundaries between different classes of excitable triangulations? Value $\varepsilon = 0.09$ corresponds to a situation when a node with 11 neighbours has at least one excited neighbour. Nodes with degree 11 are rare species in Delaunay triangulations (Fig. 6.2). However when such nodes are deprived, i.e. for $\varepsilon > 0.09$, from a chance to be excited at all by a minimal perturbation waves of excitation start to be reflected by edges of triangulation.

Excitation wave-fronts backfire when we make nodes with degree 9 non-excitable by raising excitability to over 0.11. This is because $\varepsilon = 0.11$ describes a situation of a node with 9 neighbours which has just one neighbour excited.

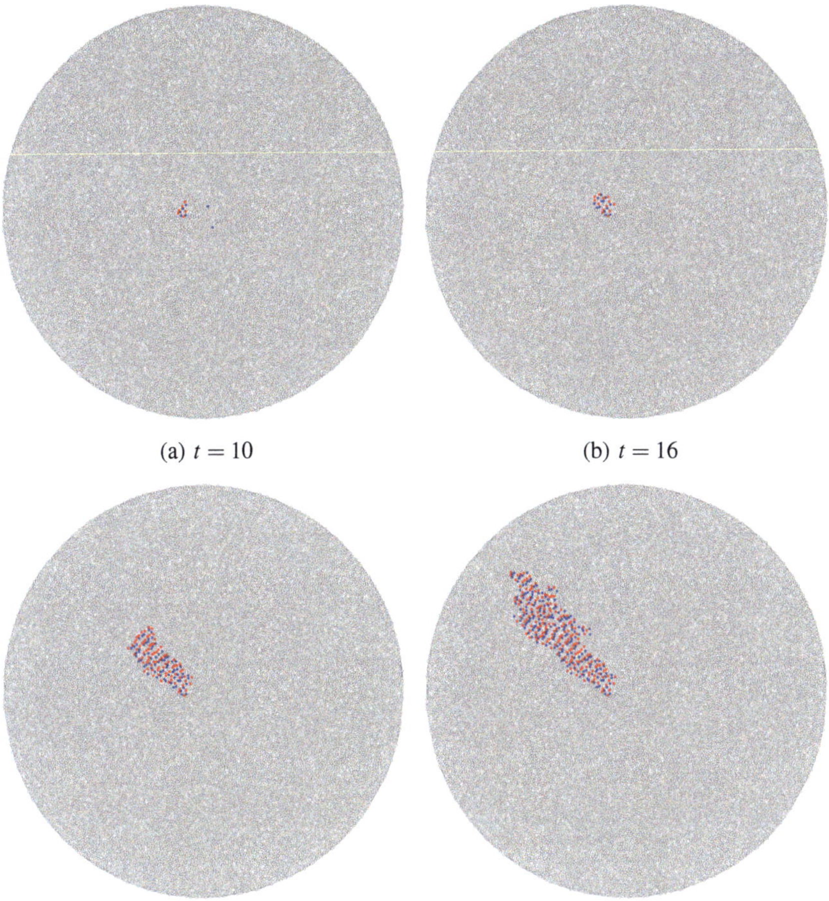

(a) $t = 10$ (b) $t = 16$

(c) $t = 40$ (d) $t = 142$

Fig. 6.11 Example of excitation dynamics for $\varepsilon = 0.17$. A local perturbation leads to localised excitations and slowly growing domains. The domain shown in (d) has stationary boundaries however configuration of excited and refractory states inside the domain is changing. Density of disc-nodes in triangulation is $\phi = 0.407$.

Localizations emerge in triangulations when $\varepsilon \geq 0.17$. In principle this value of ε corresponds to two situations: a node with six neighbours has one excited neighbour and a node with 12 neighbours has two excited neighbours. The latter situation is not important because nodes with 12 neighbours are extremely rare. Thus, we can propose that excitation becomes localised when six-neighbour nodes are activated by at least two neighbours. Two excited neighbours is a minimal requirement for existence of localizations in orthogonal automaton networks with eight-cell neighbourhoods [9]. In a regular network, or cellular automaton, a travelling localisation preserves its shape and velocity vector as long as necessary.

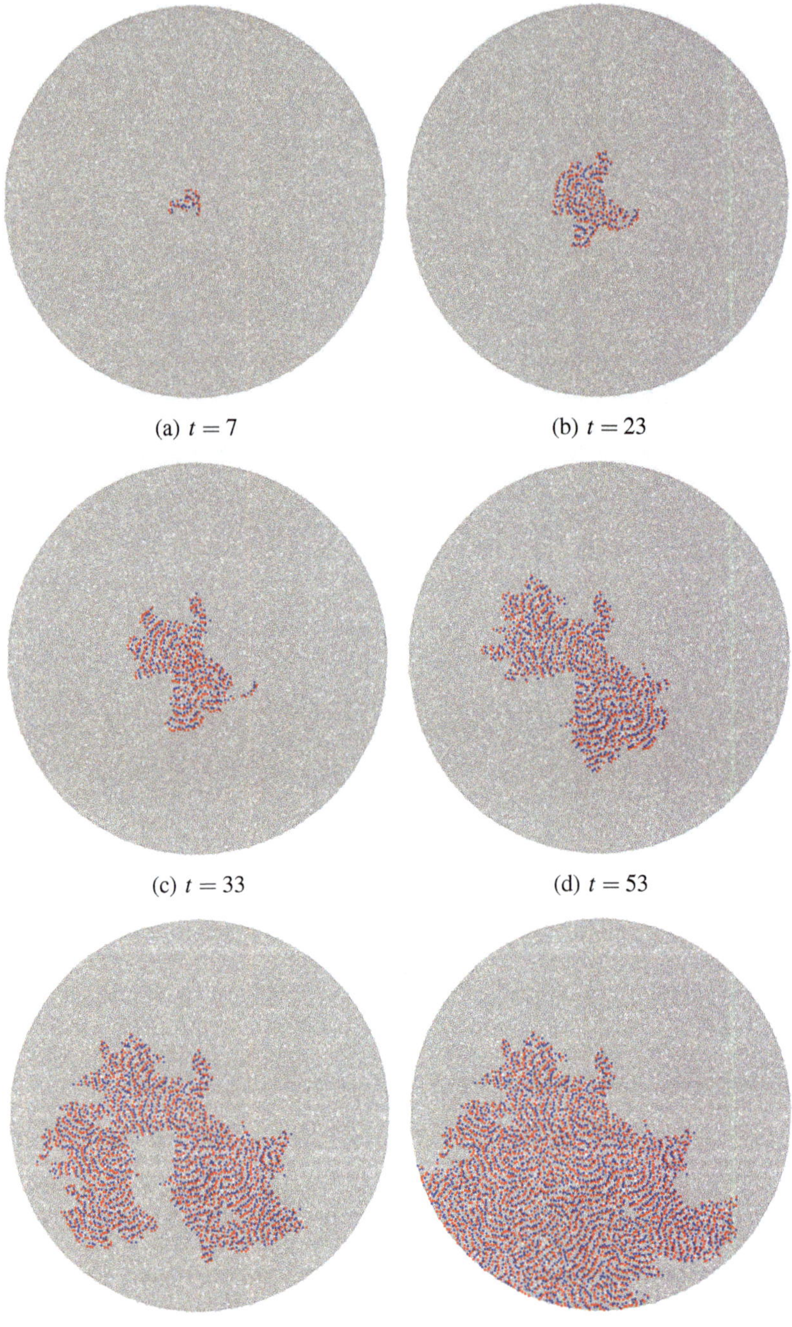

(a) $t = 7$ (b) $t = 23$

(c) $t = 33$ (d) $t = 53$

(e) $t = 119$ (f) $t = 278$

Fig. 6.12 In triangulation, density of disc-nodes is $\phi = 0.407$, excited by rule (6.2) with $\varepsilon = 0.17$ a local perturbation leads to localised excitations and slowly propagating domains, which may occupy significant number of nodes

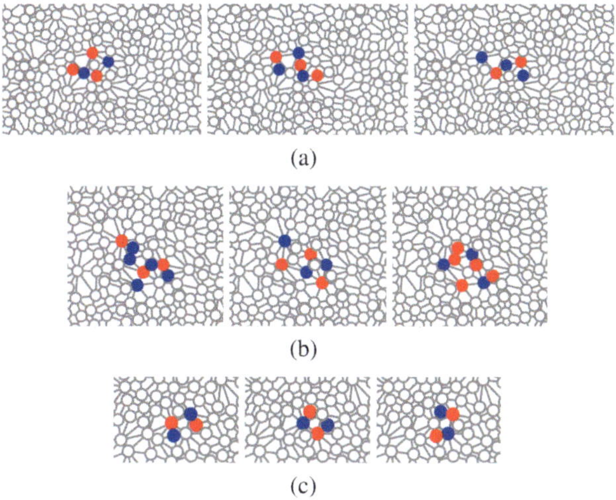

Fig. 6.13 Examples of oscillators: (a) and (b) two oscillators for excitability $\varepsilon = 0.2$, (c) minimal oscillator, $\varepsilon = 0.21$. Each sub-figure presents three consecutive states of an oscillator. Density of disc-nodes in triangulation is $\phi = 0.407$.

Delaunay triangulations are irregular therefore localised wave-fragments do not survive for a long time. These travelling localizations frequently change directions of their motion. Soon after its birth a localisation either perishes or expands, backfires and forms a slowly growing cluster of excitation (Fig. 6.12). Any excitation ceases to sustain when ε exceeds 0.25, which corresponds to one of four, two of eight, and three of twelve excited neighbours.

Does density of disc-nodes affect structure of behavioural space?

For relative excitability $0 < \varepsilon < 0.09$ decrease of density ϕ increases chances of wave-generators to be formed from a simple propagating wave-front. This possibly happens due to increased inhomogeneity of node degrees and increasing role of each particular node in excitation propagation. In case of $0.09 \leq \varepsilon < 0.11$ decrease of ϕ increases opportunities for wave-generators to form when a wave collides to edge of triangulation.

For excitability $0.11 \leq \varepsilon < 0.17$ time of backfiring and formation of generators behind wave-fronts, is proportional to density ϕ, higher the ϕ the longer it takes for a singular wave-front to backfire. And finally, chances that an excitation domain growing from a single perturbation, in triangulations with excitability $0.17 \leq \varepsilon < 0.2$, occupies the whole triangulation are inversely proportional to density of disc-nodes in the triangulation.

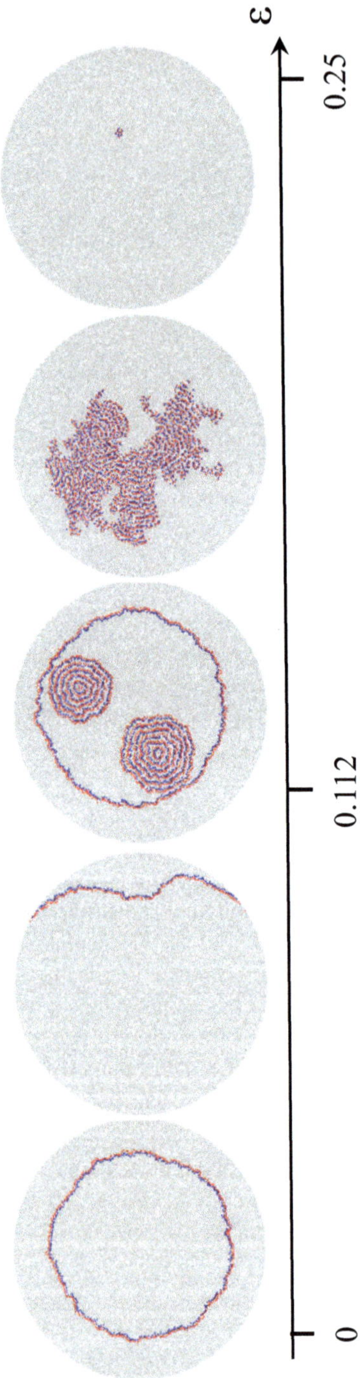

Fig. 6.14 Visualisation of basic types of excitation parameterised by relative excitability threshold ε

6.4 Summary

Excitable automata on Delaunay triangulations exhibit a range of non-trivial space-time dynamics ranging from reflection of excitation waves by edge of triangulation to branching domains of excitation guided by travelling localised excitations. Growing domains of localised excitation may be analogous to patterns of excitation in Belousov-Zhabotinsky reaction in micro-emulsion [153]. Reflection of waves by edge of triangulation supports previously published results on active waves reflection in spatially inhomogeneous non-linear media [190] and reflection of reaction-diffusion waves by no-flux boundary due to consumption of reactant between a wave and a boundary [193]. Further studies of wave speed up along edge of triangulation could be done by additional techniques of characterising topological and dynamical properties of complex networks, e.g. a diversity entropy proposed in [251]. We believe the findings will contribute towards designs of novel computing substrates in non-crystalline structure. Also, automaton interpretation of excitation dynamics on Delaunay triangulation can make a viable model of automaton-network approaches to design of nano-computing devices [146].

Chapter 7
Excitable β-Skeletons

Given a set **V** of planar points, for any two points p and q we define β-neighbourhood $U_\beta(p,q)$ as an intersection of two discs with radius $\beta|p-q|/2$ centered at points $((1-\frac{\beta}{2})p, \frac{\beta}{2}q)$ and $(\frac{\beta}{2}p, (1-\frac{\beta}{2})q)$, $\beta \geq 1$ [150, 160], see examples of the lunes in Fig. 7.1. Points p and q are connected by an edge in β-skeleton if the pair's β-neighbourhood contains no other points from **V**.

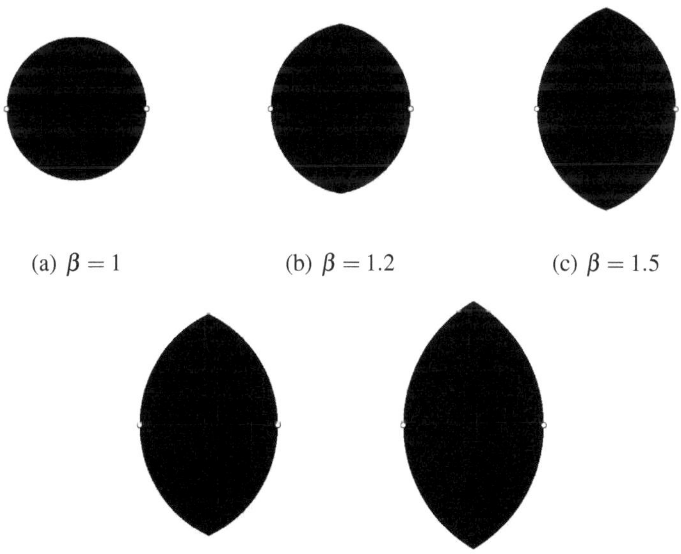

(a) $\beta = 1$ (b) $\beta = 1.2$ (c) $\beta = 1.5$

(d) $\beta = 1.7$ (e) $\beta = 2$

Fig. 7.1 Examples of the lunes between two points (small circles) for various values of β

A. Adamatzky: Reaction-Diffusion Automata, ECC 1, pp. 135–153.
springerlink.com

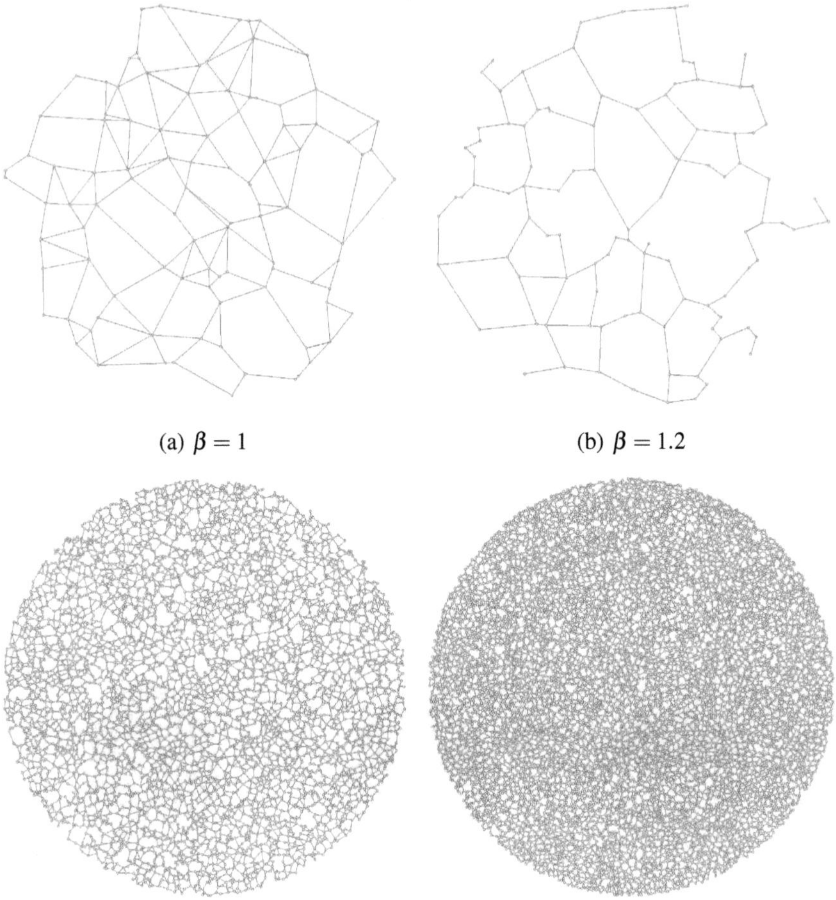

(a) $\beta = 1$ (b) $\beta = 1.2$

(c) $\beta = 1$ (d) $\beta = 1.2$

Fig. 7.2 Examples of β-skeletons: (a) $n = 100$, $\phi = 0.0027$, and $\beta = 1$, (b) $n = 100$, $\phi = 0.0027$, and $\beta = 2$, (c) $n = 5000$, $\phi = 0.136$, and $\beta = 1.5$, (b) $n = 10000$, $\phi = 0.271$, and $\beta = 1.5$

A β-skeleton is a graph $G_\beta(\mathbf{V}) = \langle \mathbf{V}, \mathbf{E}, \beta \rangle$, where nodes $\mathbf{V} \subset \mathbf{R}^2$, edges $\mathbf{E}$, and for $p, q \in \mathbf{V}$ edge $(pq) \in \mathbf{E}$ if $U_\beta(p,q) \cap \mathbf{V}/\{p,q\} = \emptyset$. Parameterisation β is monotonous: if $\beta_1 > \beta_2$ then $G_{\beta_1}(\mathbf{V}) \subset G_{\beta_2}(\mathbf{V})$ [150, 160]. A β-skeleton is Gabriel graph [191] for $\beta = 1$ and the skeleton is relative neighbourhood graph for $\beta = 2$. We consider only skeletons with $1 \leq \beta \leq 2$ because β-skeletons are non-planar for $\beta < 1$ and they are disconnected for $\beta > 2$.

We study β-skeletons of planar set of n discs. Centres of the discs form set $\mathbf{V}$. Each disc has a radius 2.5 units. The discs are randomly distributed in a large disc with radius 480 units. We undertake computational experiments with n varying from

1000 to 15000 (Fig. 7.2). Number of nodes *per se* is not as important as density of their packing therefore we refer to any particular **V** via density of node packing $\phi = 0.027, \cdots, 0.407$.

An excitable β-skeleton is defined as follows. A node $p \in \mathbf{V}$ is a finite state machine. Every node takes three states: resting ($\circ$) state, excited ($+$) state and refractory ($-$) state. A node updates its states in discrete time depending on states of its neighbours. All nodes update their states simultaneously.

We assume that a resting node excites depending on a number of excited neighbours. If a node is excited at time t the node takes refractory state at time step $t+1$, independently on states of its neighbours. Transition from refractory to resting state is also unconditional.

Let $\nu(p) = \{q \in \mathbf{V} : (pq) \in \mathbf{E}\}$ be a neighbourhood, or set of neighbours, of node p, p^t a state of node p at time step t, $\sigma^t(p)$ a number of excited neighbours of p at step t, and $d(p)$ a degree, or a number of neighbours $|\nu(p)|$, of node p. Then the node-state transition function can be defined as follows:

$$p^{t+1} = \begin{cases} +, \text{ if } C(u(p)^t) \\ -, \text{ if } p^t = + \\ \circ, \text{ if } p^t = - \end{cases} \tag{7.1}$$

We consider two versions of the excitation condition $C(u(p)^t) \in \{ \text{TRUE}, \text{FALSE} \}$:

- *Absolute excitability*: $C(u(p)^t) = \sigma^t(p) \geq \theta$, $\theta = 1, \cdots$
- *Relative excitability*: $C(u(p)^t) = (\frac{\sigma^t(p)}{d(p)} > \varepsilon), \varepsilon \in [0, 1]$.

Absolute excitability is the most common approach of defining rules in excitable discrete systems however it does not account for diversity of node degrees in disordered systems. Thus we also explore skeletons with relative excitability.

In the chapter we illustrate space-dynamics of excitation by snapshots of skeleton configurations. We analyse integral dynamics of skeletons using activity α. The activity α is a ratio of excited nodes to a total number of nodes averaged over several time steps. The activity α is measured after initial transient period, when excitation patterns are given a chance to occupy the whole skeleton (usually a hundred time steps is sufficient).

7.1 Absolutely Excitable Skeletons

In β-skeletons governed by absolute excitability rule excitation persists only for threshold $\theta = 1$.

Finding 18. *Activity α is proportional to density ϕ and inversely proportional to β.*

Figure 7.3 shows exactly how activity α depends on β. Note that maximum possible α equals 0.33, for sustainable activity, because an excited node always becomes

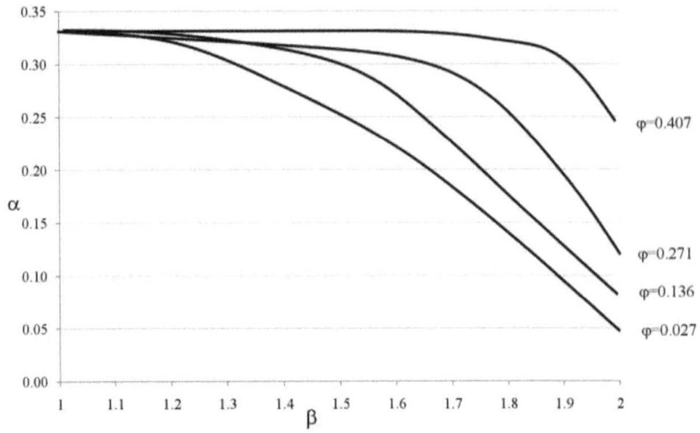

Fig. 7.3 Activity α versus β for various densities ϕ of node packing in β-skeletons with absolute excitation threshold $\theta = 1$. Skeletons are developed from initial random configuration, where a node is assigned excited state probability 0.1 and is resting otherwise.

refractory, and a refractory node always becomes resting. Thus even if the whole skeleton is active, only third of its nodes are excited at any given time step. The activity decreases polynomially with increase of β, degree of the polynomial is proportional to density ϕ of nodes in the skeleton.

Spatial counterparts of integral activity are shown in Fig. 7.4. These are examples of excitation patterns emerged after a single node of the resting skeleton was excited. In β-skeletons with high density of nodes packing $\phi = 0.2071$ and 0.407 excitation of a single node of a resting skeleton leads to formation of a generator of spiral and target waves if $1 \leq \beta < 1.4$. The most noticeable pattern of the waves is observed for $\phi = 0.407$ and $\beta = 1$. Increase of β leads to break up wave-fronts and only localised compact clusters of activity remain on the skeleton.

What values of ϕ and β cause excitation fail to span the whole graph? For what values of ϕ and β only tiny localised clusters of oscillating activity remain in a skeleton? We call values β critical if patterns of excitation change abruptly, e.g. excitation dynamics changes from target waves filling the whole skeleton to slowly growing, β_d, or even localised domains of activity, oscillators, β_o.

Finding 19. *Critical values β_d and β_o are linearly proportional to density ϕ of nodes in skeleton.*

Assuming that a skeleton fully occupied by target waves has activity level $\alpha = 0.33$ even tiniest decrease in α indicates presence of a stationary resting domain. See example in Fig. 7.4, $\phi = 0.271$ and $\beta = 1.4$. Let us assume that excitation fails to span a skeleton when $\alpha \leq 0.32$ and only bounded domains of excitation are present if $\alpha \leq 0.25$. Let β_d and β_o be critical values of β thus that if for a given β-skeleton value β exceeds β_d then the skeleton is not fully occupied by excitation, and if $\beta \geq \beta_o$ then only localised oscillators are formed. Figure 7.5 presents critical values of β computed for 25 sample configurations of β-skeletons with ϕ varying from

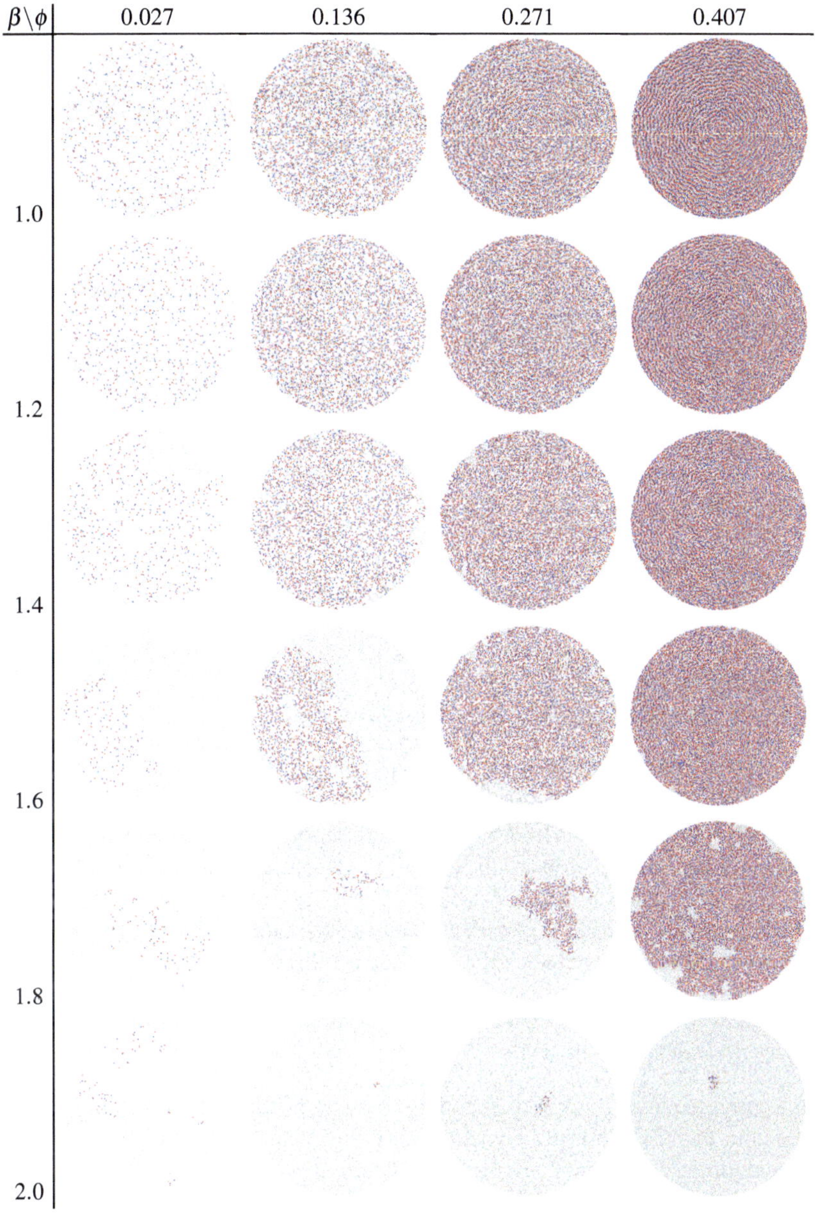

Fig. 7.4 Configurations of excitation on β-skeletons with node density $\phi = 0.027, 0.054, \cdots, 0.407$ and absolute threshold of excitation $\theta = 1$. All configurations are developed from a localised initial excitation: all nodes but one are resting, one node is excited. The configurations are snapped after 50 steps of development.

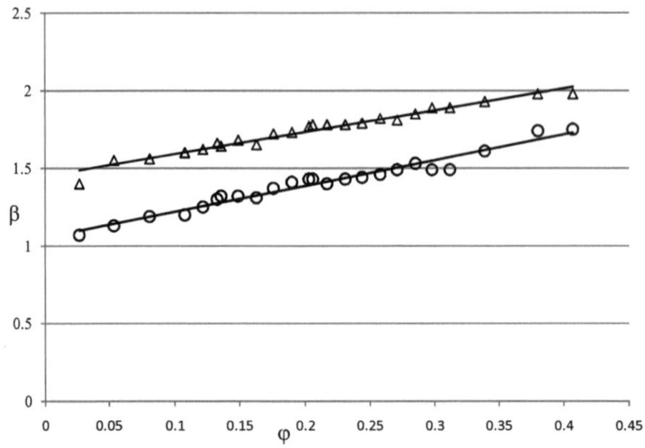

Fig. 7.5 Dependence of critical values β_d and β_o on ϕ. Data points for β_d are shown by circles, for β_a by triangles. Trend lines are $\beta_d = 1.65\phi + 1.05$, $R^2 = 0.961$ and $\beta_a = 1.42\phi + 1.45$, $R^2 = 0.964$.

0.027 to 0.407. What are structural correlates responsible for the transition from a wide-spread excitation to localised domains?

Finding 20. *Excitation ceases to occupy a whole skeleton when a mode of node degree distribution of the skeleton changes from 4 to 3.*

We found this by directly comparing β_d and integral characteristics of node degree distributions for a range of β and ϕ (Fig. 7.6a). Mode of node degree distribution is not helpful however in detecting skeletons supporting localised oscillators. As shown in Fig. 7.6a skeletons with mode 3 of degree distribution can exhibit very large and very small domains of oscillating activity. An average node degree of a skeleton gives us a bit more detailed picture of ϕ-β-induced structural transitions and thus can be used to detect when localised oscillators are formed.

Finding 21. *A skeleton exhibits only localised oscillating domains of activity when average node degree drops below 3.*

The average node degrees corresponding to onset of localised oscillations are underlined in Fig. 7.6a. Further comparison of average node degrees (Fig. 7.6a), and corresponding space-time configurations of activity (see e.g. Fig. 7.4) allows us to state the following observation. Average node degree δ of a skeleton determines space-time dynamics of excitation as follows:

- $3.9 \leq \delta$: classical wave-like dynamics, generators of target waves, ordered dynamics,
- $3.7 \leq \delta < 3.9$: generators of wave-fragments, disordered excitation activity,
- $3.0 \leq \delta < 3.7$: formation of slowly propagating domains of excitation, some loci of a skeleton remain resting,

$\phi\backslash\beta$	1.0	1.1	1.2	1.3	1.4	1.5	1.6	1.7	1.8	1.9	2.0
0.027	4	3	3	3	3	3	3	3	3	3	3
0.054	4	4	3	3	3	3	3	3	3	3	3
0.081	4	4	4	3	3	3	3	3	3	3	3
0.108	4	4	4	3	3	3	3	3	3	3	3
0.136	4	4	4	3	3	3	3	3	3	3	3
0.163	4	4	4	4	3	3	3	3	3	3	3
0.190	4	4	4	4	3	3	3	3	3	3	3
0.217	4	4	4	4	4	3	3	3	3	3	3
0.244	**5**	4	4	4	4	3	3	3	3	3	3
0.271	**5**	4	4	4	4	3	3	3	3	3	3
0.298	**5**	4	4	4	4	4	3	3	3	3	3
0.325	**5**	**5**	4	4	4	4	3	3	3	3	3
0.352	**5**	**5**	4	4	4	4	4	3	3	3	3
0.380	**5**	**5**	4	4	4	4	4	3	3	3	3
0.407	**5**	**5**	4	4	4	4	4	3	3	3	3

(a)

$\phi\backslash\beta$	1.0	1.1	1.2	1.3	1.4	1.5	1.6	1.7	1.8	1.9	2.0
0.027	4.0	3.6	3.5	3.3	3.2	3.0	2.9	2.8	2.7	2.6	2.5
0.054	4.1	3.8	3.6	3.4	3.2	3.1	2.9	2.8	2.7	2.6	2.6
0.081	4.2	3.9	3.7	3.5	3.3	3.2	3.0	2.9	2.8	2.7	2.6
0.108	4.2	4.0	3.7	3.5	3.3	3.2	3.1	2.9	2.8	2.7	2.7
0.136	4.3	4.0	3.8	3.6	3.4	3.2	3.1	3.0	2.9	2.8	2.7
0.163	4.4	4.1	3.9	3.7	3.5	3.3	3.2	3.0	2.9	2.8	2.7
0.190	4.5	4.2	3.9	3.7	3.5	3.3	3.2	3.1	2.9	2.8	2.7
0.217	4.6	4.3	4.0	3.8	3.6	3.4	3.3	3.1	3.0	2.9	2.8
0.244	4.6	4.3	4.1	3.8	3.6	3.5	3.3	3.2	3.0	2.9	2.8
0.271	4.7	4.4	4.1	3.9	3.7	3.5	3.3	3.2	3.1	2.9	2.8
0.298	4.8	4.5	4.2	4.0	3.7	3.6	3.4	3.3	3.1	3.0	2.9
0.325	4.9	4.5	4.3	4.0	3.8	3.6	3.5	3.3	3.2	3.0	2.9
0.353	4.9	4.6	4.3	4.1	3.9	3.7	3.5	3.4	3.2	3.1	2.9
0.380	5.0	4.7	4.4	4.2	4.0	3.8	3.6	3.4	3.2	3.1	3.0
0.407	5.1	4.7	4.5	4.2	4.0	3.8	3.6	3.4	3.3	3.1	3.0

(b)

Fig. 7.6 Mode (a) and average (b) node degree of β-skeletons for $\beta \in [1,2]$ and $\phi = 0.027, \cdots, 0.407$. In (b) entries corresponding to $\beta_o(\phi)$ are underlined.

- $\delta \leq 3.0$: only small localised stationary domains of oscillating activity are formed.

The proposed parameterisation is visualised in Fig. 7.7.

7.2 Relatively Excitable Skeletons

Here we consider behaviour of skeletons with node density $\phi = 0.407$. Skeletons with density $\phi = 0.407$ (15,000 disc-nodes) show a wide representation of structural properties because they exhibit widest range of modes of node degree distributions (Fig. 7.6). In computational experiments we perturbed a resting skeleton with small excitation: every node is assigned an excited state with probability 0.1. We allowed the skeleton to develop its excitation patterns for 100 iterations and recorded activity α then. Examples of integral activity of skeletons for $\beta = 1$, 1.5 and 2 are shown in

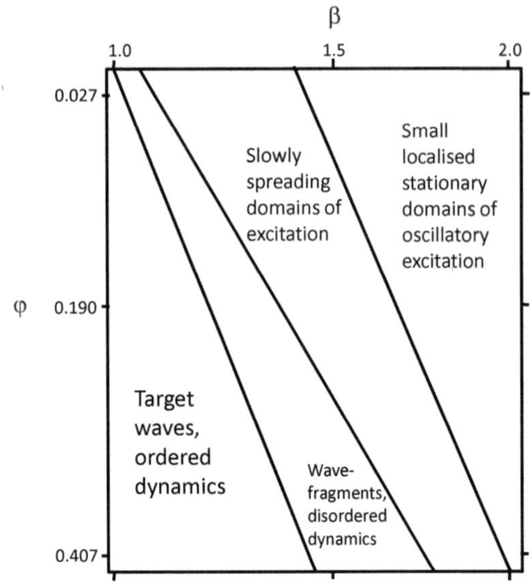

Fig. 7.7 Parameterisaton of a space-time dynamic of β-skeleton with absolute excitation threshold by node density ϕ and β

Fig. 7.8. We see that drastic changes in activity levels occur when ε changes from 0.199 to 2 (first drop in activity level is most clear in Fig. 7.8a), and when ε changes from 0.249 to 0.25 (second drop in activity level, see e.g. Fig. 7.8b).

More detailed description of how α depends on β is provided in Fig. 7.9. For $\varepsilon \in [0, 0.2[$ activity level α is not changed with increase of β, $1 \leq \beta \leq 1.7$. However as soon as β reaches 1.8 activity level slightly decreases. The activity level drops down by one third when $\beta = 2$ Fig. 7.9. For $\varepsilon \in [0.25, 0.33[$ activity monotonously increases with increase of β. For $\varepsilon \in [0.2, 0.249[$ activity α grows with growth of β till $\beta = 1.6$. The activity remains unchanged for $\beta = 1.7$ and then decreases with further increase of β. Let us discuss space-time dynamics of these β-skeletons for selected values of β.

Gabriel Graph, $\beta = 1$

For $\varepsilon \in [0, 0.166]$ skeletons exhibit classical properties of discrete excitable media. A single excitation causes formation of a spiral wave or a generator of spiral or target waves. Successions of circular waves fill the whole skeleton (Fig. 7.10ab). As soon as ε exceed 0.166 excitation wave-fronts break up into separate wave fragments. Thus branching domains of excitation activity are formed (Fig. 7.10c).

Only tiny oscillating activity domains emerge for $\varepsilon \geq 0.2$ (Fig. 7.10d). Life-cycles of three oscillators are shown in Fig. 7.11. Left oscillator in Fig. 7.11 is a

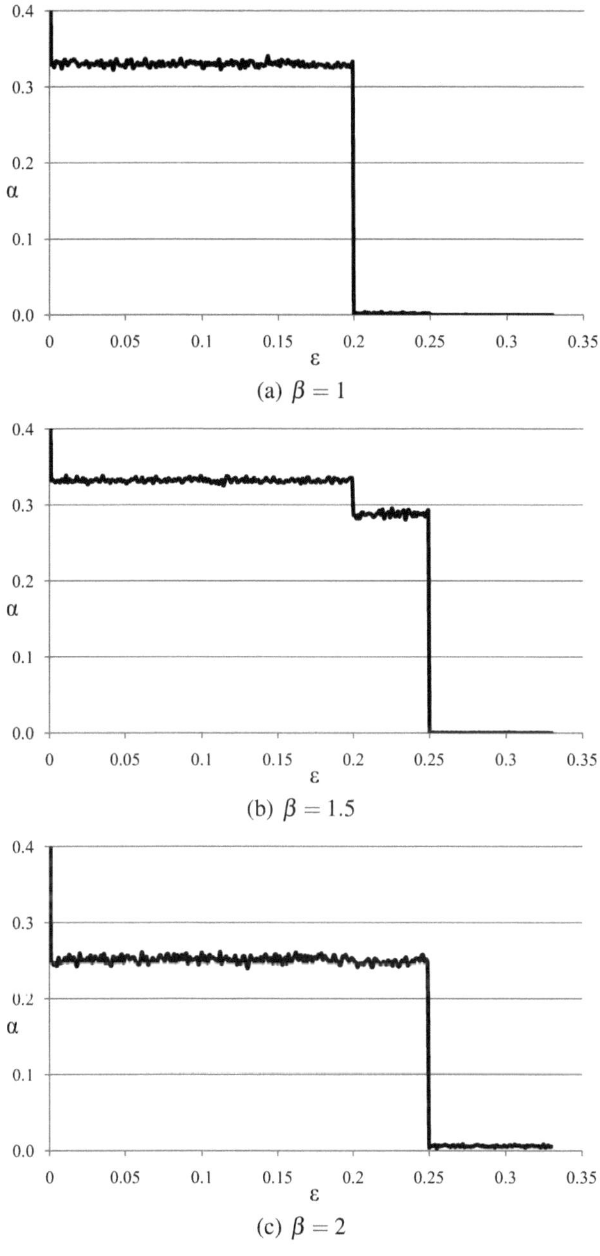

Fig. 7.8 Activity α versus relative excitation threshold ε of β-skeletons for (a) $\beta = 1$, (b) $\beta = 1.5$, (c) $\beta = 2$

$\beta\backslash\varepsilon$	$\varepsilon \in [0,0.2[$	$\varepsilon \in [0.2,0.25[$	$\varepsilon \in [0.25,0.33[$
1	0.33	0.002	0
1.1	0.33	0.01	0
1.2	0.33	0.04	0
1.3	0.33	0.15	0
1.4	0.33	0.23	0.00007
1.5	0.33	0.29	0.00010
1.6	0.33	0.31	0.00015
1.7	0.33	0.31	0.00020
1.8	0.32	0.305	0.00150
1.9	0.295	0.285	0.00300
2	0.24	0.245	0.01

Fig. 7.9 Dependence of activity level α of β-skeletons on β and ε. Initially resting skeleton is set to a random configuration of excitation by exciting each node with probability 0.1. Values of α are measured 100 iterations after initial excitation of skeletons.

typical one. Excitation wave, consisting of one excited state and one refractory state, runs along a cycle of five nodes. The oscillator thus has period five (Fig. 7.11a–f). The oscillator conserves number of non-resting states.

Oscillator in the top part of snapshots in Fig. 7.11 has period three. It changes its state between configurations with two excited and one refractory and one excited and two refractory states. The activation does not cycle but is exchanged between several nodes. First one node a excites its two neighbours. These excited nodes convey excitation to two other nodes (their neighbours), which in turn excite the original node a (Fig. 7.11a–d).

Oscillator shown in right part of snapshots in Fig. 7.11 consists of four excited and four refractory states, the number of the non-resting states is conserved. The oscillator has period three (Fig. 7.11a–d). This oscillator has changes its size by shifting between compressed (Fig. 7.11c), intermediate (Fig. 7.11a) and fully expanded (Fig. 7.11b) configuration.

Skeleton becomes non-excitable, i.e. no excitation persists, when ε exceeds 0.25.

$\beta = 1.5$

Propagating domains of excitation are circularly shaped however they do not show any pronounced wave-fronts, the excitation patterns are rather disordered for $\varepsilon \in [0,0.199]$ (Fig. 7.12ab). With ε exceeding 0.199 activity domains change their shapes form circular to branching, tree-like propagating domains (Fig. 7.12c). Size of such branching domains decreases with increase of ε. When $\varepsilon \geq 0.25$ all domains cease propagating and just tiny clusters, oscillators, of activity are formed (Fig. 7.12d). A typical oscillator is a cycle of three or four nodes around which a quasi-one-dimensional excitation propagates. Sometimes the running excitation

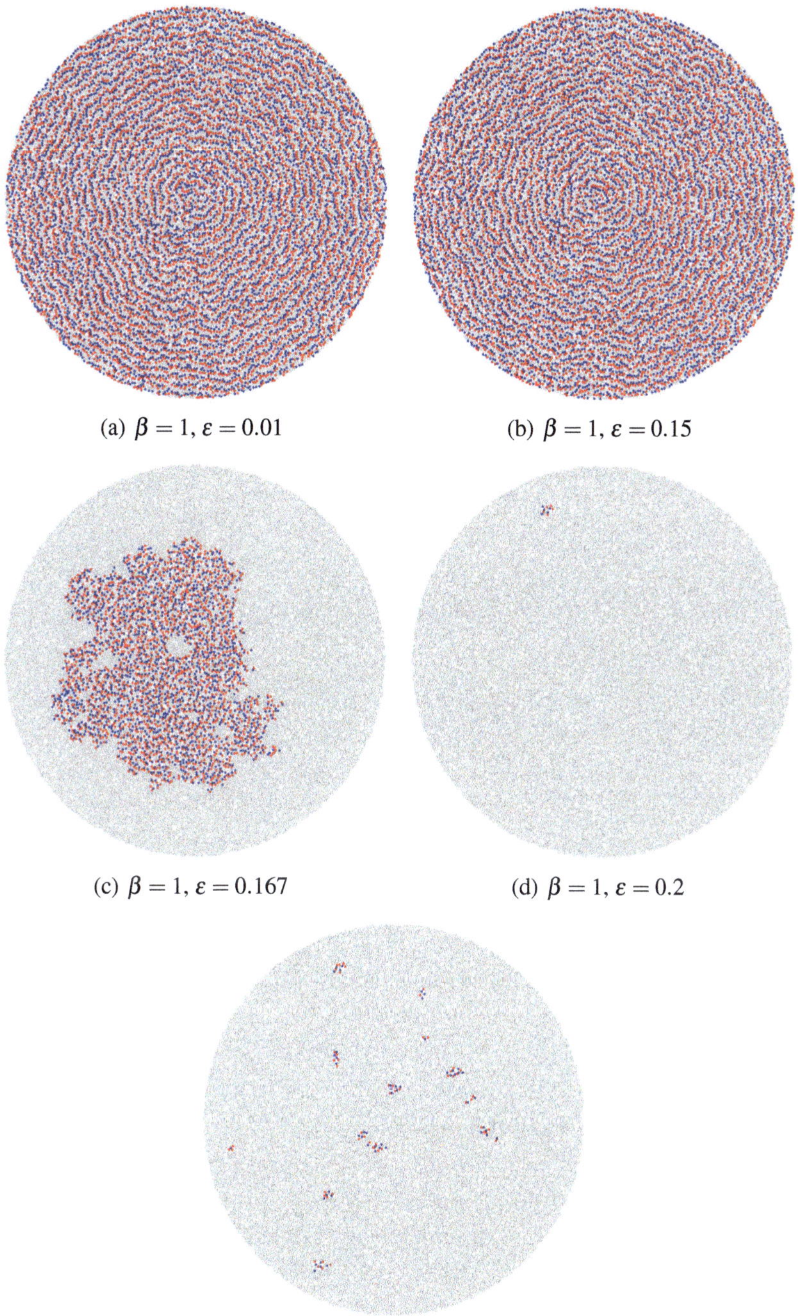

(a) $\beta = 1, \varepsilon = 0.01$

(b) $\beta = 1, \varepsilon = 0.15$

(c) $\beta = 1, \varepsilon = 0.167$

(d) $\beta = 1, \varepsilon = 0.2$

(e) $\beta = 1, \varepsilon = 0.24$

Fig. 7.10 Snapshots of excitation dynamics in β-skeletons ($\beta = 1$) with relative excitation threshold. (a)–(d) singular initial excitation, (e) random initial excitation, a node is excited with probability 0.1.

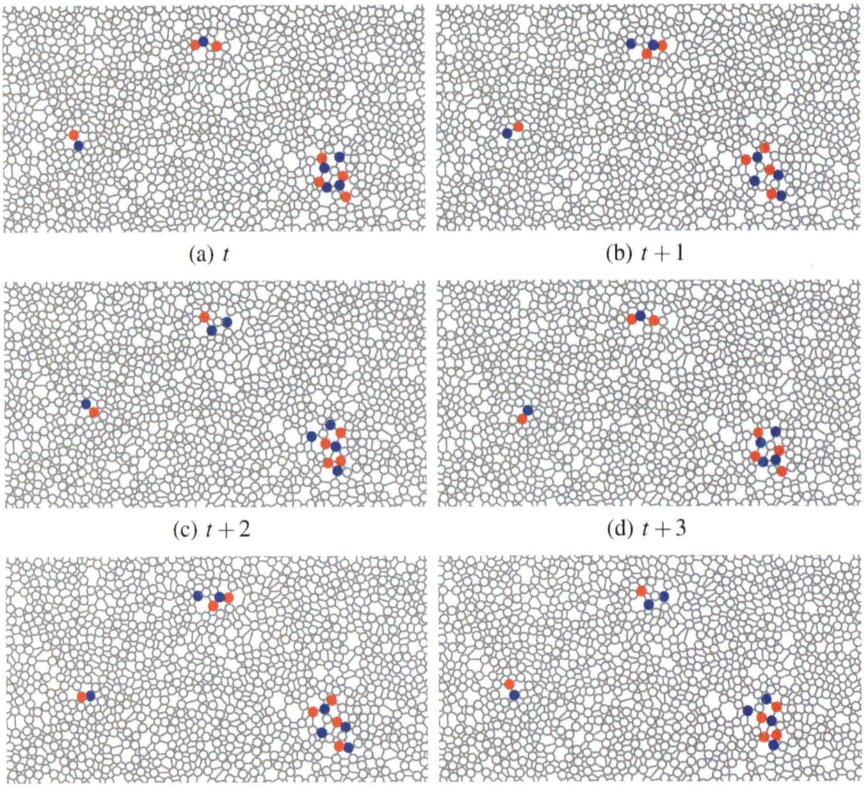

(a) t (b) $t+1$ (c) $t+2$ (d) $t+3$ (e) $t+4$ (f) $t+5$

Fig. 7.11 Examples of oscillators in β-skeletons ($\beta = 1$) with relative excitation threshold $\varepsilon = 0.24$

sends 'sparks' of activity to lateral nodes; these sparks extinguish in few steps of development. No excitation persists in skeletons for $\varepsilon > 0.33$.

Relative Neighbourhood Graph, $\beta = 2$

Single excitation gives birth to slowly propagating irregularly-shaped domains of excitation activity (Fig. 7.13a). The domains originated form a single excitation never span the whole skeleton. Sizes of domains decrease with increase of ε (Fig. 7.13bc).

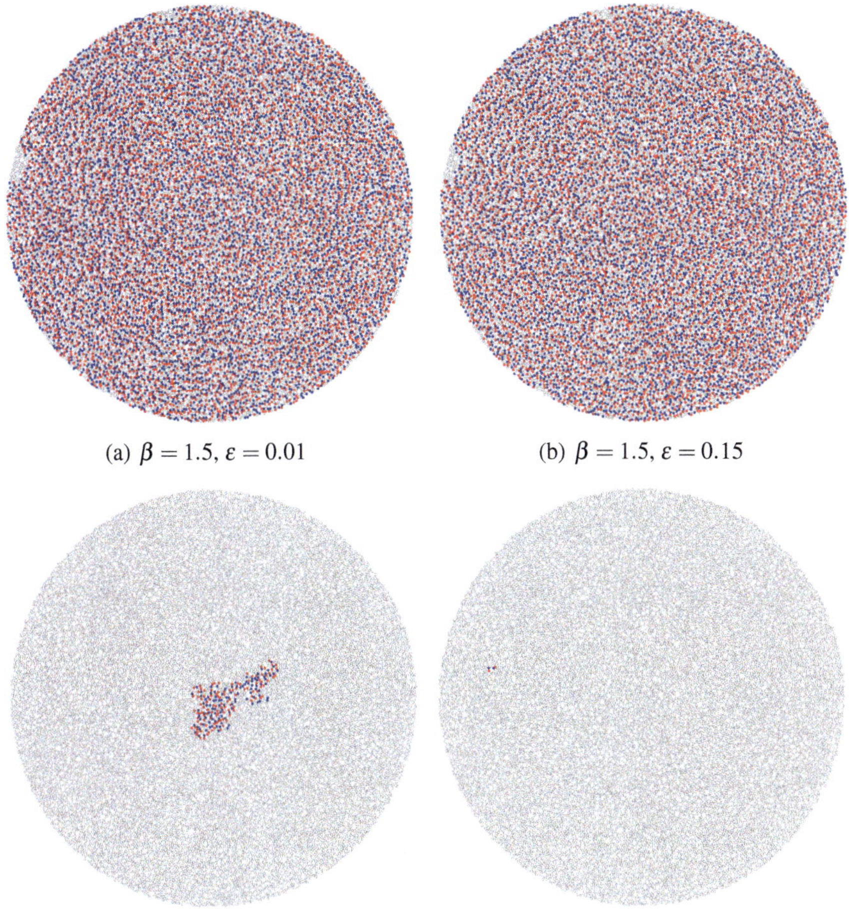

(a) $\beta = 1.5$, $\varepsilon = 0.01$ (b) $\beta = 1.5$, $\varepsilon = 0.15$

(c) $\beta = 1.5$, $\varepsilon = 0.2$ (d) $\beta = 1.5$, $\varepsilon = 0.3$

Fig. 7.12 Snapshots of excitation dynamics in β-skeletons ($\beta = 1.5$) with relative excitation threshold

Only tiny clusters of excitation are formed for $\varepsilon \in [0.25, 0.333[$ (Fig. 7.13d). A life-cycle of most typical oscillator is shown in Fig. 7.14. This oscillator has period seven. Its core structure is a singleton-excitation (accompanied by a refractory tail) running around seven-node cycle, anti-clockwise. At certain moments of its life-cycle the excitation excites few nodes neighbouring to the 'core cycle'. Thus sparks of excitation are formed (Fig. 7.14c–f).

No excitation persists for $\varepsilon \geq 0.333$.

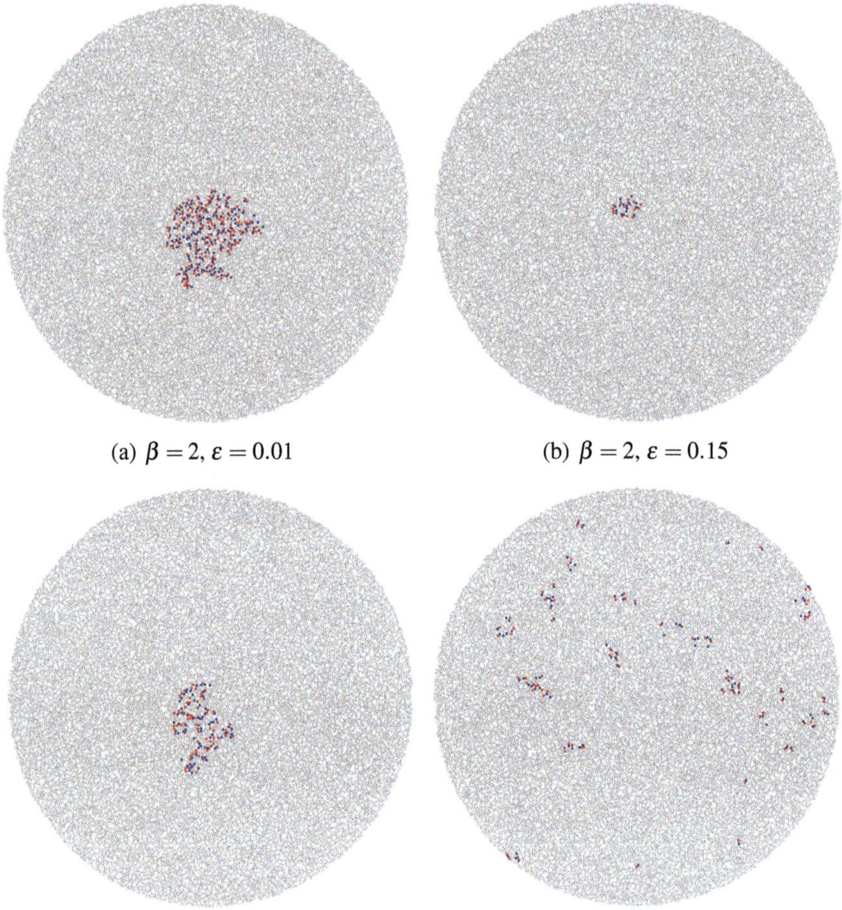

(a) $\beta = 2, \varepsilon = 0.01$ (b) $\beta = 2, \varepsilon = 0.15$

(c) $\beta = 2, \varepsilon = 0.2$ (d) $\beta = 2, \varepsilon = 0.3$

Fig. 7.13 Snapshots of excitation dynamics in β-skeletons ($\beta = 2$) with relative excitation threshold

Finding 22. *Excitable beta-skeletons with relative threshold of excitation exhibit the following classes of space-time dynamics: (1) spiral and target waves similar to classical excitation waves, (2) omnidirectional propagating patterns of disordered excitation, (3) localised domains of excitations, (4) branching domains, and, (5) tiny oscillators.*

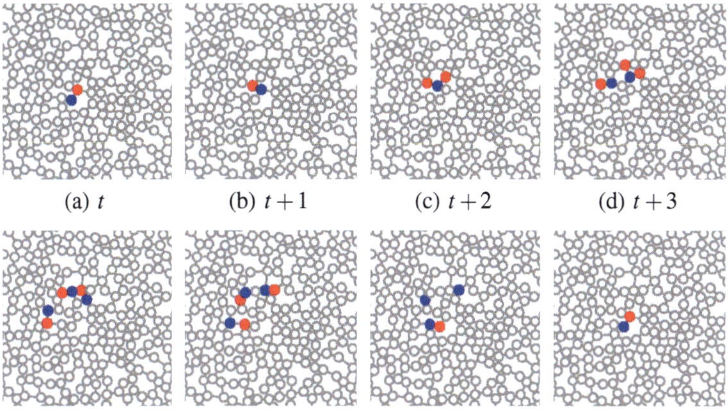

(a) t (b) $t+1$ (c) $t+2$ (d) $t+3$

(e) $t+4$ (f) $t+5$ (g) $t+6$ (h) $t+7$

Fig. 7.14 Examples of oscillators in β-skeletons ($\beta = 2$) with relative excitation threshold $\varepsilon = 0.29$

Position of the classes on $\beta - \varepsilon$-plane is shown in Fig. 7.15. There are sharp boundaries between classes, characterised by abrupt changes in morphology of excitation patterns. Critical values of relative excitability threshold are $\varepsilon = 0.167, 0.2, 0.25$ and 0.33. These thresholds reflect situations when a resting node excites if at least one of six ($\varepsilon = 0.166$), one of five ($\varepsilon = 0.2$), one of four ($\varepsilon = 0.25$) and one of three ($\varepsilon = 0.33$) neighbours is excited. Excitability threshold $\varepsilon = 0.166$ is critical only for β-skeletons with β very close to 1. This is because such skeletons have mode 5 of node degree distribution and average node degree 5.1 (Fig. 7.6, $\phi = 0.407$), i.e. nodes with six neighbours are second (after nodes with five neighbours) dominating nodes on the graph. For similar reasons $\varepsilon = 0.2$ is not critical threshold value for β-skeletons with β close to 2 because over half of the nodes in such skeletons have three neighbours, and nodes with two or four neighbours are second dominants in these skeletons.

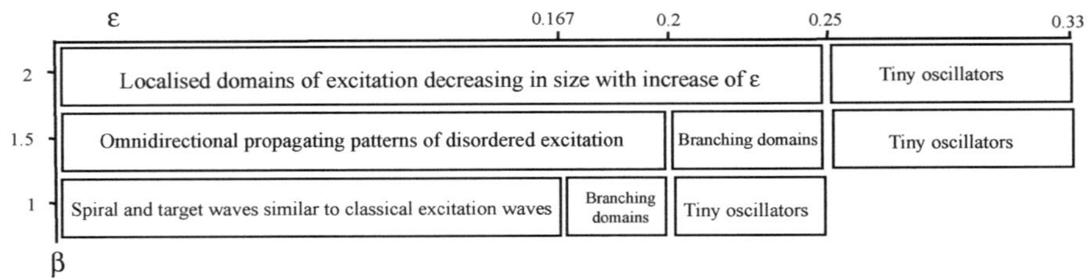

Fig. 7.15 Parameterisaton of space-time dynamics of β-skeletons with relative excitation threshold by β and ε

7.3 Stability of Localised Oscillators

Finding 23. *Localised excitations, or oscillators, are stable under changes of relative excitability threshold.*

The finding is illustrated for $\beta = 1$ and $\beta = 2$ in Fig. 7.16. We choose values of threshold ε at the edge of excitability — $\varepsilon = 0.204$ for $\beta = 1$ and $\varepsilon = 0.25$ for $\beta = 2$ — and perturb β-skeleton with a random configuration of excitations. Few localised oscillators emerge (Fig. 7.16ad). Then we decrease excitability thresholds — dynamically, without resetting the skeleton to resting state — to $\varepsilon = 0.199$ ($\beta = 1$) and $\varepsilon = 0.249$ ($\beta = 2$). The oscillators start to generate excitation wave-fragments. The wave-fragments emitted by the oscillators fill the skeletons (Fig. 7.16be). When we reduce excitability thresholds back to their original levels 0.204 and 0.25 traveling wave-fragments disappear but oscillators remain untouched, they stay where they were before relative excitability threshold changed (Fig. 7.16cf).

7.4 Summary

Excitation dynamics of smoothly parameterised family of planar proximity graphs — β-skeletons — show diverse modalities. Overall level of activity in an absolutely excitable β-skeleton (a resting node excites if number of its neighbours exceeds certain threshold) is proportional to node packing density and inversely proportional to β. Space-time dynamics of absolutely excitable β-skeletons can be classified as follows:

- spiral and target waves,
- disordered dynamics with irregularly traveling wave-fragments,
- slowly spreading domains of excitation, and
- small localised stationary domains of oscillatory activity.

Transition of a skeleton between these classes is controlled by changing β and density of node packing. Both β and density affect distribution of node degrees, thus the classes can be parameterised using just one parameter — average node degree.

There are five classes of space-time activity of the relatively excitable (a resting node excites if a ratio of its excited neighbours to a total number of neighbours exceeds certain threshold) β-skeletons with tightly packed nodes:

- spiral and target waves,
- omnidirectional propagating patterns of disordered excitation,
- localised domains of excitations,
- branching domains of activity, and
- tiny oscillators.

The classes are controlled by β and the relative excitability threshold ε. Tiny oscillators usually emerge in skeletons with just below edge excitability. The oscillators are

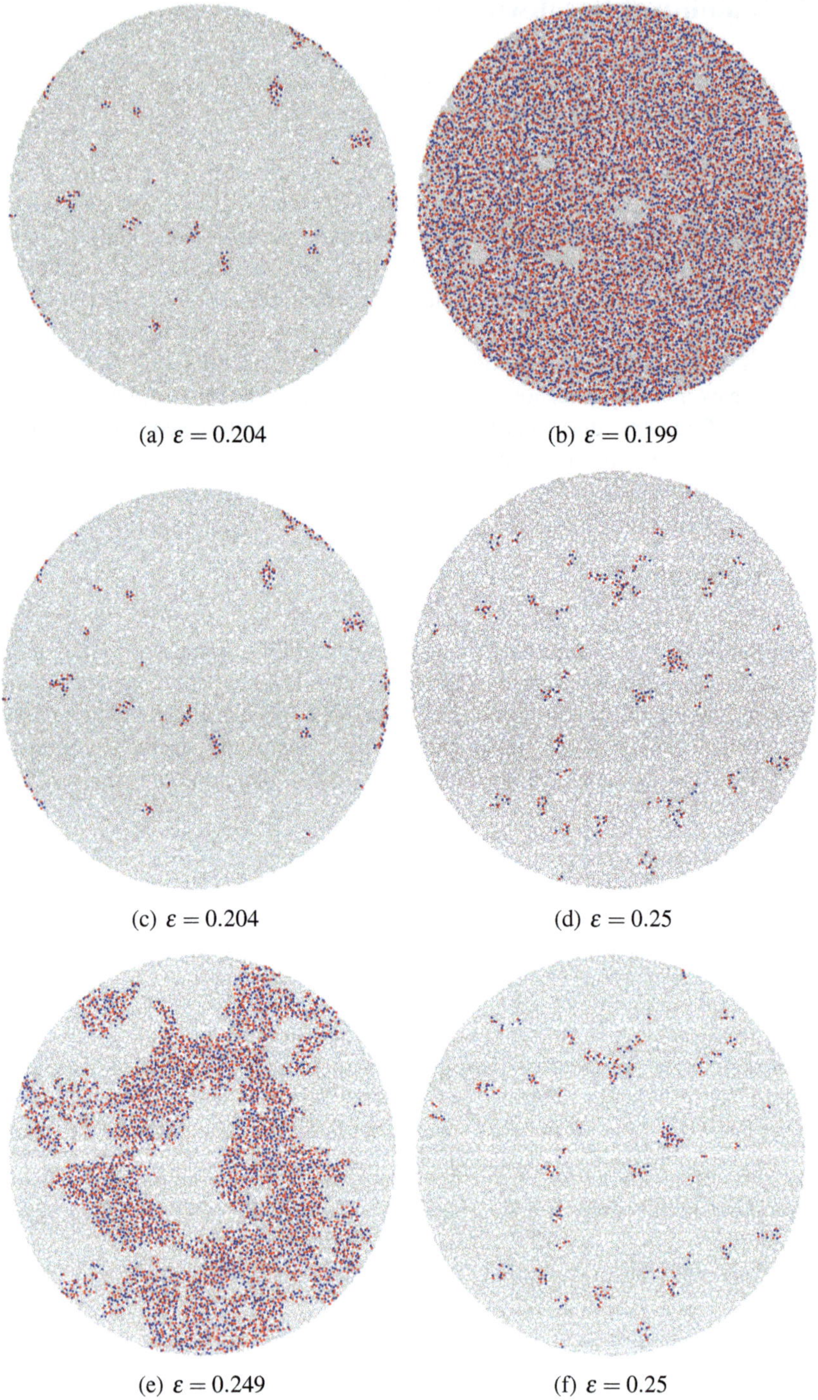

(a) $\varepsilon = 0.204$ (b) $\varepsilon = 0.199$

(c) $\varepsilon = 0.204$ (d) $\varepsilon = 0.25$

(e) $\varepsilon = 0.249$ (f) $\varepsilon = 0.25$

Fig. 7.16 Stability of oscillators under changes of ε: (a)–(c) $\beta = 1$, (d)–(f) $\beta = 2$

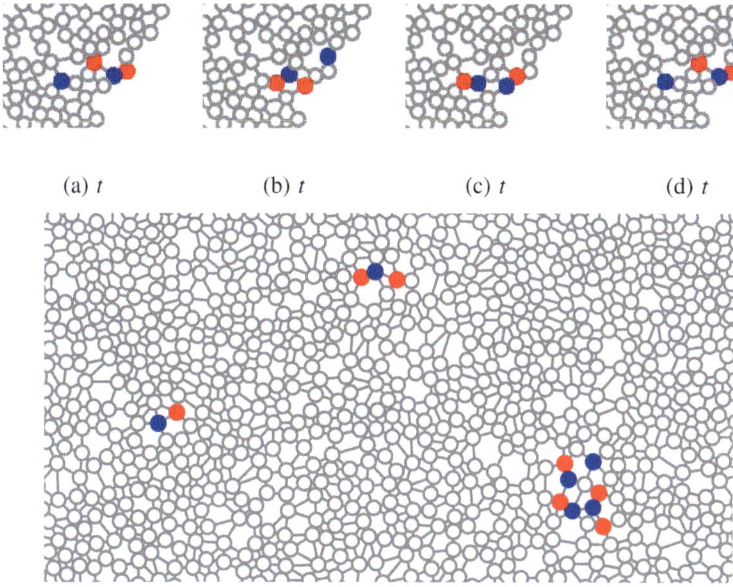

(a) t (b) t (c) t (d) t

(e) $t+6$

Fig. 7.17 Examples of an oscillator in β-skeletons ($\beta = 1.5$) with relative excitation threshold $\varepsilon = 0.32$

stable under relative excitability threshold. Decreasing threshold can increase number of excited loci around the oscillators but can not modify oscillators themselves. As soon as we increase the threshold all excitation activity but original oscillators disappear. The class of branching domains is a transitional class occupying a part of $\beta - \varepsilon$-space between full excitability and non-excitability. With regards to propagating patterns, waves are pronounced for $\beta = 1$, and are alike classical excitation waves in discrete media. Increase of β to 1.5 causes wave-fronts to break up into separate wave-fragments, see e.g. Fig. 7.17.

The ideas and findings of the chapters will find their applications in analysis of spreading crowd activities, e.g. riots, on a city's streets; distributed containment of traffic jams on motorways networks; study of foraging patterns of social insects and myxomycetes; control of excitation dynamics and propagation of defects in soft matter; design and simulation of conglomerates of simple excitable elements.

Chapter 8
Evolving Localizations in Reaction-Diffusion Automata

We consider a ternary-state totalistic hexagonal cellular automaton, where a cell updates its state depending on just the numbers, not positions, of different cell-states in its neighbourhoods. One cell-state, S, is a dedicated substrate state: a cell in state S, whose neighbourhood is filled only with states S, does not change its state (S is a an analogue of quiescent state in cellular automaton models). Two other states, A and B, are assigned to be reactants. The cell-state transition rule can be written as follows:

$$x^{t+1} = f(\sigma_A(x)^t, \sigma_B(x)^t, \sigma_S(x)^t),$$

where $\sigma_p(x)^t$ is the number of cell x's neighbours with cell-state $p \in \{A, B, S\}$ at time step t. Every cell x, assuming infinite or periodic boundaries, has seven neighbours: the central cell x and its six closest neighbours.

To give a compact representation of the cell-state transition rule, we adopt the formalism of [15], and represent the cell-state transition rule as a matrix $\mathbf{M} = (M_{ij})$, where $0 \leq i \leq j \leq 7$, $0 \leq i + j \leq 7$, and $M_{ij} \in \{A, B, S\}$. The output state of each neighbourhood is given by the row-index i (the number of neighbours in cell-state A) and column-index j (the number of neighbours in cell-state B). We do not have to count the number of neighbours in cell-state S, because it is given by $7 - (i + j)$. A cell with a neighbourhood represented by indexes i and j will update its state to cell-state M_{ij} which can be read off the matrix. In terms of the cell-state transition function this can be presented as follows: $x^{t+1} = M_{\sigma_A(x)^t \sigma_B(x)^t}$.

How do matrix $\mathbf{M}$ entries correspond to phenomena in reaction-diffusion chemical systems? The entries $M_{i0} = A$ and $M_{0j} = B$, $i > 0$, symbolise the diffusion of reactants A and B. Entries $M_{ij} = p$, $p \in \{A, B, S\}$, can be interpreted as a quasi-chemical reaction

$$iA + jB \longrightarrow p,$$

where i molecules of species A react with j molecules of species B to produce species p [17].

A. Adamatzky: Reaction-Diffusion Automata, ECC 1, pp. 155–162.
springerlink.com

8.1 Breeding Glider-Supporting Rules

Emmanuel Sapin [222–225] discovered cellular automata which support gliders (mobile localisations) using evolutionary computation techniques. Sapin's evolutionary algorithm maintains a population of patterns, initially generated at random, that evolves according genetic operations: the rules of selection, recombination, mutation, and survival. A shared 'environment' is used to determine the fitness or performance of each individual in the population. The fittest individuals are more likely to be selected for reproduction through recombination and mutation. The search space is a set of cell-state transition rules, where each rule represents a unique cellular automaton. An individual is an automaton coded as a bit-string representing values of a cell at the next generation for each neighbourhood state. A string is composed of two substrings. The first substring represents the neighbourhood states used by gliders. The second substring is initialised at random. The search space contains 3^{36} possible cell-state transition rules.

A fitness function is computed as follows. Random configurations of cells are developed by the tested automaton. They allowed to be evolved by cellular automata and checked for presence of gliders. A value of the fitness function isa number of gliders detected divided by the total number of cells. The 36 bits of each individual are initialised at random. The mutation function consists of mutating one bit among 36. The recombination is a random point crossover. An elitist strategy in which the best individual of population is kept is used. The value of the fitness function and the generation of the best rule are memorised. If after ten new generations the algorithm has not found a better rule the algorithm stops.

8.2 Likehood of Gliders

From the set of evolved functions we calculated[1] a set of matrices $F^z = (F_{ij})$, where i is the number of neighbours in state A, and j is the number of neighbours in state B, $z \in \{A, B, S\}$, which reflect likelihood for any particular transition $\langle \sigma_A, \sigma_B \rangle \to z$. A matrix $F^{\#}$ which indicates a likelihood of any particular state transition to be redundant, i.e. not necessary for supporting gliders. Exact structures of the matrices are shown in Fig. 8.1.

Just a few examples of mobile localizations discovered are shown in Fig. 8.2. Most typical localizations are gliders, which change their configuration every second time step (Fig. 8.2ab), and puffer trains — travelling localisations leaving trail of oscillators or breathers behind themselves (Fig. 8.2cd). The evolutionary algorithm employed was tuned to select transition functions supporting localisations, not selecting rules with manageable dynamics of localisations. Therefore, those who will start experimenting with matrices presented in Fig. 8.1 may be disappointed

[1] The probabilistic matrices are calculated by Emmanuel Sapin, in the framework of EPSRC project EP/E005241/1

$$\mathbf{F}^S = \begin{bmatrix} 1 & .63 & .46 & .22 & .34 & .14 & .1 & .06 \\ .66 & .23 & .21 & .41 & .21 & .09 & .08 & \\ .53 & .26 & .36 & .25 & .14 & .09 & & \\ .29 & .3 & .29 & .11 & .06 & & & \\ .21 & .25 & .17 & .08 & & & & \\ .18 & .11 & .07 & & & & & \\ .15 & .09 & & & & & & \\ .04 & & & & & & & \end{bmatrix} \qquad \mathbf{F}^A = \begin{bmatrix} 0 & .09 & .14 & .13 & 0 & 0 & 0 & 0 \\ .14 & .43 & .25 & 0 & 0 & 0 & 0 & \\ .12 & .27 & .23 & 0 & 0 & 0 & & \\ .19 & .16 & 0 & 0 & 0 & & & \\ .15 & 0 & 0 & 0 & & & & \\ 0 & 0 & 0 & & & & & \\ 0 & 0 & & & & & & \\ 0 & & & & & & & \end{bmatrix}$$

(a) (b)

$$\mathbf{F}^B = \begin{bmatrix} 0 & .24 & .16 & .13 & 0 & 0 & 0 & 0 \\ .16 & .25 & .22 & 0 & 0 & 0 & 0 & \\ .2 & .24 & 0 & 0 & 0 & 0 & & \\ .15 & .05 & 0 & 0 & 0 & & & \\ .07 & 0 & 0 & 0 & & & & \\ 0 & 0 & 0 & & & & & \\ 0 & 0 & & & & & & \\ 0 & & & & & & & \end{bmatrix} \qquad \mathbf{F}^\# = \begin{bmatrix} 0 & .04 & .24 & .52 & .66 & .86 & .9 & .94 \\ .04 & .09 & .32 & .59 & .79 & .91 & .92 & \\ .15 & .23 & .41 & .75 & .86 & .91 & & \\ .37 & .49 & .71 & .89 & .94 & & & \\ .57 & .75 & .83 & .92 & & & & \\ .82 & .89 & .93 & & & & & \\ .85 & .91 & & & & & & \\ .96 & & & & & & & \end{bmatrix}$$

(c) (d)

Fig. 8.1 Glider-likelihood matrices

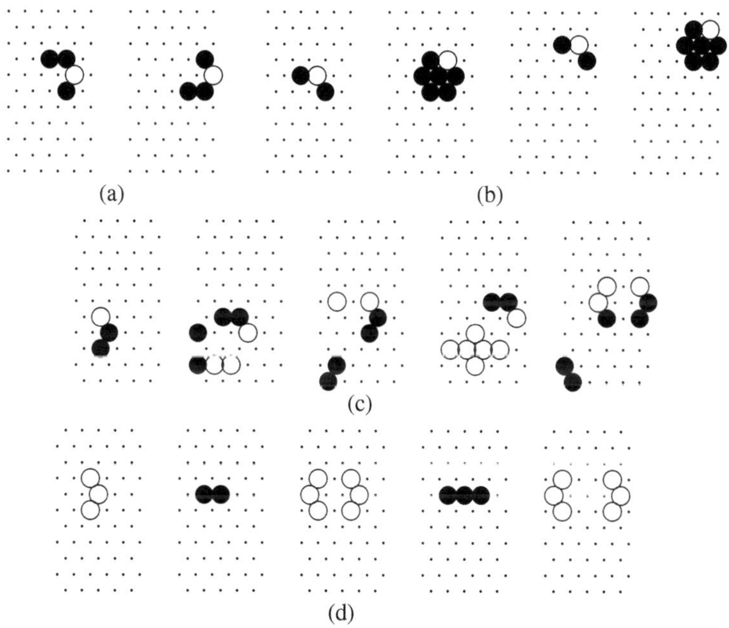

Fig. 8.2 Examples of localizations observed in rules presented by glider-likelihood matrices. State A is shown by circle, state B by solid disc. (a) glider propagating east; (b) glider propagating north-east; (c) puffer train propagating north-east; (d) puffer train propagating east.

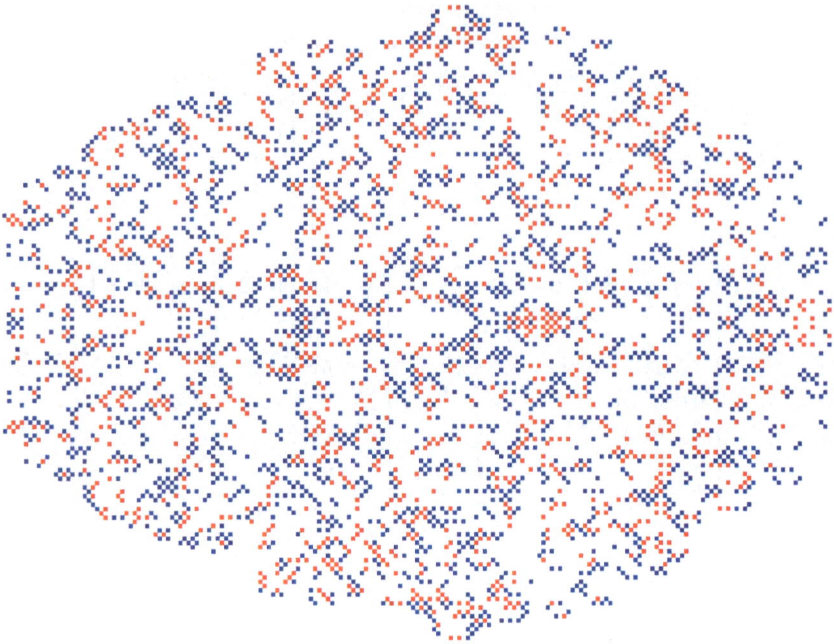

Fig. 8.3 An example of a typical configuration generated by glider-supporting rules. Gliders multiply during collisions, so the whole space is filled by visually disorganised but in reality highly ordered pattern.

to see that, in most cases, development of an automaton from initial random configurations leads to disorderly looking configurations, even if the patch of initial stimulation was small enough. This is because gliders inhabit such spaces in abundance, they interact one with another, produce more gliders in result of their interaction, and populations of swarming gliders look like quasi-chaotic patterns for an untrained eye (Fig. 8.3).

8.3 Quasi-chemical Reaction

In order to characterise sets of possible quasi-chemical reactions, which lead to the formation of stationary and mobile perturbations of reagent concentrations, we provide isoline representation of the glider likelihood matrices in Fig. 8.4. Based on the distributions shown in Fig. 8.4 we can speculate that

Finding 24. *A probability, or rate, of dissociation of reactants decreases with the increase of number of the reactant molecules in each one local site of the medium.*

See topology of isolines in Fig. 8.4a.

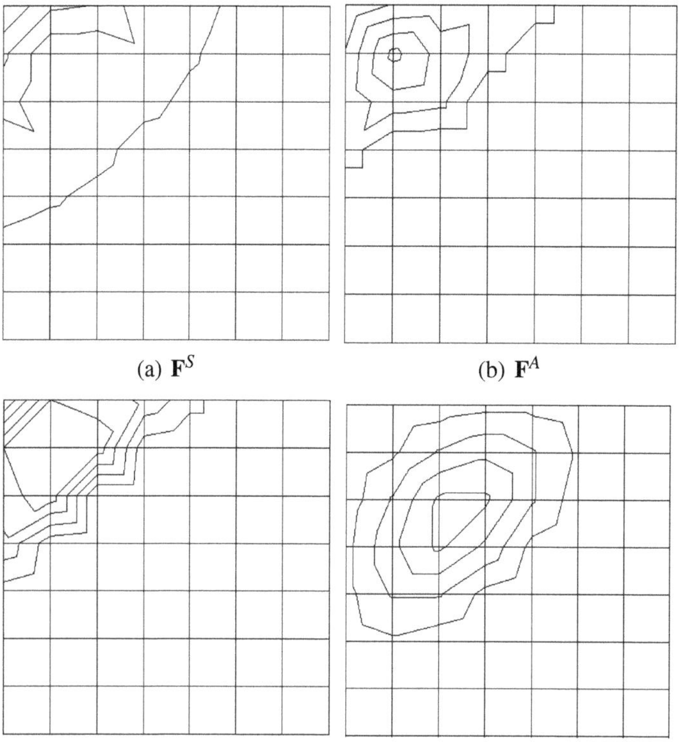

(a) $\mathbf{F}^S$ (b) $\mathbf{F}^A$ (c) $\mathbf{F}^B$ (d) $\mathbf{F}^\#$

Fig. 8.4 Isolines representation for glider likelihood matrices. Number of states of reactant *A* increases from top left corner to bottom left, number of states of reactant *B* increases from top left corner to top right one. In each case there is a single elevation. Approximate locations of elevations are F^S_{00}, F^A_{11}, F^B_{11}, and $F^\#_{22}$.

Finding 25. *Reactions leading to production of reactants A and B are most likely to involve not more then four (usually between one and three) molecules of the reactants.*

Distributions shown in Fig. 8.4ab reach their zero levels — no reaction take place — for the sites which satisfy the condition: Sum of molecules *A* and molecules *B* does not exceed four.

Finding 26. *Reactant B is more likely to be produced during quasi-chemical reactions derived from* **F***-matrices.*

See Fig. 8.4ab to compare area of elevations.

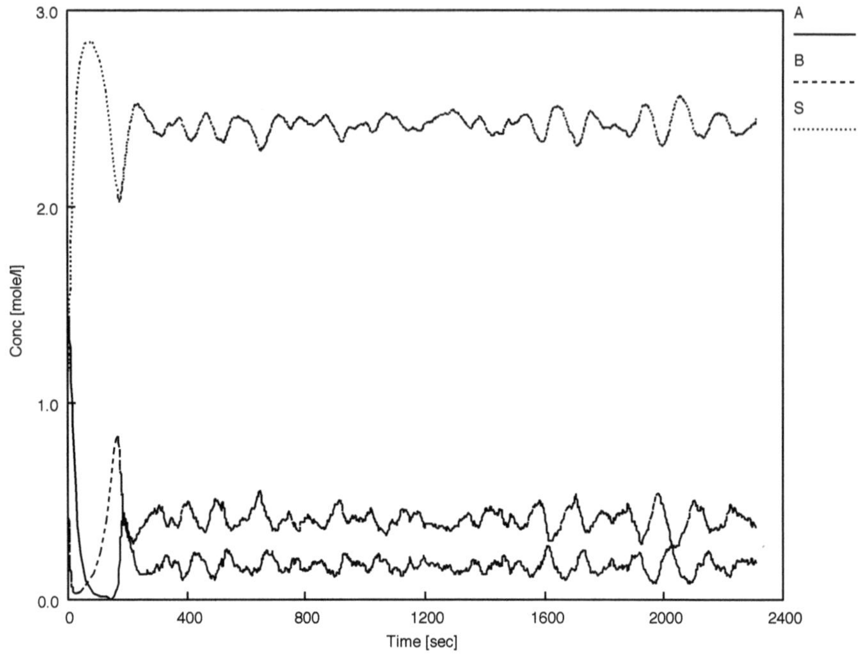

Fig. 8.5 Concentration dynamics of A, B and S species in a well-stirred reactor. Simulated on CKS (Chemical Kinetics Simulator, IBM) with the following parameters: pressure and temperature are kept constant, volume is not tracked. 100K particles were simulated. Initial concentration of each species is 1 mole/l.

We generalise glider likelihood matrices to a set of abstract, quasi-chemical, reactions as follows:

$$\begin{array}{ll} A+B \xrightarrow{1.} 2A & A+2B \xrightarrow{0.4} 2A+B \\ 2A+B \xrightarrow{0.1} 3A & 2A+2B \xrightarrow{0.01} 3A+B \\ 3A+B \xrightarrow{0.01} 4A & A+B \xrightarrow{1.} 2B \\ A+2B \xrightarrow{0.1} 3B & 2A+B \xrightarrow{0.05} A+2B \\ B+S \xrightarrow{0.01.} 2B & A \xrightarrow{0.054} S \\ B \xrightarrow{0.0015} S & \end{array} \quad (8.1)$$

A typical scenario of how the system (8.1) behaves in a well-stirred reactor is shown in Fig. 8.5. The reaction scheme developed represents an oscillatory chemical system, where concentration of substrate is significantly higher than concentrations of reactants A and B. This indeed conforms with the nature of spreading localizations and pulsating behaviour of generators of localizations, or glider guns. The presence of both types of reactants in a localisation is essential for the majority of propagating localisations, which is identified by the similarity of the dynamic of concentration profiles of reactants A and B shown in Fig. 8.5.

8.4 Reductions of Transitions Functions

What would be the behaviour of automata governed by the most likely cell-state transition rules? In order to find an answer we decided to reduce matrices of glider likelihoods to their most articulated forms. First, we have made matrices more symmetrical by the following procedure. If $|F'_{ij} - F''_{ji}| < \varepsilon$ then we update entries (ij) and (ji) as follows: $F'_{ij} = F''_{ji} = \left\lfloor \frac{1}{2}(F'_{ij} + F''_{ji}) \right\rfloor$. We do this because we assume that in evolutionary experiments reactants A and B have the same diffusion coefficients.

Then we perform the following operation. If $|F^A_{ij} - F^B_{ij}| > 0.2$ and $|F^A_{ij} - F^S_{ij}| > 0.2$, or $F^B_{ij} = 0$ and $F^S_{ij} = 0$ then $M^A_{ij} = 1$. We also discard entries of $(0,5)$ and $(1,4)$ of the matrix M^B as being negligibly small. After these operation, we got the following matrix

$$\mathbf{R} = \begin{bmatrix} 0 & 0 & 0 & \{0,1,2\} & 0 & 0 & 0 & 0 \\ 0 & 1 & \{0,1,2\} & 0 & 0 & 0 & 0 & \\ 0 & \{0,1,2\} & \{0,1\} & 0 & 0 & 0 & & \\ \{0,1,2\} & \{0,1\} & 0 & 0 & 0 & & & \\ \{0,1\} & 0 & 0 & 0 & & & & \\ 0 & 0 & 0 & & & & & \\ 0 & 0 & & & & & & \\ 0 & & & & & & & \end{bmatrix}$$

where R_{ij} is a set of all possible states that a cell x can take if its neighbourhood $u(x)$ has i cells in state A and j cells in state B. We found that

Finding 27. *Starting its development from a random configuration a cellular automaton governed by a local transition functions expressed by matrix* **R** *exhibits still and breathing localizations.*

Most common localizations observed in cellular automata with local transition rules from the class **R** are shown in Fig. 8.6. Minimal weight localizations are still triples of cells in state A or B, or triples switching between A and B (Fig. 8.6a). These are followed by compact still patterns of four cells in A or B states, see examples in Fig. 8.6bc, switching compact patterns of four non-resting states Fig. 8.6d; and, wish-bone shaped clusters of A and B states Fig. 8.6e–h. The next heavier common localizations are the switching patterns of six Fig. 8.6i and eighth Fig. 8.6j non-resting states, and, the breathing localisation Fig. 8.6k oscillating between 12 and 13 non-resting states.

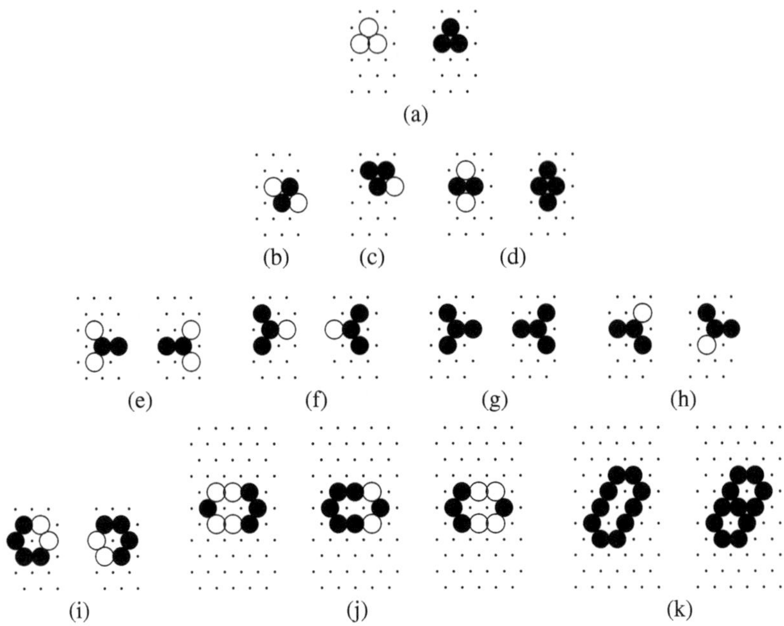

Fig. 8.6 Stationary localizations typically observed in automata governed by cell-state transition rules from class **R**. State *A* is shown by a circle, state *B* by a solid disc.

8.5 Summary

By hybridised the paradigm of reaction-diffusion cellular automata [15, 17] with evolutionary techniques of breeding glider-supporting rules [222–225] we statistically evaluate the set of all possible totalistic cell-state transition functions which support mobile and stationary localizations and calculated exact structures of glider likelihood matrices and interpreted them in terms of abstract reactions. We shown that quasi-chemical systems derived from glider-supporting rules exhibit classical dynamics of excitable chemical systems. By reducing glider likelihood matrices to their strong components one can obtain a set of local transition rules that exhibit exclusively stationary localizations.

Part II
Population Dynamics

Chapter 9
Population Dynamics in Automata

Let us consider a two-dimensional hexagonal cellular automaton, every cell of which takes two states: species a and species b, and updates its state in discrete time indexautomaton!population depending on its own state and just the numbers of cell-states of its six neighbours. Let $\sigma_a^t(x)$ and $\sigma_b^t(x)$ be sums of cell x's neighbours in state a and b, respectively, at time step t. Then the cell-state transition function will be as follows:

$$x^{t+1} = f(x^t, \sigma_a^t(x), \sigma_b^t(x)). \tag{9.1}$$

A cell in state a updates its state depending on a number of neighbours in state b, a cell in state b updates its state depending on a number of neighbours in state a. We represent the rule (9.1) by threshold conditions for a cell to keep its state, i.e. we consider only conditions of the transitions $x^t = a \rightarrow x^{t+1} = a$ and $x^t = b \rightarrow x^{t+1} = b$. The conditions of the cell-state transitions $a \rightarrow a$ and $b \rightarrow b$ are as follows:

	transition	
interaction	$a \rightarrow a$	$b \rightarrow b$
mutualism	$\sigma_b > \theta$	$\sigma_a > \delta$
commensalism	$\sigma_b > \theta$	
parasitism	$\sigma_b < \theta$	$\sigma_a > \delta$
amensalism	$\sigma_b < \theta$	
neutralism		
competition	$\sigma_b < \theta$	$\sigma_a < \delta$

where $0 \leq \theta, \delta \leq 6$. If no conditions are attached to a transition, the transition takes place by default (unconditionally).

Values of parameters θ and δ indicate a strength of an influence of one species to another. Smaller values correspond to higher strength because they imply that less amount of one species is required to influence another species. Higher values of the parameters mean weaker strength of inter-species interactions.

For example, consider a parasitic interaction with parameters $\theta = 3$ and $\delta = 1$. A cell in state a at time step t takes the same state a at time step $t+1$ if it has at least

A. Adamatzky: Reaction-Diffusion Automata, ECC 1, pp. 165–182.
springerlink.com

three neighbours in state b, otherwise the cell takes state b. A cell in state b at time step t takes the same state b if it has at least one neighbour in state a.

In case of amensalism, a cell in state a at time step t will remain in state a at time step $t+1$ if not more then θ of its neighbours are in state b; otherwise the cell takes state b. A cell in state b will remain in the state b indefinitely.

We employ the two following integral parameters to characterise space-time dynamics of studied cellular automata: neighbourhood frequency and cell-update frequency.

Let $\nu(u)$ be a frequency of occurrence of neighbourhood configuration $u \in 0, 1^7)$ (the neighbourhood includes central cell) in configuration of a cellular automata, and η be an average distance of each frequency from the frequency of neighbourhood configurations in a random configuration, $\eta = \sqrt{(\sum_{u \in \{0,1\}^7} \nu(u) - \frac{128}{1})}$. For a random configuration, where each cell-state occurs with probability $\frac{1}{2}$, the value of η is 0; configuration where cells are in the same state (solid fill) is characterised by $\eta = 1$. The parameter η characterises a completeness of local representation in the system. When $\eta = 0$ all possible configurations of neighbourhood are presented in the single configuration, and thus the configuration is locally complete. A configuration becomes more and more incomplete when η tends to 1.

The parameter ϕ is an average cell-state update frequency, the value measured in t steps indicates a probability for a cell to change its state during any given step of evolution. The parameter ϕ may be thought of as a degree of local visible activity: the more ϕ the often cells change their states. We stress that this only applies to the visible activity because each cell of a cellular automaton updates its state at every step of the automaton development, the cell might not change its state however.

To compare morphological characteristics of configurations emerged in development of the cellular automata models, we initiated the systems with random pattern of species a and b and then recorded configurations after a transient period. The transient period is assumed to be completed when the automaton reaches its global fixed point of evolution (still configuration) or a cycle (when succession of configuration is looped). Exemplar configurations, recorded after the transient period are shown in Figs. 9.1 – 9.16.

9.1 Mutualism

Spatial dynamics of cellular automata imitating mutualistic interactions, in scenarios when both species are equally represented in initial configurations, is characterised by still or switching patters (Fig. 9.1). Size of the same species domains increases when dependency of one species increases relatively to dependency of other species, see e.g. increase of θ while δ is fixed in Fig. 9.1. When dependencies of both species increase simultaneously (e.g. the same increases in values of θ and δ, Fig. 9.1) then spatial configurations of the automata change their morphological characteristics from fine-grained (see $\theta = 1$ and $\delta = 1$, Fig. 9.1) to pronounced switching clusters (see $\theta = 3$ and $\delta = 3$, Fig. 9.1) and then return back to fine-granular configurations (see $\theta = 5$ and $\delta = 5$, Fig. 9.1).

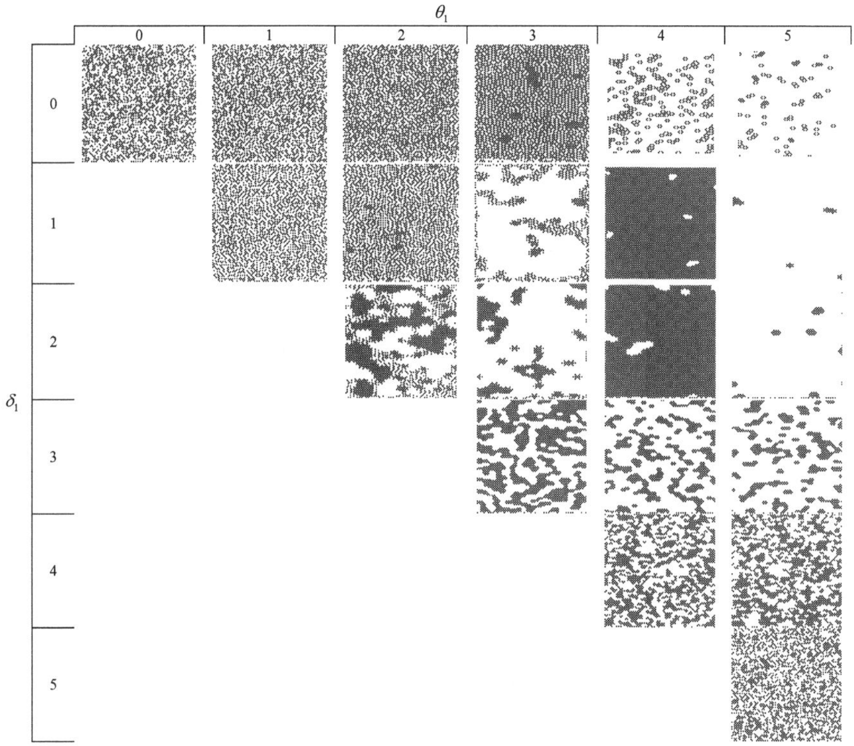

Fig. 9.1 Configurations of cellular automaton imitating mutualistic interactions for different values of dependency parameters θ and δ. In each case automaton started evolution in a configuration where states a and b assigned to cells with the same probability 0.5. States a shown by white pixels, states b by black pixels.

Typical examples of switching configuration for parameters $\delta = 2, \theta = 3$ are shown in Fig. 9.2ab. One can observe compact clusters of one species on the lattice homogeneously filled with other species (Fig. 9.2a), and strips of elongated domains (Fig. 9.2a). Mechanics of the switching is attributed to the absence of a no-species states: when cell in state a lacks enough neighbours in state b, the cell switches to state b.

In a cellular automaton imitating mutualism, small concentration of one species, e.g. a, gives rise to traveling target waves on the lattice filled with another species b, see example in Fig. 9.3a–e. Cells not occupied by propagating wave patterns switch synchronously between b and a states. When wave-fronts of propagating patterns collide they stop their propagation and a quasi-stationary structure is formed (Fig. 9.3f).

As to integral characteristics of cellular automaton dynamics we can see (Fig. 9.4a) that local representation becomes incomplete and less and less rich when influence of one species to another decreases while influence of other species

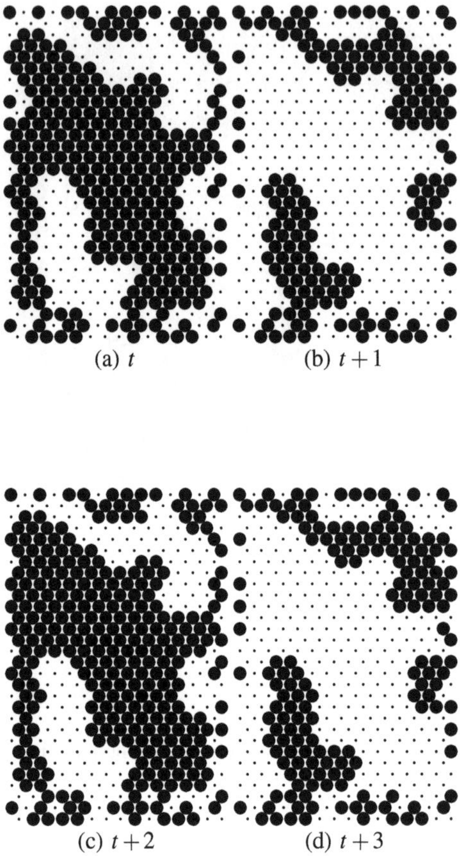

Fig. 9.2 Examples of switching patterns in mutualistic model, $\delta = 2$, $\theta = 3$

remains unchanged. There is just slight decrease in completeness when influences of both species decrease simultaneously, see Fig. 9.4aη. Activity of the cellular automata simulating the interacting systems increases with decrease of influences between species, see Fig. 9.4aϕ.

9.2 Commensalism and Amensalism

Cell-state transition functions related to commensalism — we consider the case when species a benefits from presence of species b, but species b are not affected by species a — do not support any propagating patterns. This is because once a cell takes a state b it stays in the state b forever. As a consequence of this, an amount

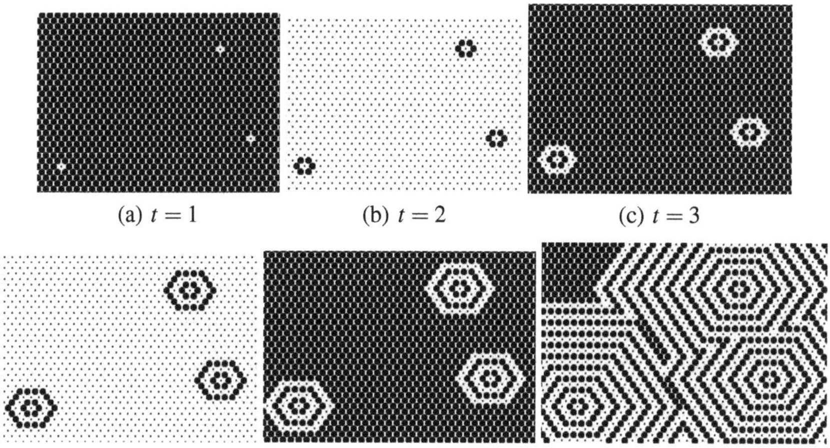

(a) $t=1$ (b) $t=2$ (c) $t=3$

(d) $t=4$ (e) $t=5$ (f) $t=22$

Fig. 9.3 Propagating patterns in cellular automaton imitating mutualistic interactions: (a)–(f) $\delta=0$, $\theta=0$, filled with a, with few, probability 0.001, cells in state b at initial stage of wave propagation

of cells in state a can either decrease or remain unchanged in the development of commensalistic cellular automaton. With increase of parameter θ a number of cells in state a decreases, as illustrated in Fig. 9.5.

In amensalistic interactions species a suffers, up to a degree θ, from the presence of species b. When value θ is small (Fig. 9.5b) then species a quickly extinct and species b propagate on the lattice till the wholes space becomes filled with b. With increasing θ localized domains of a emerge and then dominate the space (Fig. 9.5b).

By weakening links between species one can control morphology. Thus, a decrease of influence of one species to another leads to increase of incompleteness of configuration morphology in models of commensalism but increase of completeness of morphology in cellular automata imitating amensalism Fig. 9.4bd.

9.3 Parasitism

In parasitic interaction, particularly for strong dependencies between host species a and parasite species b, parasites propagate on the lattice filled with hosts, see e.g. Fig. 9.6.

When dependencies between species become weaker propagation of parasites gets less trivial. Thus for $\delta=2$ and $\theta=2$ and an initial configuration of lattice filled with hosts a and just few parasites b, small domains of parasites stay still while bigger domains give rise to spreading patterns (Fig. 9.7). The propagating

$\eta_{mutualism}$	θ = 0	1	2	3	4	5
δ = 0	0.02	0.04	0.1	0.2	0.45	0.85
1		0.06	0.11	0.7	0.93	0.95
2			0.27	0.69	0.88	0.95
3				0.37	0.48	0.54
4					0.16	0.16
5						0.03

$\phi_{mutualism}$	θ = 0	1	2	3	4	5
δ = 0	0	0	0	0.03	0.62	0.9
1		0	0.02	0.82	0.98	0.99
2			0.67	0.95	0.98	0.99
3				0.99	0.99	0.99
4					0.99	0.99
5						0.99

(a) mutualism

$\eta_{commensalism}$	θ = 0	1	2	3	4	5
	0.01	0.04	0.09	0.29	0.66	0.93

$\phi_{commensalism}$	θ = 0	1	2	3	4	5
	0	0	0	0	0	0

(b) commensalism

$\eta_{parasitism}$	θ = 1	2	3	4	5	6
δ = 0	0.26	0.24	0.20	0.14	0.03	0.01
1	0.18	0.15	0.11	0.09	0.05	0.04
2	0.13	0.11	0.07	0.15	0.1	0.09
3	0.1	0.07	0.48	0.21	0.2	0.26
4	0.04	0.03	0.59	0.3	0.43	0.62
5	0.016	0.07	0.73	0.49	0.64	0.87

$\phi_{parasitism}$	θ = 1	2	3	4	5	6
δ = 0	0.47	0.45	0.37	0.07	0.03	0.01
1	0.66	0.57	0.47	0.05	0.02	0.002
2	0.77	0.66	0.42	0.01	0.005	0.001
3	0.85	0.74	0.004	0.008	0.002	0.004
4	0.97	0.89	0.008	0.002	0.003	0.001
5	0.99	1	0.008	0.002	0.001	0.001

(c) parasitism

$\eta_{amensalism}$	θ = 1	2	3	4	5	6
	1	1	1	0.84	0.13	0.02

$\phi_{amensalism}$	θ = 1	2	3	4	5	6
	0	0	0	0	0.0	0

(d) amensalisms

$\eta_{competition}$	θ = 1	2	3	4	5	6
δ = 1	0.02	0.06	1	1	1	1
2		0.06	1	1	1	1
3			0.3	1	1	1
4				0.38	0.55	0.78
5					0.16	0.13
6						0.03

$\phi_{competition}$	θ = 1	2	3	4	5	6
δ = 1	1	1	0	0	0	0
2		1	0	0	0	0
3			0.3	0	0	0
4				0	0	0
5					0	0
6						0

(e) competition

Fig. 9.4 Integral characteristics of cellular automata imitating two-species interactions η is a completeness of local representation in configurations and ϕ is a cell-update frequency

patterns exhibit distinctive wave fronts, which is due to parasites' need to have a 'comfortable' concentration of hosts around them.

By weakening further dependency parameters δ and θ we shift character of a spatio-temporal dynamic towards formation of slowly expanding irregular clusters of parasites with short-living mobile localizations of parasitic activity propagating between them (Fig. 9.8).

In parametric portrait Fig. 9.9 we observe that the increase in dependency parameters is reflected by the following transitions in spatio-temporal dynamics: from waves of parasites to still patterns of parasites, exchanging with each other mobile localizations, to still patterns.

In situations of equal densities of hosts and parasites in initial configuration one can observe the following:

- small domains switching between few-site groups of parasites to domains of parasites spanning significant part of the lattice, see Fig. 9.10a;

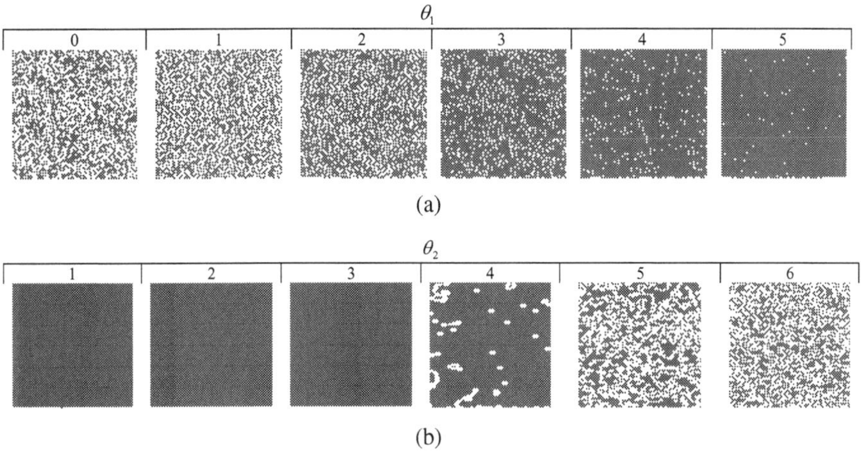

Fig. 9.5 Configurations of cellular automaton imitating (a) commensalism, different values of dependency parameter θ, (b) amensalism, different values of dependency parameter θ. In each case automaton starts its development in a random configuration where states a and b assigned to cells with the same probability 0.5. States a are shown by white pixels, states b by black pixels.

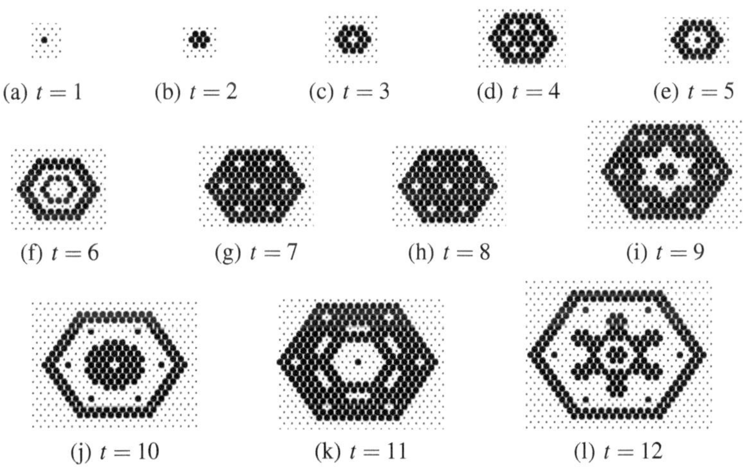

Fig. 9.6 Propagation of parasites in a lattice filled with hosts, parasitism $\delta = 1$ and $\theta = 1$, initial density of parasites is 0.001

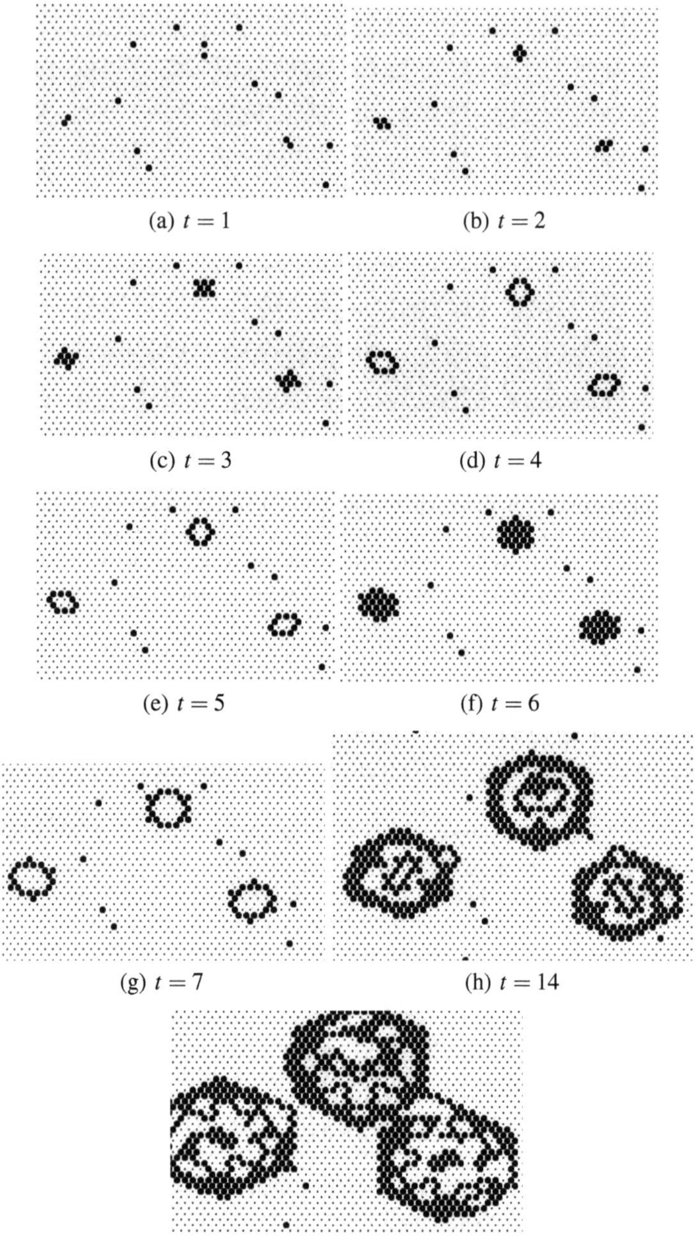

(a) $t = 1$ (b) $t = 2$

(c) $t = 3$ (d) $t = 4$

(e) $t = 5$ (f) $t = 6$

(g) $t = 7$ (h) $t = 14$

(i) $t = 18$

Fig. 9.7 Propagation of parasites in a space filled with hosts for minimal dependencies between hosts and parasites: $\delta = 2$ and $\theta = 2$, initially cells were assigned parasite states with probability 0.01, the rest are hosts

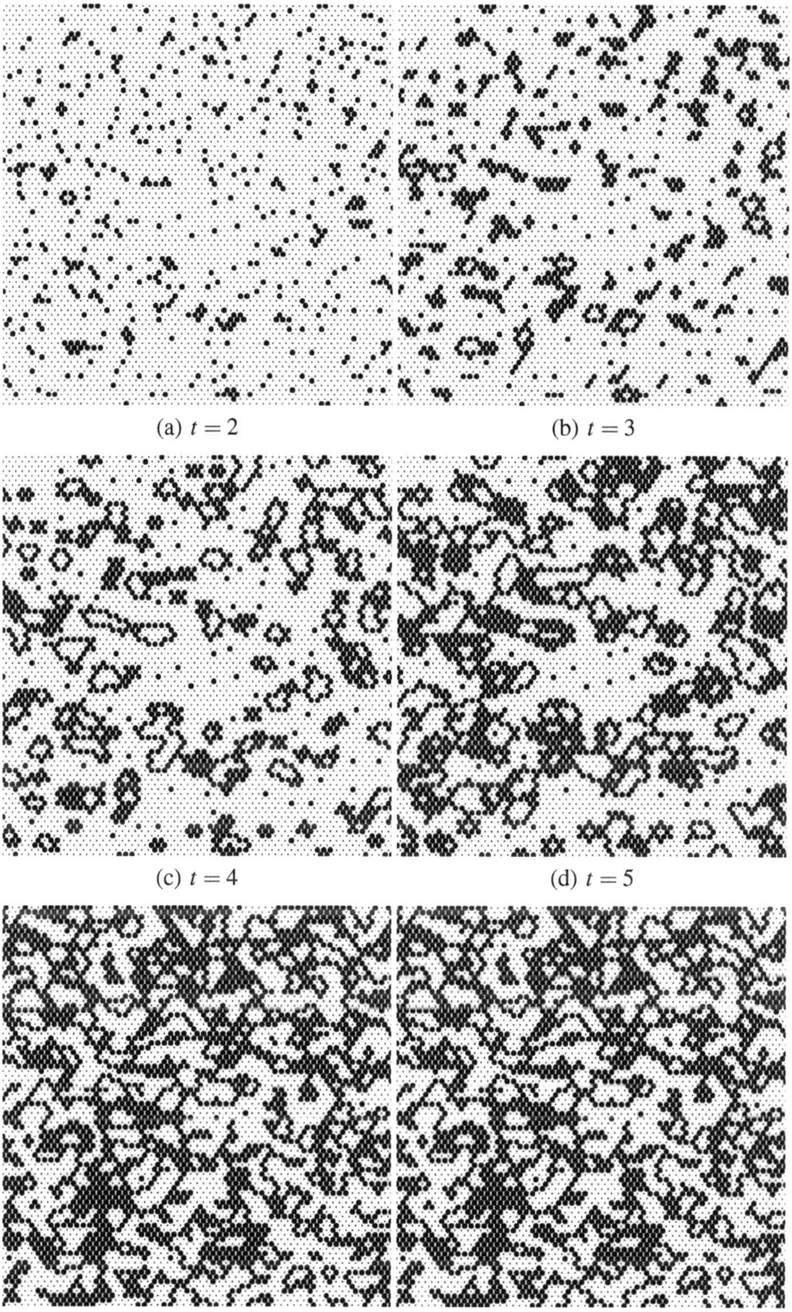

(a) $t = 2$ (b) $t = 3$

(c) $t = 4$ (d) $t = 5$

(e) $t = 29$ (f) $t = 30$

Fig. 9.8 Example of spatio-temporal dynamics in parasite-host cellular automaton for weak dependencies between host a and parasite b, $\delta = 2$ and $\theta = 3$; initial density of parasites is 0.1

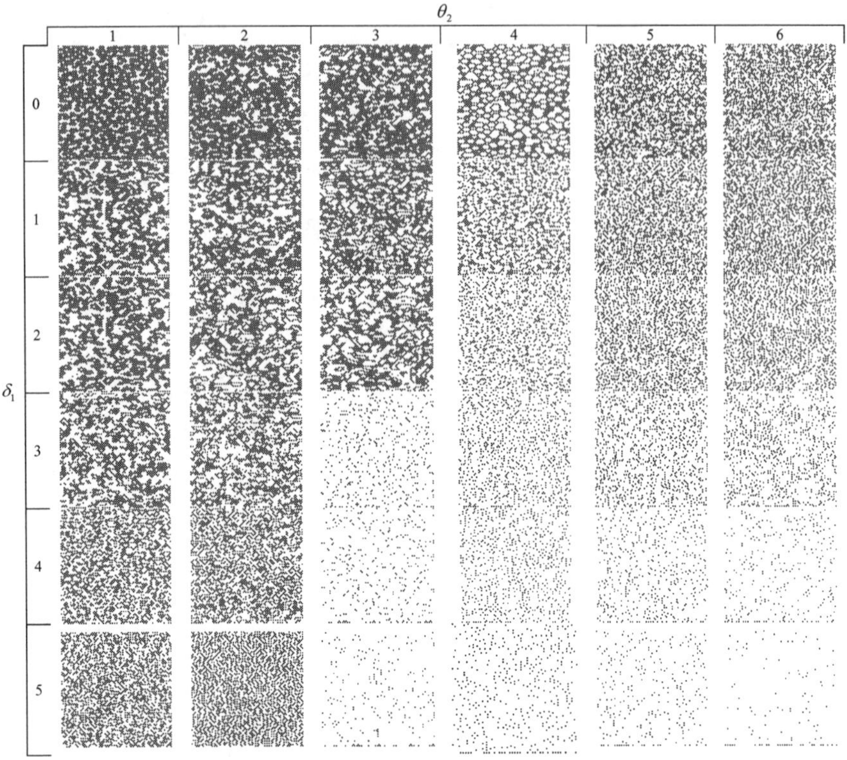

Fig. 9.9 Configurations of cellular automaton imitating parasitic interactions for different values of dependency parameters θ and δ. In each case automaton starts its development in a configuration where states a and b assigned to cells with the same probability 0.5. States a shown by white pixels, states b by black pixels.

- branching clusters of parasites, 'generating' singletons of parasites, Fig. 9.10b;
- networks of parasites penetrating hosts (each network consists of relatively ordered domains of parasites, nodes of the network exhibits dynamical links with their closest neighbouwrs), Fig. 9.11a;
- combination of disordered clusters of parasites and loci of highly ordered groups of parasites, Fig. 9.11b.

In parasitic interactions, when influence of one species stay fixed while influence of another species decreases, we observe increase of local completeness of morphological representation (Fig. 9.4cη). Simultaneous decrease in species influences first leads to increase of local completeness, followed by decrease of the completeness. The turning point is at mid-range influence, degrees of influence between two and three.

Activity ϕ decreases with decreasing influence of the parasite species on host species. The activity increasing with decreasing of the benefits the parasites species

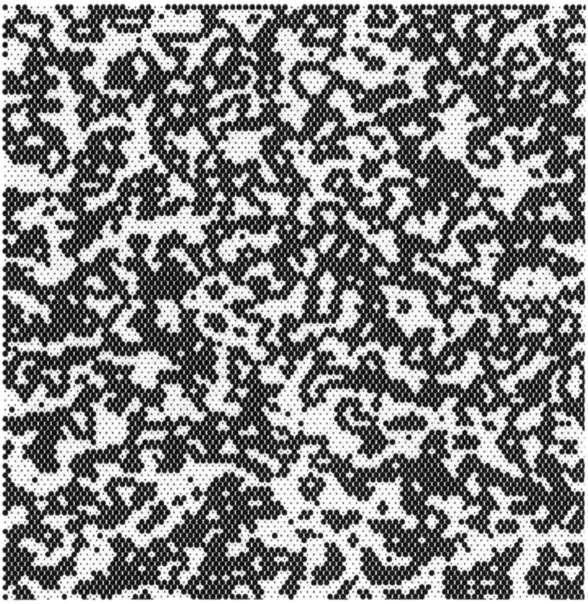

(a) $t = 2$

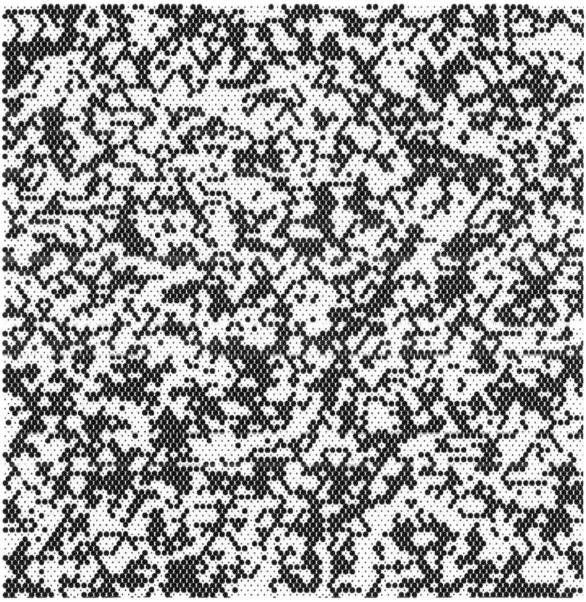

(b) $t = 2$

Fig. 9.10 Examples of structures formed in parasite-host systems (a) $\delta = 1$ and $\theta = 1$; (b) $\delta = 2$ and $\theta = 3$; initial density of parasites is 0.5. Left configuration represent automaton configuration at time step t, right configuration at time step $t + 1$.

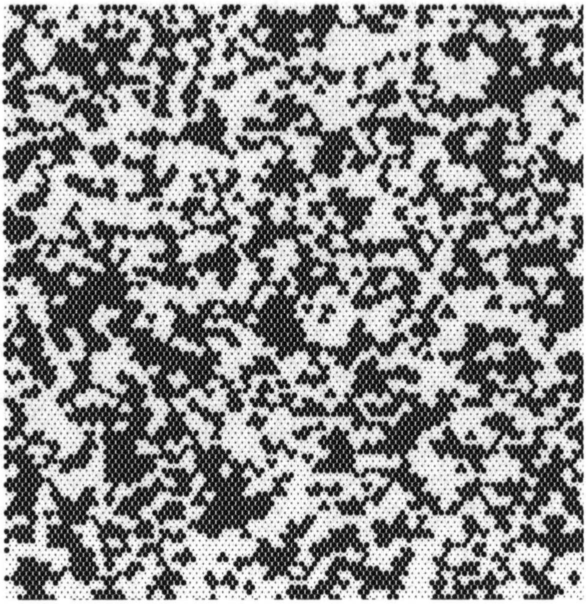

(a) $t = 2$

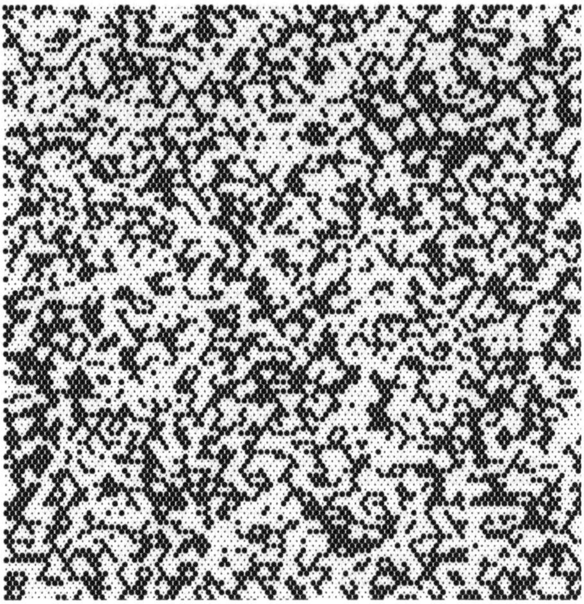

(b) $t = 2$

Fig. 9.11 Examples of structures formed in parasite-host systems (a) $\delta = 0$ and $\theta = 4$; (b) $\delta = 3$ and $\theta = 2$. initial density of parasites is 0.5. Left configuration represent automaton configuration at time step t, right configuration at time step $t+1$.

receive from the host species (Fig. 9.4cϕ). Simultaneous decrease of influences of both host and parasite species leads to decrease in the activity.

9.4 Competition

Cellular automaton imitating competition interaction exhibit wave patterns, similar to that observed in excitable media. If competition between species is strong, e.g. $\theta = 1$ and $\delta = 1$ (Fig. 9.12), and one of the species is sparsely distributed in the automaton initial configuration then target waves are formed and propagate (Fig. 9.12).

In scenario of dissimilar dependencies between species, e.g. $\delta = 1$ and $\theta = 2$, propagating target waves give rise to stationary domains when waves collide with each other (Fig. 9.13). The still domains are filled with the species which is lesser affected by other species, and thus eventually wins competition.

Propagating 'solid' (filled with one species) domains is yet another feature of the cellular automata imitating competition. This is typical for scenarios where there is a significant difference between dependencies of each species, e.g. $\theta = 5$ and $\delta = 2$ (Fig. 9.14). There we can see that certain configurations of one species in the lattice filled with another species form domains which increase in size and ultimately fill the whole lattice.

When almost all cells of automaton lattice take the same state apart of few cells which take different state, then breathing, oscillating and switching patterns can be observed (Fig. 9.15). This is particularly visible for the following values of dependencies: $\delta = 2$ and $\theta > 1$ or $\delta > 1$ and $\theta = 2$. The structures formed are similar to switching patters in Conway's Game of Life and to stationary waves in cellular automaton models of reaction-diffusion systems.

Momentary configurations, recorded after transient period, of cellular automata imitating competing species are shown in Fig. 9.16. The automata developed from initial random configurations, where both species have been represented equally. Multiplicity of wave sources makes target waves almost unrecognisable. Therefore we can only concentrate on general characteristics of the configurations morphology. We see that the increase in competing dependencies lead to multiplication of large one-species domains into many smaller domains, which further multiply to singletons (one or two cells in one state in the 'ocean' of other states).

In cellular automata imitating competition activity level is high (almost every cell changes its state every time step) only when degrees of influence between species are high, namely for $\theta, \delta \in \{1,2\}$ and also when $\theta = \delta = 3$ (Fig. 9.4e). The activity deceases for weak influences between species.

Morphologies of configurations representing competition are only complete when degrees of inter-species influences have relatively similar values, see diagonal $\theta = \delta$ in Fig. 9.4e. In this case with decrease of influences (corresponding to simultaneous growth of θ and δ) completeness decreases till $\theta = \delta = 4$ and then start to increase.

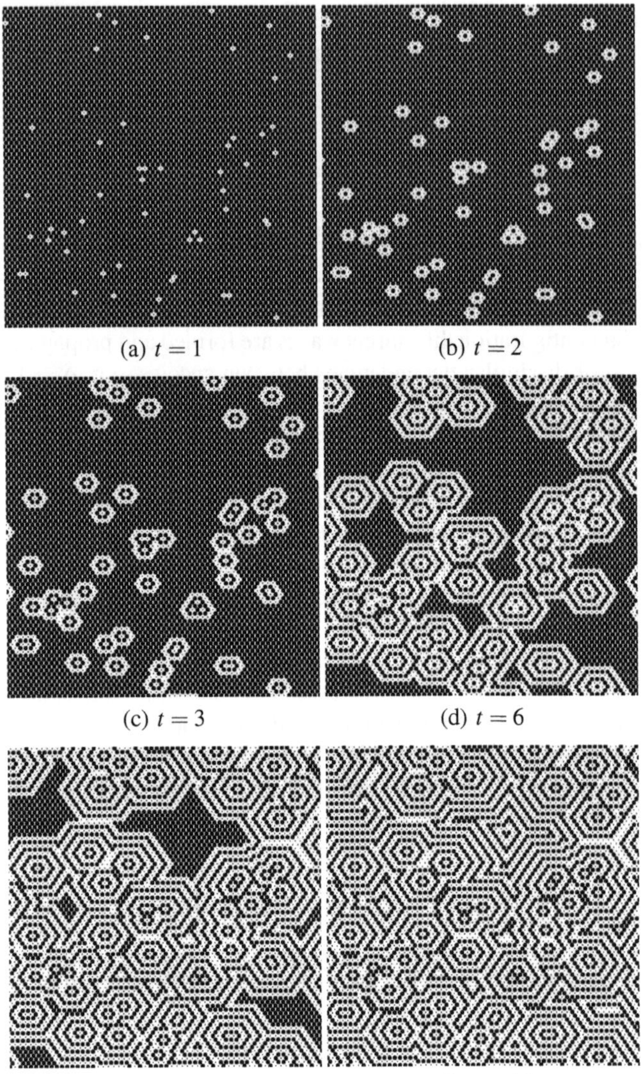

(a) $t = 1$ (b) $t = 2$ (c) $t = 3$ (d) $t = 6$ (e) $t = 9$ (f) $t = 15$

Fig. 9.12 Propagation and interaction of wave-patterns on the lattice with competing species a and b, $\theta = 1$ and $\delta = 1$, initially almost every cell takes state b and few (probability 0.001) cells take state a

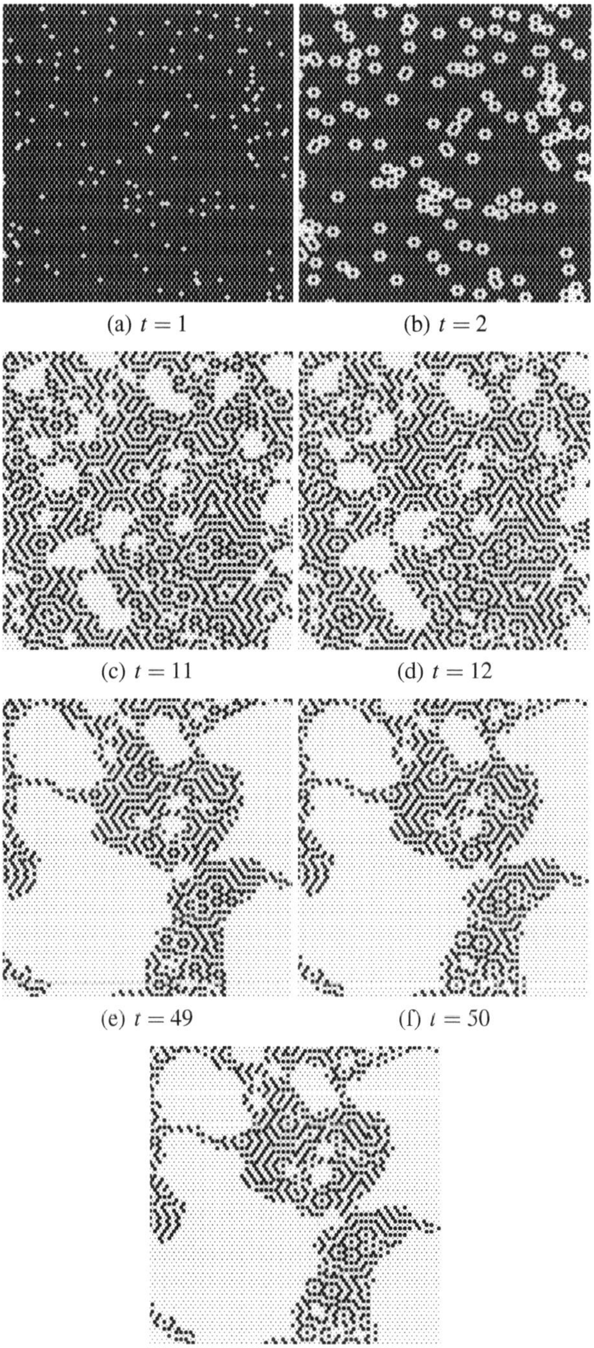

(a) $t = 1$ (b) $t = 2$

(c) $t = 11$ (d) $t = 12$

(e) $t = 49$ (f) $t = 50$

(g) $t = 51$

Fig. 9.13 Formation of growing domains of species a during interaction of waves, $\theta = 1$ and $\delta = 3$, initially almost every cell takes state b and few (probability 0.3) cells take state a

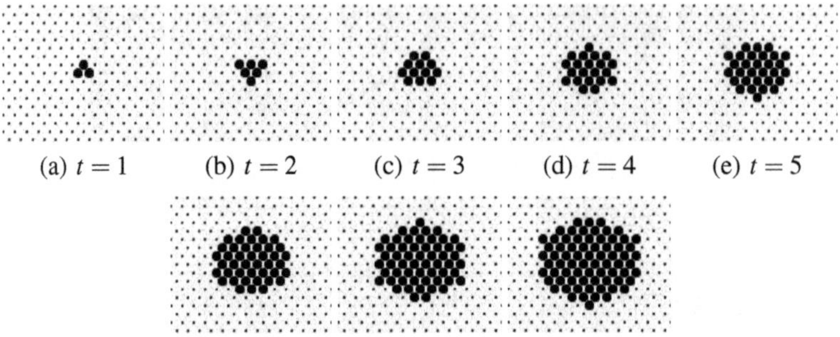

(a) $t=1$ (b) $t=2$ (c) $t=3$ (d) $t=4$ (e) $t=5$

(f) $t=6$ (g) $t=7$ (h) $t=8$

Fig. 9.14 Another way for stationary domains of species a to form: the still and 'solid' domain of a states grows from a tiny cluster of cells in the state a, $\theta=5$ and $\delta=2$

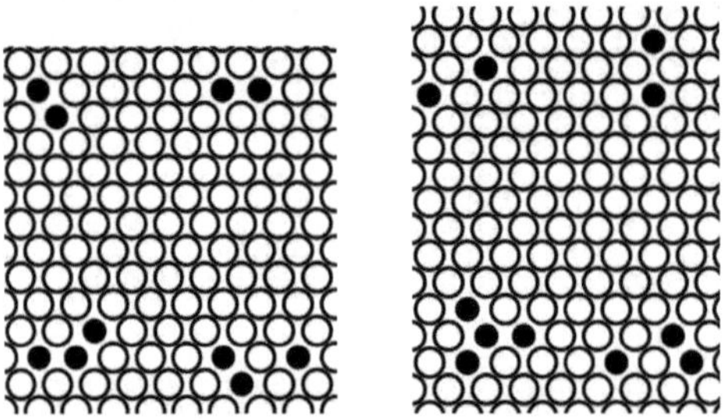

(a) $t=1$ (b) $t=2$

Fig. 9.15 Typical oscillators in competing systems for $\theta=2$ and $\delta=5$. Similar oscillators can be for the parameters $\theta=3$ and $\delta=2$; $\theta=3$ and $\delta=2$; $\theta=5$ and $\delta=2$.

9.5 Summary

Finding 28. *The proposed hierarchy of complexity of two-species interactions, as inferred from analysis of cellular automata models, will be as follows:* $\{$*commensalism, amensalism*$\} \prec$ *competition* $\prec$ *parasitism* $\prec$ *mutualism.*

The complexity of these systems show that a full biological understanding of symbiotic associations requires examining the direct and indirect interactions of symbionts in their ecological community context. In other cases, however, such as the human gut microbial community (with a long-tailed rarefaction curve of involved species), it is difficult to assess the number of symbiotic partners and their interaction strengths. Basic spatial processes of symbiotic interaction, as assessed by our model, are necessarily blurred in excessively complex natural systems, which usually involve more than two partners. Yet, the simple model of interactive excitation could serve as a primer to investigate more complex systems. The knowledge of basic configurations of symbiotic excitation might also help to design nature-inspired

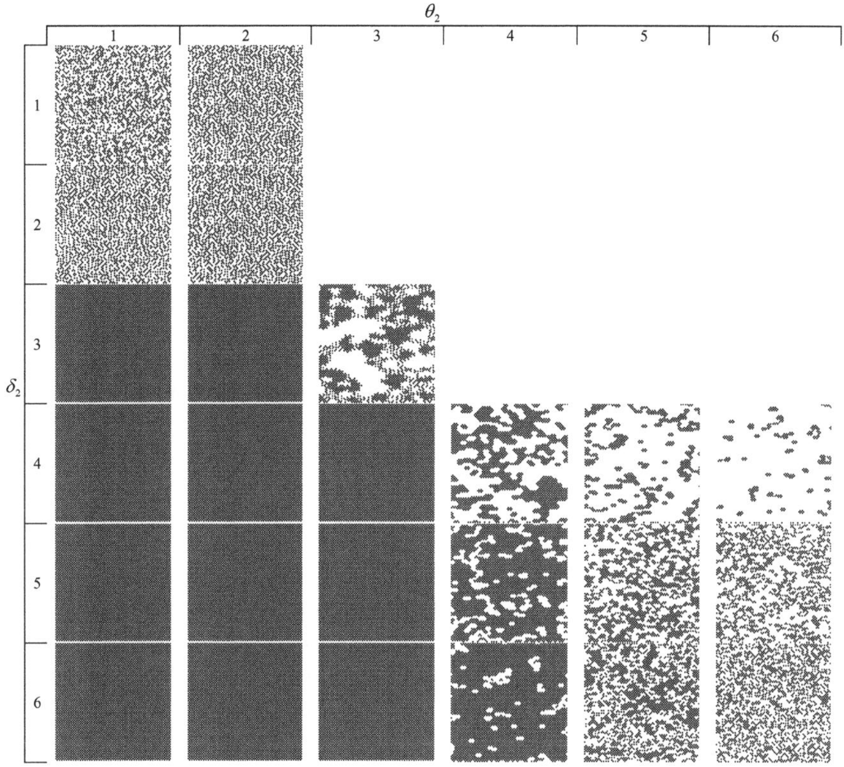

Fig. 9.16 Configurations of cellular automaton imitating competition interactions for different values of dependency parameters θ and δ. In each case automaton started evolution in a configuration where states a and b assigned to cells with the same probability 0.5. States a shown by white pixels, states b by black pixels.

and massively parallel computing approaches based on (variable numbers of) symbiotic/interacting agents.

What is a complexity hierarchy of inter-species interactions? Neutralism is out of question in our setup because cells do not change their states from the beginning of the systems' development. Amensalism and commensalism have lowest complexities: cell-state update frequencies are minimal, transient periods are short, no non-trivial propagating patterns observed. Competition occupies the next above bottom level of complexity. Cellular automata models of competing species do exhibit target waves and growing domains however they are characterised by low values of local completeness and minimal frequency of cell-state updates.

Parasitism and mutualism are top candidates for the most complex inter-species interactions. Cell-state update frequencies and local completeness of spatial configurations are in overall higher in automaton models of mutualistic interactions than in parasitic ones. Both types of interaction support formation and propagation of wave patterns, domains and switching patterns. Cellular automata imitating mutualism exhibit wider spectrum of breathing domains than automata imitating parasitism. These findings persuade us to conclude that mutualistic interactions are more complex than parasitic interactions.

We aimed to present a minimal model of inter-species interactions, therefore we did not include an empty space or a substrate (which would be a third cell-state) in the cellular automaton model. Because the model is reduced to the minimal number of parameters, it is actually difficult to find similar configurations in natural interactions. For example, the consequence of the lack of an 'empty' space in the model implies instantaneous occupation of space by species a (b) if species b (a) 'dies'. This may lead to somewhat paradoxical interpretation of interaction rules, e.g. in case of mutualism we have the following: species a and b benefit from presence of each other in their vicinities, however they also benefit from extinction of other species. Such a shift in interpretation of results is not entirely nonsense. In real-world the relationships can be combined, channeled via different routes, and transformed when applied in a complex networks of interaction species. Even the most simple cases of bacterial interactions are certainly more complex, as they do not restrict on two dimensions and frequently involve additional mechanisms, such as diffusive quorum-sensing signals. Symbioses among organisms with different level of organisation can be much more complex. Ant gardens are among of the best examples: there one can find commensalism (relationships between ants and bacteria), amensalim (between bacteria and parasitic fungi) and one more commensalism (between bacteria and cultivated fungi) [220]. The leaf-cutter ant symbiosis is similarly complex with five recognised members [174], including a parasitism between fungi, amensalism between bacteria and parasitic fungi, and mutualism between ants and fungi. The strength of one interaction may here also vary with the abundance of a partner influencing another interaction [175]. Another example — a transformation of preying to mutualism is provided in [58, 106]. They discuss that preys of American Alligator do actually benefit not suffer from the alligators' predation because for any particular species A, alligators not only consume organisms of the species A but also other species, which involved in competitive relationships with the species A.

Chapter 10
Automaton Mechanics of Mutualism

We simulate mutualistic relationships in a two-dimensional hexagonal cellular automaton. Every cell takes three states: 0, 1 and 2. States 1 and 2 represent species species '1' and species '2'. State 0 represents an 'empty space', or a substrate. Two processes must be simulated: propagation of species and survival of species. A cell of the hexagonal cellular automaton can be occupied exclusively by 'empty space', state 0, or by one of the species, states 1 or 2. A cell x in 'empty state' at time t ($x^t = 0$) becomes occupied by species 1 at time $t+1$ if it has more neighbours in state 1 than in state 2 ($\sigma_1^t > \sigma_2^t$, $\sigma_i^t = |\{y \in u(x) : y^t = i\}|$, $i \in \{1,2\}$) but enough neighbours in state 2 to support species 1 depending on them ($\sigma_2^t > \theta_{01}$), or it has equal amount of both species ($\sigma_1^t = \sigma_2^t$) but there are more species 2 to support species 1 than species 1 to support species 2 ($\sigma_2^t - \theta_{01} > \sigma_1^t - \theta_{02}$). Propagation of species 2 can be discussed similarly.

A cell in state 1 (2) remains in state 1 (2) if number of its neighbours in state 2 (1) exceeds specified threshold, $\sigma_2^t > \theta_{11}$ ($\sigma_1^t > \theta_{22}$). There are no cell-state transitions $1 \to 2$ and $2 \to 1$. See summary of the cell-state transitions in Fig. 10.1.

The thresholds θ_{01} and θ_{02} are propagation, or sustainability, parameters. The thresholds θ_{11} and θ_{22} are survivability parameters. In general, $1 \leq \theta_{01}, \theta_{02}, \theta_{11}, \theta_{22} \leq 6$, however no activity persists on the lattice for values higher than 3, so in the chapter we consider only $1 \leq \theta_{01}, \theta_{02}, \theta_{11}, \theta_{22} \leq 3$.

Sometimes we are addressing a cell-state rule as rule $R(\theta_{01}, \theta_{02}, \theta_{11}, \theta_{22})$.

x^t / x^{t+1}	0	1	2
0	$\sigma_1^t + \sigma_2^t = 0$	$A \vee (\neg A \wedge B)$	$C \vee (\neg C \wedge D)$
1	$\sigma_2^t < \theta_{11}$	$\sigma_2^t \geq \theta_{11}$	never
2	$\sigma_1^t < \theta_{22}$	never	$\sigma_1^t \geq \theta_{22}$

Fig. 10.1 Cell-state transition table imitating mutualistic interactions. The conditions A, B, C and D are as follows: $A = (\sigma_1^t > \sigma_2^t) \wedge (\sigma_2^t > \theta_{01}) \vee (\sigma_1^t + \sigma_2^t > 0) \wedge (\sigma_2^t - \theta_{01} > \sigma_1^t - \theta_{02})$, $C = (\sigma_2^t > \sigma_1^t) \wedge (\sigma_1^t > \theta_{02}) \vee (\sigma_1^t + \sigma_2^t > 0) \wedge (\sigma_2^t - \theta_{01} < \sigma_1^t - \theta_{02})$, $B = (\texttt{random}(1) < 0.5)$, $D = (\texttt{random}(1) \geq 0.5)$, and `random` generates a random real number between 0 and 1.

A. Adamatzky: Reaction-Diffusion Automata, ECC 1, pp. 183–194.
springerlink.com

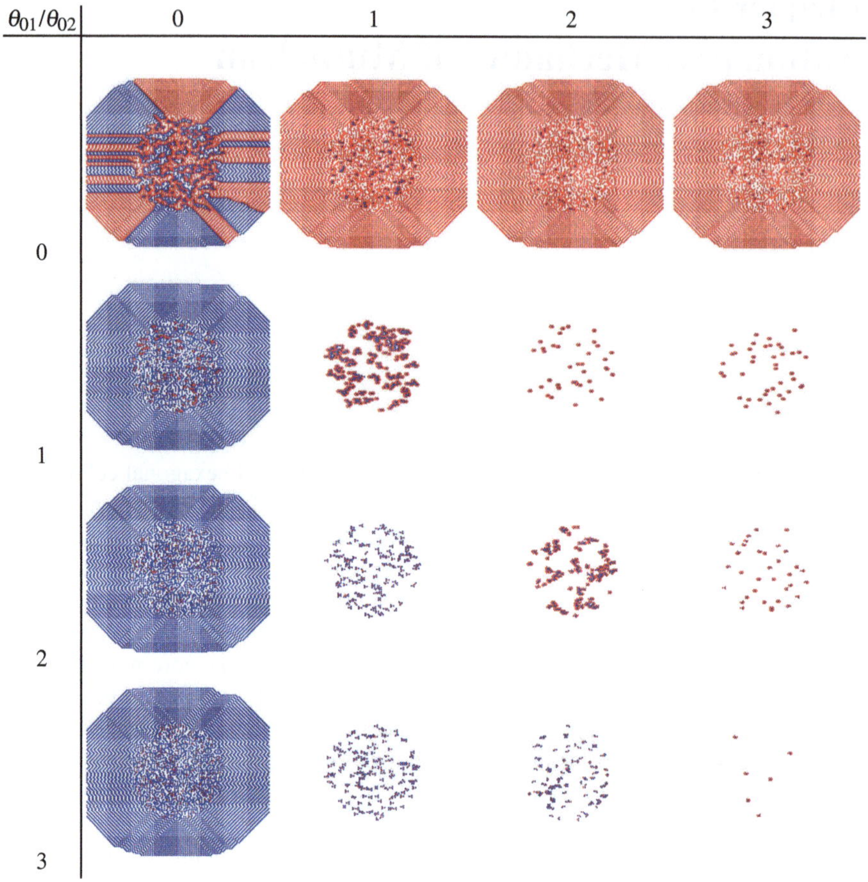

Fig. 10.2 Samples of configurations parameterised by values of θ_{01}, rows, and θ_{02}, columns. Each entry $[\alpha, \beta]$ in the table represents a sample configuration for parameters $\theta_{01} = \alpha$, $\theta_{02} = \beta$, $\theta_{11} = 1$ and $\theta_{22} = 2$. Lattice is 200×200. Initially cells not further then 50 cells from the lattice centre are assigned states 1 or 2 with probability 0.1 each.

10.1 Phenomenology

We have analysed development of automata from random initial configurations for all values of four thresholds $1 \leq \theta_{01}, \theta_{02}, \theta_{11}, \theta_{22} \leq 3$. We found that when one of the species 1 (2) does not depend on another species to propagate $\theta_{02} = 0$ ($\theta_{01} = 0$) then this species 1 (2) propagates unlimitedly while other species remains confined to the zone of initial inoculation. See few selected samples in Fig. 10.2.

Dynamics of unlimited propagation depends on parameters $\theta 01$ and θ_{02}. Consider for example, $\theta_{01} = 0$ and $\theta_{02} < 4$. Species 1 propagates on the lattice because they do not need species 2 to invade new space. If $\theta_{11} = 0$ then lattice becomes filled

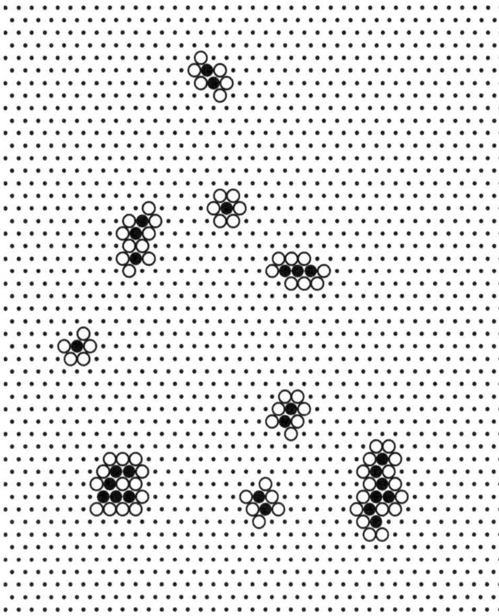

Fig. 10.3 Stationary still localizations in rule $R(2221)$. State '1' is shown by solid disc, state '2' by circle, state '0' by dot.

with 'solid' pattern of species 1, once a site occupied by 1 it always remains in the state 1. For $\theta_{11} > 0$ we observe structures similar to target waves because species 1 can not remain at the same site for more then one step of discrete time t, it dies, or vacates the site, at time $t+1$, however the site becomes occupied again by species 1 at time step $t+1$.

For the parameters $\theta_{02} > 0$ and $\theta_{01} > 0$ localisations — compact groups of non-quiescent (not equal 0) states — are observed. The number of localisations in configurations, recorded after transient period, decreases with increase of $\theta_{02} + \theta_{01}$. Principle localisations are discussed in next section.

10.2 Localisations in Mutualistic Systems

In exhaustive experiments we discovered four principle types of localisations: stationary still localizations, stationary oscillating localizations, propagating quasi-one-dimensional localizations, or worms, and sub-linearly growing domains. 'Stationary' means that the pattern does not travel along the lattice, 'still' means that the pattern does not change its structure during automaton development.

Stationary still localisation is a compact groups of cells in non-quiescent (non-'0') states. Diameter of such group usually does not exceed 1-3 cells. Typically one species occupies central position in the localisation and it is surrounded by

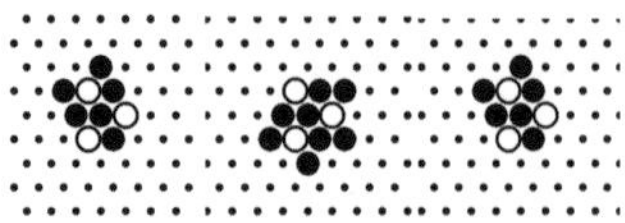

(a) t (b) $t+1$ (c) $t+2$

Fig. 10.4 Oscillators in rule $R(2323)$

sites occupied by another species. Thus in Fig. 10.3 we see sites of species 1 (solid discs) surrounded by several sites with species 2 (circles). Species 1 survive because they have more than two neighbouring sites with species 2 ($\theta_{02} > 2$) while less dependent species 2 are supported by being in relation with just one site occupied by species 1 ($\theta_{01} > 1$). Several elementary localizations can be linked together to form an extended cluster of localizations.

Stationary oscillator is compact group of cells in non-quiescent states, which does not move along the lattice unlimitedly. The group can be translated cyclically around some fixed point. All oscillators observed in space-time dynamics of mutualistic populations are can be classified as flip-flops, which rotate on a fixed degree around stationary centre, e.g. oscillators in Fig. 10.4 and Fig. 10.5, and breathing localisations, which has stationary core, e.g. sites in state 1 in Fig. 10.6, and switching/breathing halo, e.g. sites in state 2 in Fig. 10.6.

Some rules support only still localizations, e.g. rule $R(2221)$ (Fig. 10.3), some only oscillators, e.g. rule $R(2323)$ (Fig. 10.4), others support both still and oscillating localizations, e.g. rule $R(1113)$ (Fig. 10.6).

Worms are quasi-one-dimensional propagating localizations, which usually consists of two parallel chains of non-quiescent states, one chain is formed by sites with species 1, another chain by sites with species 2. All worms observed in our experiments have only two growing tips. For example, see in Fig. 10.7 snapshots of a single worm growing in rule $R(1111)$ automaton. The worm has two growing tips which are two sites: one occupied by state 1, another by state 2. For neighbouring quiescent cells numbers of non-quiescent states 1 and 2 is equal, so the states 1 and 2 ‘diffuse’ randomly. Species 1 and 2 can not survive without each other, therefore when one of the species advances, it must then ‘wait’ for a neighbouring empty site to be occupied by another species. Due to randomisation involved the growing tips do not propagate directly but rather implement a kind of random walk on the lattice (Fig 10.8a).

When initial density of species 1 and 2 is high enough, many worms are born (Fig 10.8b), they propagate and collide with each other. When a growing tip collides to a body of a worm it stops propagating. Thus forming a porous stationary core with free growing tips propagating centrifugally (Fig 10.8b).

For some rules, e.g. rule $R(3300)$ (Fig. 10.9), worms are short-living. In these rules worms stop propagating when their growing tips are blocked by the following conditions. Every resting cell x, $x^t = 0$, closest to non-resting cells of the worm's tip has $(\sigma_1^t > \sigma_2^t) \wedge (\sigma_2^t < \theta_{01})$, i.e. species 1 do not propagate, and also $(\sigma_2^t > \sigma_1^t) \wedge (\sigma_1^t < \theta_{02})$, i.e. species 2 do not propagate as well.

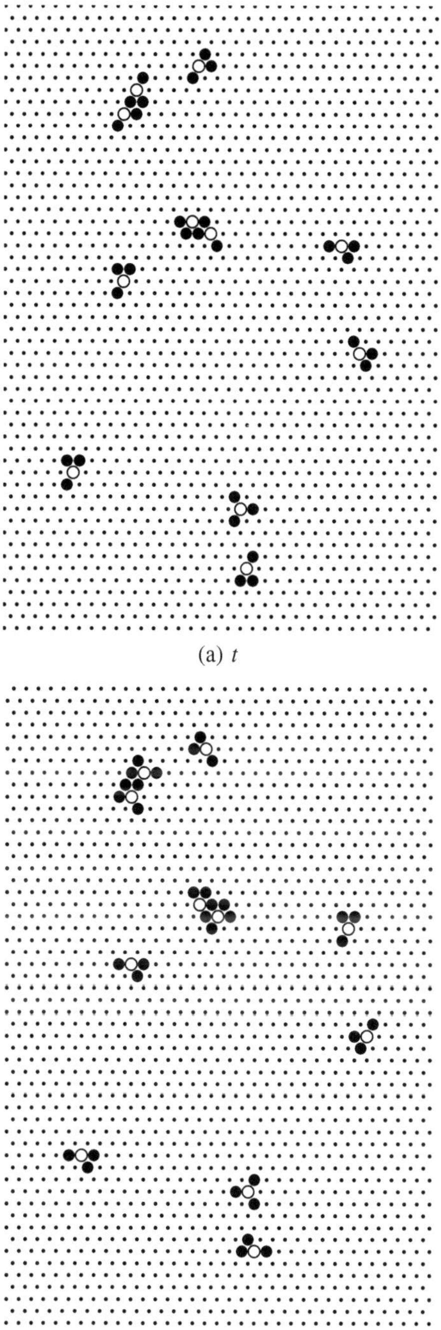

(a) t

(b) $t+1$

Fig. 10.5 Oscillators in rule $R(1223)$

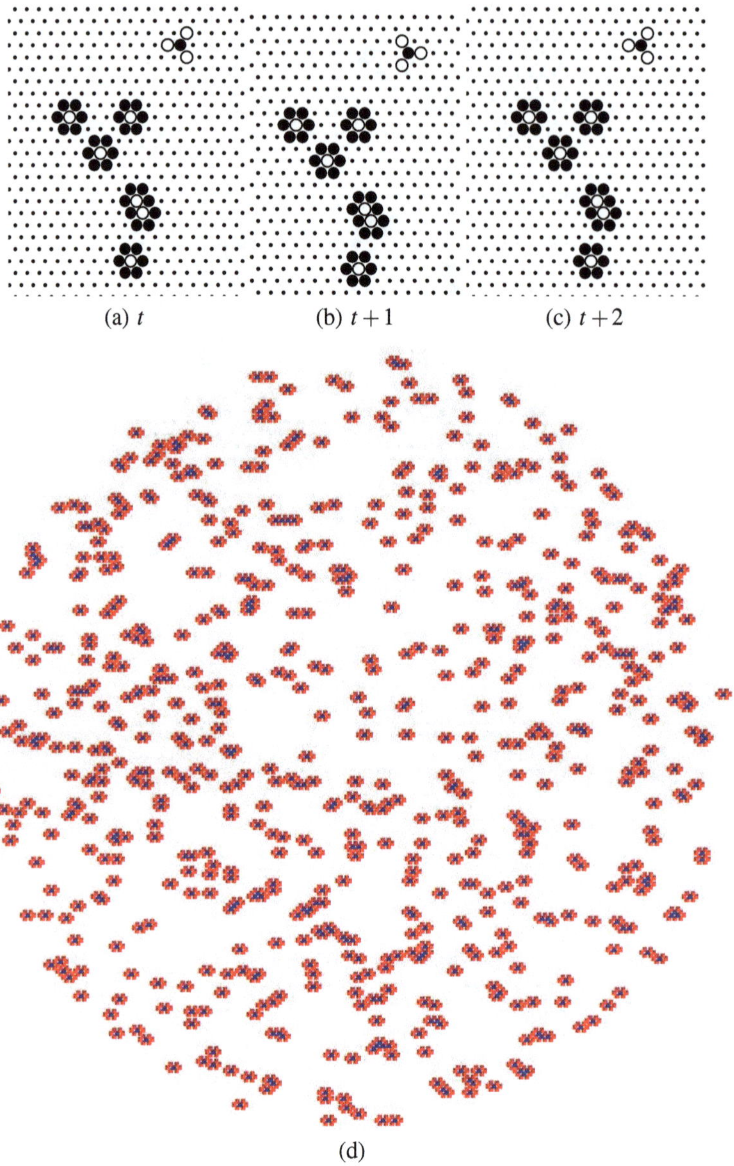

Fig. 10.6 Still localizations and oscillators in rule $R(1113)$. (a)-(c) three consecutive configurations exhibiting still localizations: sites with species 2 surrounded by sites with species 1, and stationary oscillators: sites with species surrounded by sites with species 2. (d) zoomed out configuration.

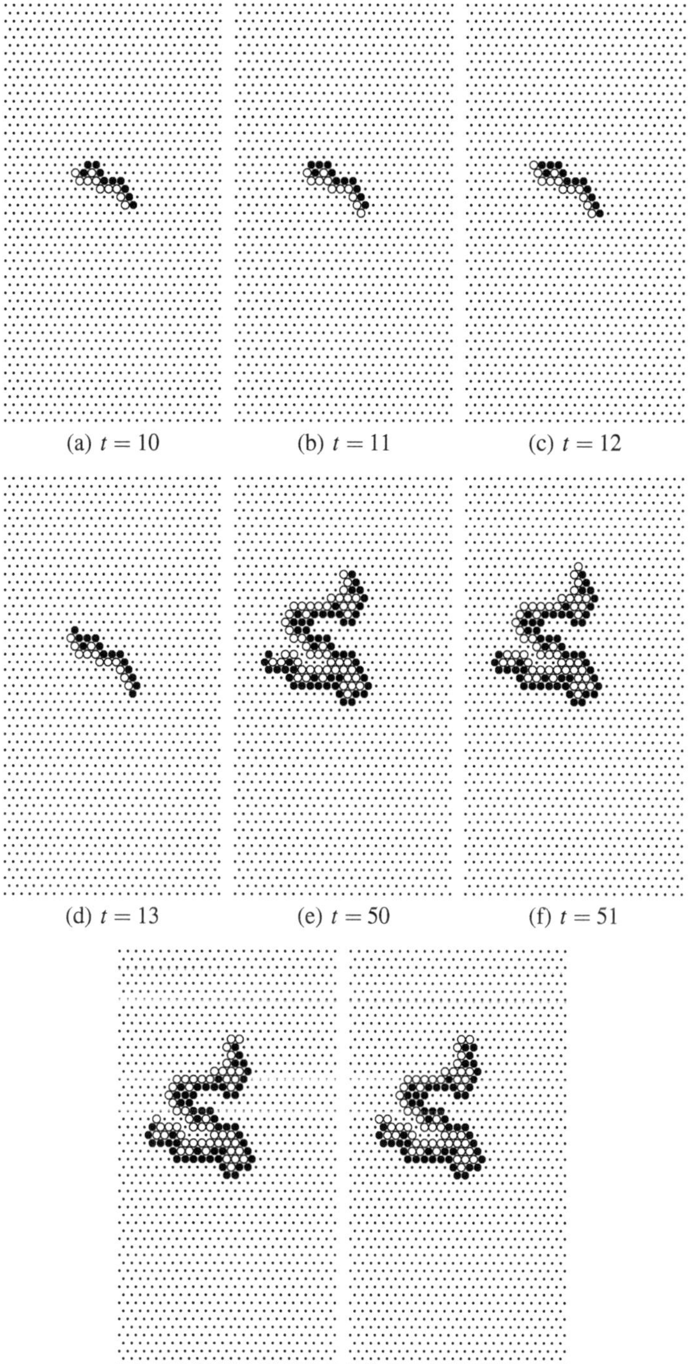

(a) $t = 10$ (b) $t = 11$ (c) $t = 12$

(d) $t = 13$ (e) $t = 50$ (f) $t = 51$

(g) $t = 52$ (h) $t = 53$

Fig. 10.7 Propagation of long-life worms in rule $R(1111)$

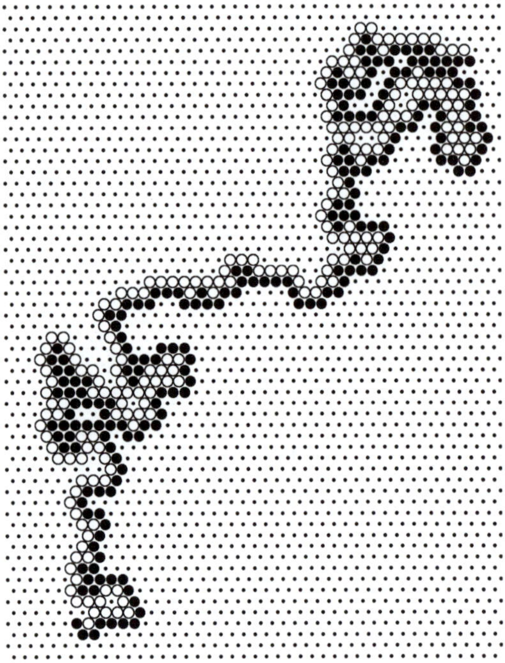

(a)

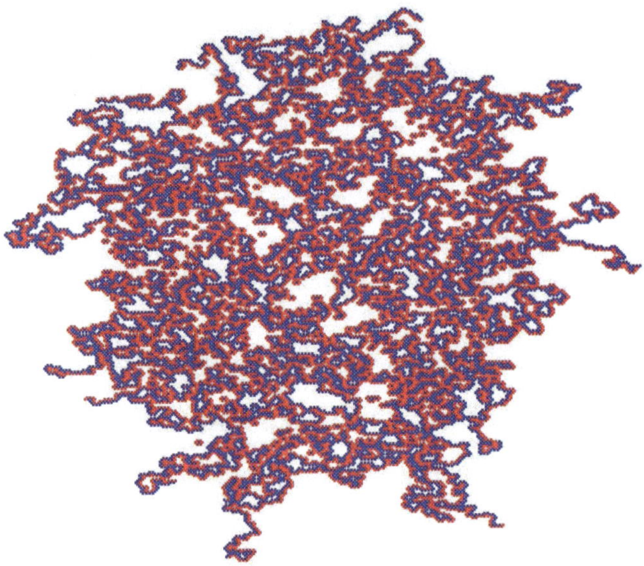

(b)

Fig. 10.8 Long-life worms in rule $R(1111)$. (a) a single worm grown for 197 steps of discrete time. (b) Zoomed out configuration of rule $R(1111)$ automaton, which starts its initial development in a random configuration, where all cells closer then 150 sites to the centre of the lattice are assigned states 1 or 2 with probability 0.5.

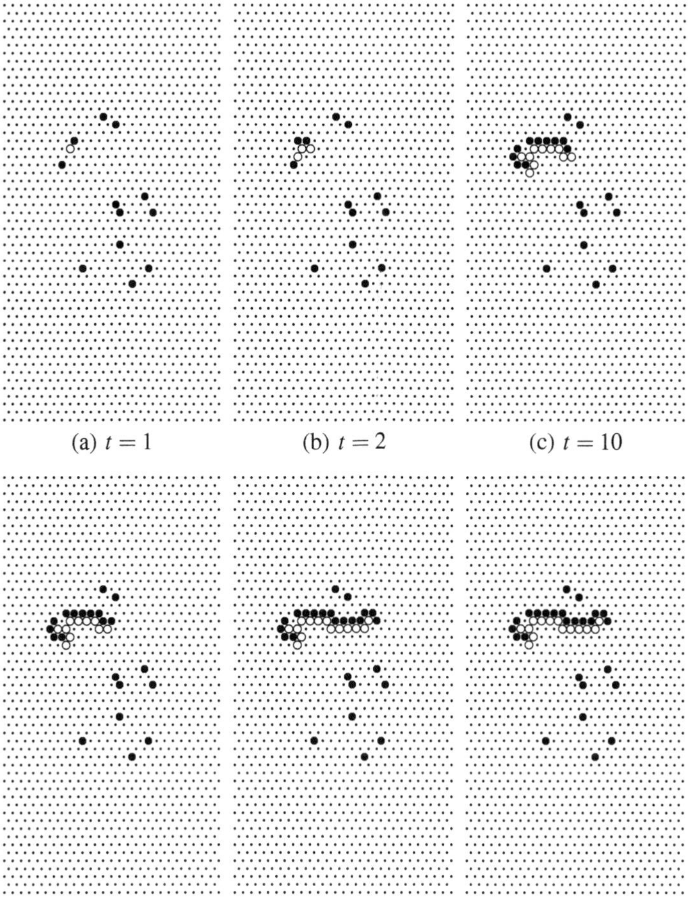

(a) $t = 1$ (b) $t = 2$ (c) $t = 10$

(d) $t = 11$ (e) $t = 35$ (f) $t = 36$

Fig. 10.9 Propagation of short-life worms in rule $R(3300)$

The last class of localizations are very slowly growing stripy domains (Fig. 10.10), found in space-time dynamics of cellular automata governed by rules $R(2222)$ and $R(3322)$. They are not proper localizations, because their size is constantly increasing. However it does increase very slow: from computational experiments we estimated that a radius of a circle around a stripy domain grows as $0.05 \cdot t$, where t is a number of time steps from the beginning of the pattern development. That is the growth rate is almost two orders smaller then typical propagation rate of usual cellular automaton patterns (e.g. excitation waves or gliders in Conway's Game of Life). So comparing to typical cellular automaton growing patterns the stripy domains behave as localizations.

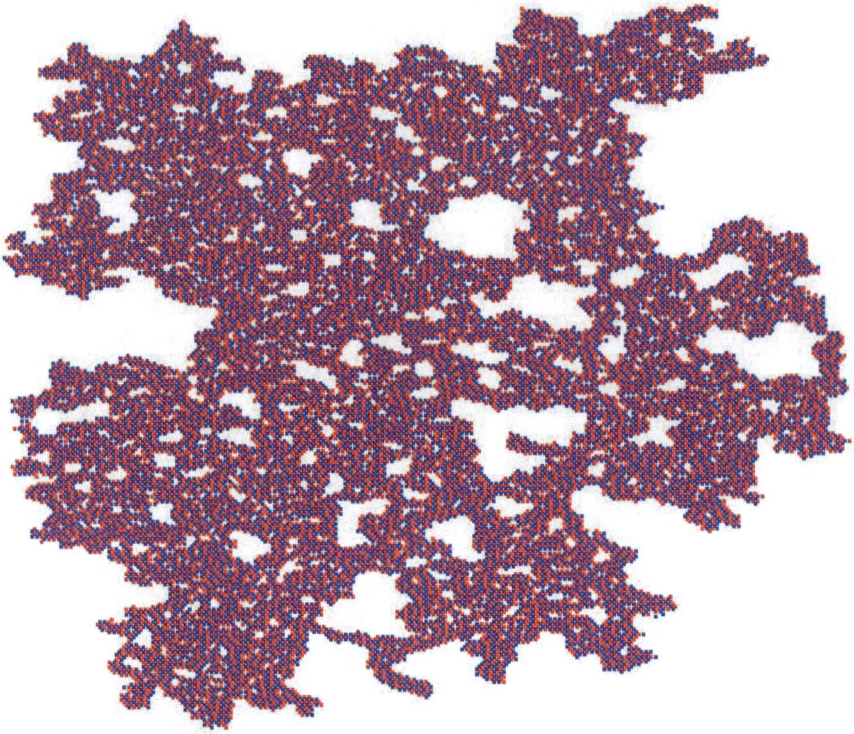

Fig. 10.10 A snapshot of a sub-linearly growing stripy domains, rule $R(2222)$, taken at 1171th step of automaton development from initial random configuration

The domains are stripy, i.e. consist of intermittent chains of 1 and 2 states, because each species needs at least two or three sites occupied by other species to propagate and at least two sites occupied by other species to survive. Thus having a chain of states 1, the chain will be covered by on both sides by states 2, which will be covered by states 1 and so on. See mechanics of the pattern formation in Fig. 10.11 and 10.12. Growth of the patterns is slow due to involved of randomization, see conditions *B* and *C* in cell-state transition table (Fig. 10.1). When an empty site accept species 1 or 2 at random, the species may not stay there if survivability criteria not met, then the site become empty again and may take of the species at random. For example, see how species 2 propagate to the North tip of the growing pattern in Fig. 10.12. At first empty site is occupied by species 2 (also because there are neighbouring species 1), however species 1 disappear because their neighbourhoods do not meet survivability criteria; and the site just occupied by species 1 becomes empty again (Fig. 10.12).

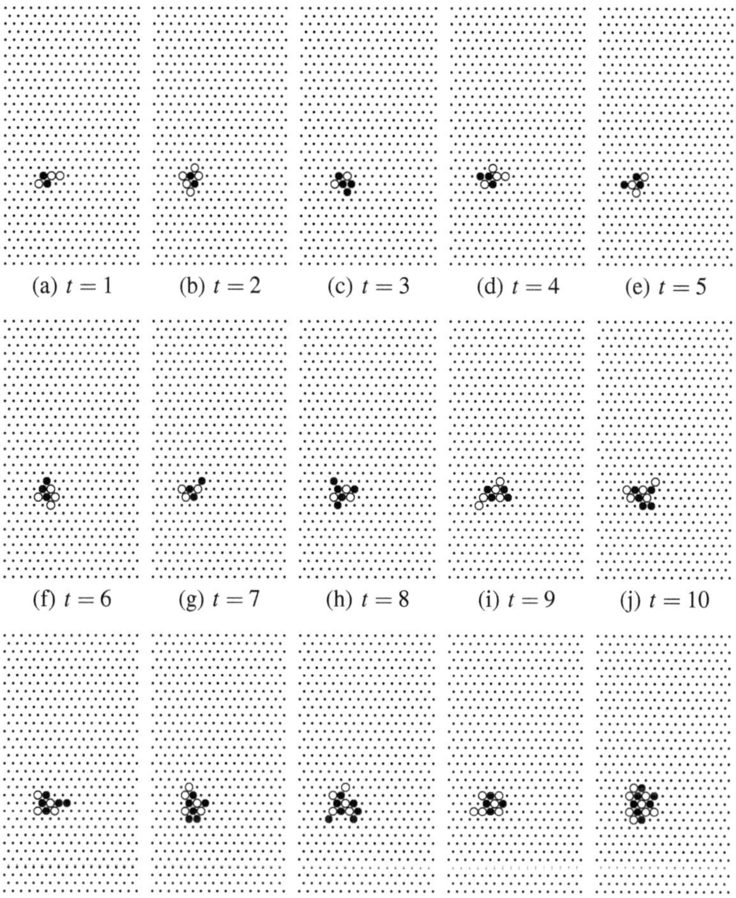

(a) $t = 1$ (b) $t = 2$ (c) $t = 3$ (d) $t = 4$ (e) $t = 5$

(f) $t = 6$ (g) $t = 7$ (h) $t = 8$ (i) $t = 9$ (j) $t = 10$

(k) $t = 11$ (l) $t = 12$ (m) $t = 13$ (n) $t = 14$ (o) $t = 15$

Fig. 10.11 Mechanics of stripy pattern formation in rule $R(2222)$: $t = 1, \ldots, 15$

10.3 Summary

A minimal cellular automaton imitating a spatially-extended system with two species in mutualistic relationships exhibit rich dynamics of pattern formation, stationary and propagating localisations. Four types of localisations are most common:

- stationary still localisations,
- stationary oscillators,
- quasi-one-dimensional propagating domains, and
- sub-linearly growing stripy domains.

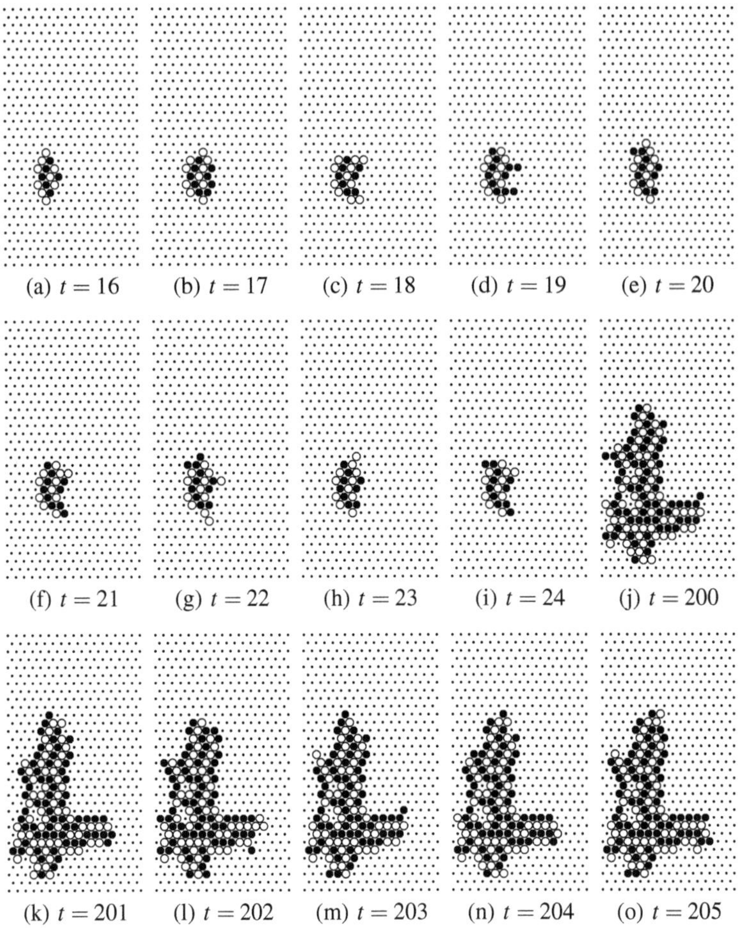

Fig. 10.12 Mechanics of stripy pattern formation in rule $R(2222)$: $t = 1,\dots,15$ and $t = 200,\dots,205$

Space-time dynamic of mutualistic populations parameterised by propagation dependencies is shown in Fig. 10.13. There we see that when propagation dependencies increase or when dependence of one species is higher then dependence of another species the stationary localizations change from propagating worms to oscillating localizations to stationary localizations. Most remarkable varieties of localizations emerge for lower but above zero propagation dependencies. The very similar correlation takes place between spatial patterns and survivability parameters, i.e. transitions from worms to oscillators to still localizations.

θ_{01}/θ_{02}	1	2	3
1	Propagating worms: $R(1100)$, $R(1101)$, $R(1111)$; Very slowly-growing worms/domains: $R(1122)$; Still and oscillating domains for other values of parameters.	All clusters are non-propagating. Majority are stationary still clusters. Only for $R(1222)$, $R(1223)$, $R(1233)$ oscillating clusters.	Mostly stationary still clusters, oscillating clusters only for $R(1322)$, $R(1323)$, $R(1333)$.
2	—	Worms propagate but freeze quickly: $R(2200)$, $R(2201)$, $R(2202)$; For $R(2202)$ life-time of worms is really short and no worms appear for $R(2203)$; for $R(2211)$ worms freeze at once and form still clusters. For $R(2212)$ and $R(2213)$ still clusters, $R(2222)$ slowly growing stripy domains.	Mostly still stationary clusters, oscillating only for $R(2322)$, $R(2323)$, for $R(2333)$ all activity extinguish all dying.
3	—	—	For $R(3300)$ worms propagate but freeze quickly; for $R(3301)$–$R(3303)$ number and life-time of warms decrease, till no worms at all for $R(3303)$; $R(3311)$ long-living worms; for $R(3312)$ to $R(3313)$ worms quickly freeze or die. For $R(3322)$ worm-like large clusters grow. There is no sustained activity for $R(3323)$ and $R(3333)$ no sustained activity.

Fig. 10.13 Characteristics of space-time dynamics for $\theta_{01} > 0$, rows, and $\theta_{02} > 0$, columns. Each entry $[\alpha, \beta]$ provides a brief description of space-time dynamics of a sample configuration for parameters $\theta_{01} = \alpha$, $\theta_{02} = \beta$, $0 \leq \theta_{11} \leq 3$ and $0 \leq \theta_{22} = 2 \leq 3$.

Part III
Computation with Excitation

Chapter 11
Voronoi Automata

Let **P** be a nonempty finite set of planar points. A planar Voronoi diagram [253] of the set **P** is a partition of the plane into such regions, that for any element of **P**, a region corresponding to a unique point p contains all those points of the plane which are closer to p than to any other node of **P**. A unique region $vor(p) = \{z \in \mathbf{R}^2 : d(p,z) < d(p,m) \forall m \in \mathbf{R}^2, m \neq z\}$ assigned to point p is called a Voronoi cell of the point p [211]. The boundary $\partial vor(p)$ of the Voronoi cell of a point p is built of segments of bisectors separating pairs of geographically closest points of the given planar set **P**. A union of $VD(\mathbf{P}) = \cup_{p \in \mathbf{P}} \partial vor(p)$ all boundaries of the Voronoi cells determines the planar Voronoi diagram [211]. A variety of Voronoi diagrams and algorithms of their construction can be found in [162, 206].

Approximation of Voronoi diagrams with propagating patterns is based on time-to-distance transformation: to approximation a bisector separating planar points p and q we initiate growing patterns at p and q. The pattern travel the same distance from the sites of origination before they meet each other, The loci where the waves meet indicate sites of the computed bisector [2, 5]. Precipitating reaction-diffusion chemical media are proved to be an ideal computing substrate for approximation of the planar Voronoi diagram [14, 88, 243]. A Voronoi diagram can be approximated in a two-reagent medium. One reagent α is saturated in the substrate, drops of another reagent β are applied to the sites corresponding to planar points to be separated by bisectors. The reagent β diffuses in the substrate and reacts with reagent α. Colored precipitate is produced in the reaction between α and β. When two or more waves of diffusing α meet, no precipitate is formed [88]. Thus uncolored loci of the reaction-diffusion medium represent bisectors of the computed diagram. A range of chemical precipitating processors is designed and working prototypes are tested in laboratory conditions [86–88, 91, 243].

11.1 Voronoi Automata

A Voronoi automaton is a tuple $\mathscr{V} = \langle \mathbf{V}(\mathbf{P}), \mathbf{Q}, \mathbf{N}, u, f \rangle$, where **P** is a finite planar set, $\mathbf{V}(\mathbf{P}) = \{V(p) : p \in \mathbf{P}\}$, **Q** is finite set, **N** is a set of natural numbers and u :

A. Adamatzky: Reaction-Diffusion Automata, ECC 1, pp. 199–208.
springerlink.com

$\mathbf{V}(\mathbf{P}) \to \mathbf{V}(\mathbf{P})^k$ is second-order neighbourhood, $0 < k < |\mathbf{P}|$, and $f : \mathbf{Q}^k \to \mathbf{Q}$ is a cell-state transition function.

Excitable-precipitating Voronoi automata studied in present chapter are specified as follows. Cell state set has for elements, $\mathbf{Q} = \{\circ, +, -, \#\}$. Thus we assign three excitation-related states — resting ($\circ$), excited ($+$) and refractory ($-$) — to cells, and one precipitate state #. Cells update their state in discrete time. A state of cell $V(x)$ are time step $t \in \mathbf{N}$ is denoted as $V(x)^t$. All cells update their states in parallel using the same cell-state transition function.

Let $w(p)$, $p \in \mathbf{P}$, be a first-order neighbourhood of a Voronoi cell $Vor(p)$ from $\mathbf{V}(\mathbf{P})$: $w(p) = \{V(q) \in \mathbf{V}(\mathbf{P}) : \partial V(p) \cap \partial V(q)\}$, i.e. a set of Voronoi cells which have common edges with $V(x)$. A second-order neighbourhood $u(p)$ is a set of neighbours of first-order neighbours of $V(p)$: $u(p) = \bigcap_{V(q) \in w(V(p))} w(V(q)$. Examples of Voronoi cell neighbourhoods are shown in Fig. 11.1a. Distributions of numbers of immediate and second neighbours and sizes of second-order neighbourhood u are given in Fig 11.1b. Predominant number, over 40%, of cells have six first-order (immediate) neighbours. Almost half of the cells have 13 or 14 second-order neighbours. For any cell size of second-order neighbourhood u is a sum of first-order neighbours and second-order neighbours. For Voronoi automata with high-density packing of 15K cells, half of the cells have either 19, or 20 or 21 neighbours. If we consider only immediate neighbourhood then $\mathcal{V}$ can be, in principle, seen as a slightly distorted, hexagonal lattice. However, by adopting second-order neighbourhood we are moving to a random structure with higher than hexagonal lattice coordinate number (a node in hexagonal lattice has 18-node second-order neighbourhood while dominating neighbourhood sizes in $\mathcal{V}$ are 19, 20 and 21).

Transition form excited to refractory state is unconditional, i.e. takes place with regards to states of a cell's neighbours. A resting cell excites if it has at least one excited neighbour. In this particular model we take a refractory state and precipitate states are absorbing: one a cell takes either of these two state it does not update its state any longer.

Neighbourhood sizes may differ between Voronoi cells therefore we use a state transition function, where a cell updates its state depending on a relative excitation in its neighbourhood. We assume that precipitation occurs in a resting cell x when a ratio $\sigma^t(x)/\nu(x)$ of excited neighbours $\sigma^t(x)$ to a total number of neighbours $\nu(x)$ exceeds some threshold $\eta \in [0,1]$. Let $\sigma(V(x)^t)$ be a number of excited cells in the cell $V(x)$'s second-order neighbourhood, $\sigma(V(x)^t) = \sum_{V(y) \in u(V(x))} |\{V(y) : V(y)^t = +\}|$ then a cell updates its state by the following rule:

$$V(x)^{t+1} = \begin{cases} \#, \text{ if } V(x)^t = \circ \text{ and } \sigma^t(x)/\nu(x) \geq \eta \\ +, \text{ if } V(x)^t = \circ \text{ and } \sigma^t(x)/\nu(x) < \eta \text{ and } \sigma^t(x) > 1 \\ -, \text{ if } V(x)^t = + \\ \circ, \text{ otherwise} \end{cases}$$

In computational experiments we construct set $\mathbf{P}$ by filling a disc-container of radius 480 units with up to 15K points. The points are packed at random but there is at least 5 units distance between any two points. Voronoi diagram $\mathcal{V}$ — on

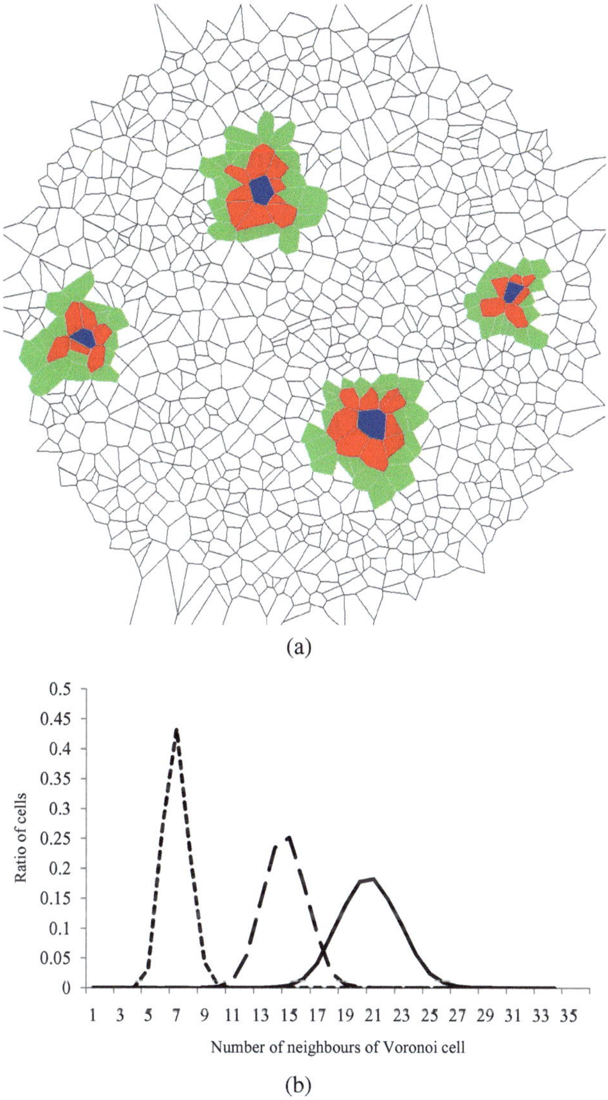

Fig. 11.1 Structure and properties of neighbourhoods. (a) Example of Voronoi tessellation with second-order neighbourhood structures highlighted for four cells. In each case central cell p is filled with blue (black in grey-scale reproduction) colour, first-order neighbours of $w(p)$ are filled with red (dark-grey) colour, and neighbours of $w(p)$ are filled with green (light-grey) colour. For each blue (black) Voronoi cell a second-order neighbourhood is a set of red (dark-grey) and green (light grey) Voronoi cells. (b) Distribution of first order, or immediate, neighbours (dotted line), second order neighbours (dashed line) and size of second order neighbourhood (solid line) in Voronoi automata of 15K cells.

which Voronoi automaton is constructed — is calculated by a classical sweepline algorithm [111].

11.2 Constructing Voronoi Diagram on Voronoi Automata

Let **B** be a set of planar points on which a Voronoi diagram $\mathfrak{V}$ on $\mathcal{V}$, $\eta = 0.4$, is approximated. We project **B** onto $\mathcal{V}$ and excite cells of $\mathcal{V}$ which are closer than 9 units to points of **B**. Excitation waves spread on $\mathcal{V}$. The excitation waves collide and precipitation occurs nearby sites of the waves' collisions. Configuration of cells in precipitate state represents edges of $\mathfrak{V}(\mathbf{B})$. An example of excitation-precipitation dynamics in $\mathcal{V}$ which approximates $\mathfrak{V}(\mathbf{B})$, where **B** is a planar set of seven points, is shown in Fig. 11.2. A scheme of the Voronoi diagram computed is shown in Fig. 11.6a.

Even in a single example shown in Fig. 11.2 it is clear that $\mathcal{V}$ does not calculate $\mathfrak{V}(\mathbf{B})$ precisely, it rather approximates it with discrete domains of precipitation, and also introduces some noise (sites of precipitation not coinciding with bisectors of $\mathfrak{V}(\mathbf{V})$). Only threshold of precipitation η can be varied in our model. How are a degree of approximation and noise to signal ratio depends on value of η? To answer we undertook a series of computational experiments, illustrated – for some values of η — in Fig. 11.3.

For a set **B** of seven points we constructed 'classical' planar Voronoi diagram $\mathfrak{V}(\mathbf{B})$. Then for $\eta = 0.2 \ldots 0.525$ we excited $\mathcal{V}$ with **B**, waited till $\mathcal{V}$ reaches its stationary configuration, where all cells are in either precipitate or refractory state, and compared the configurations of $\mathcal{V}$ and image of $\mathfrak{V}(\mathbf{B})$. For each η we recorded a fraction ρ of precipitate-cells in a final configuration of $\mathcal{V}(\eta)$ to a total number of precipitate-cells in the same configuration (Fig. 11.4a).

We found that while density ν, calculated as a ratio of a number of precipitate-cells in stationary configuration $\mathcal{V}(\eta)$ to a number of precipitate-cells in configuration of $\mathcal{V}(0.2)$, exponentially decreases with increase of precipitation threshold η (Fig. 11.4b) degree of approximation ρ shows more intriguing behaviour (Fig. 11.4a). From $\eta = 0.2$ till $\eta = 0.41$ 'signal-to-noise' ratio $\rho(\eta)$ grows almost exponentially. Trend of $\rho(\eta)$ undergoes S-type transition between $\eta = 0.41$ and $\eta = 0.453$ with one pick $\rho(0.44)$ and two cavities $\rho(0.43)$ and $\rho(0.452)$. $\mathcal{V}$ with precipitation threshold in the zone around $\eta = 0.4$, onset of 'strange' behaviour of $\rho(\eta)$), demonstrates best approximation of $\mathfrak{V}(\mathbf{B})$, as recognised by unaided eye.

11.3 Arbitrary-Shaped Planar Objects and Contours

The Voronoi automaton $\mathcal{V}$ copes well with data set **O** of planar finite shapes. To approximate diagram $\mathfrak{V}$ of arbitrary planar shapes we project the objects of **O**

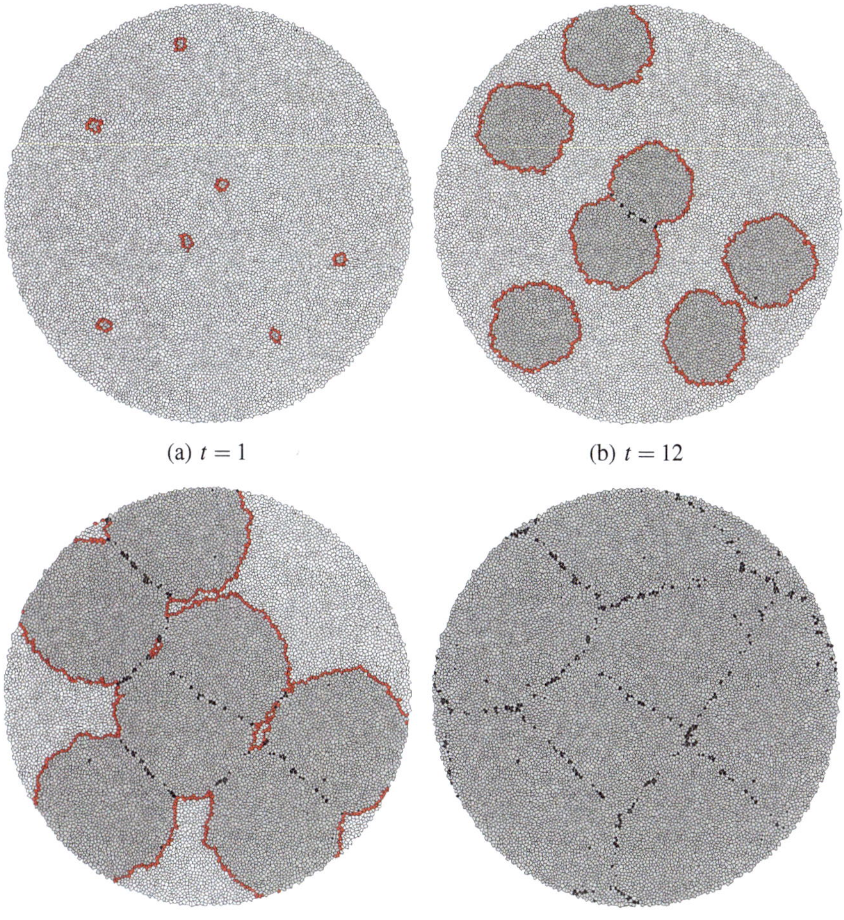

(a) $t = 1$ (b) $t = 12$ (c) $t = 20$ (d) $t = 44$

Fig. 11.2 Approximation of Voronoi diagram $\mathfrak{V}(\mathbf{B})$ on Voronoi automaton $\mathcal{V}$, $\eta = 0.4$. Points of **B** are projected onto $\mathcal{V}$ at time step $t = 0$. Seven Voronoi cells of $\mathcal{V}$ are excited and generated quasi-circular excitation waves. Configuration of $\mathcal{V}$ at time step $t = 1$, when wave have just started to develop, is shown in (a). The waves propagate outwards their initial stimulation sites and covert Voronoi cells they are occupying into refractory states (bc). When two or more waves collide precipitation occurs. By 44th step of the automaton development excitation becomes extinct but domains of precipitate-cells represent edges of the approximated Voronoi diagram $\mathcal{V}$. Resting cells are blank, excited cells red (dark grey), refractory cells grey, and precipitate cells black.

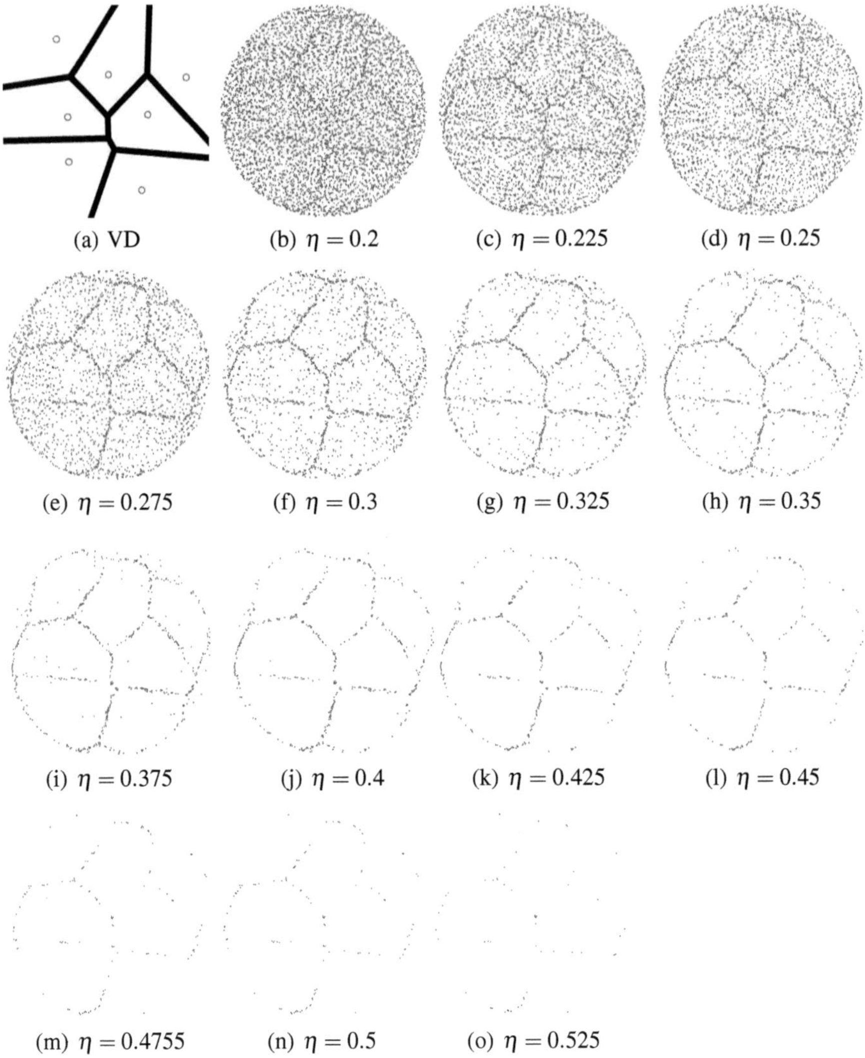

Fig. 11.3 Quality of approximation depends on η. (a) Voronoi diagram $\mathfrak{V}(\mathbf{B})$ of seven planar points set **B** constructed by Fortuna's sweepline algorithm, (b)–(o) stationary configurations of precipitate in automaton $\mathcal{V}$ excited by **B** for different $\eta = 0.2 \ldots 0.525$. In snapshots (b)–(o) black pixels symbolise Voronoi cells of $\mathcal{V}$ in precipitate state, cells in refractory states are blank.

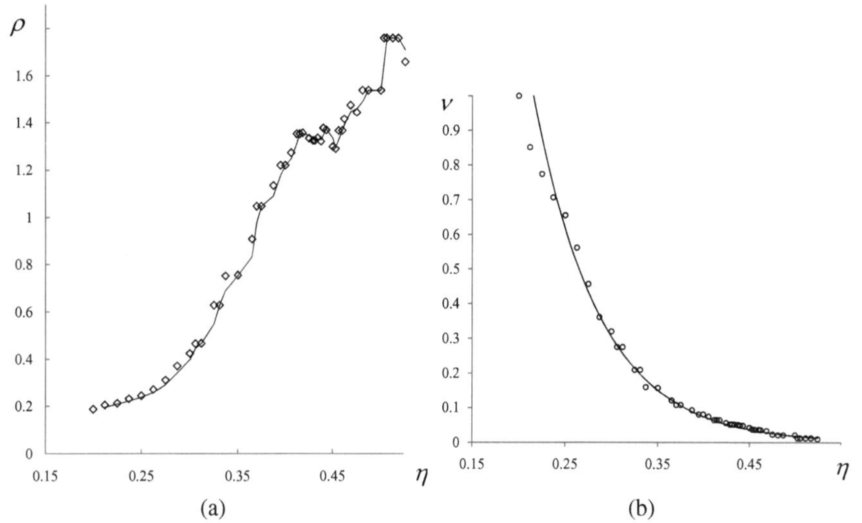

Fig. 11.4 Characterisation of approximation of $\mathfrak{V}(\mathbf{B})$ by automaton $\mathscr{V}$: (a) ρ vs η and (b) v vs η. Markers (rhombs and circles) show actual values of $v(\eta)$ and $\rho(\eta)$, while solid lines are moving average trend line purely for eye guidance.

(Fig. 11.5a) onto $\mathscr{V}$ in such a manner that at time $t = 0$ all cells of $\mathscr{V}$ covered (coinciding with pixels of) by shapes from **O** becomes excited (Fig. 11.5b). Waves of excitation travel across $\mathscr{V}$ and provoke precipitation during their collisions (Fig. 11.5c–e). When all cells take refractory or precipitate state the automaton's configuration becomes stationary. The spatial distribution of precipitate represents Voronoi diagram of data shapes (see Fig. 11.5f and Fig. 11.6b). A precipitate-enhancement of original position of date shape is a byproduct of $\mathscr{V}$'s development.

A skeleton of a planar contour is a set of centres of bitangent circles lying inside the contour [62]. Blum's grass-fire algorithm for computing skeleton employs propagating patterns [53, 54, 62]: to compute a skeleton we set a contour on fire and let the fire spread, quench points where the advancing fire-fronts collide represent the skeleton. Most known algorithms of skeletonisation are based on Blum's approach: simulation of grass-fire [171], distance transform [216], analytical construction of medial axis and topological thinning [41, 208].

To approximate a skeleton of a shape **S** (Fig. 11.7a) we project **S** onto $\mathscr{V}$. Voronoi cells of $\mathscr{V}$ corresponding to black pixels of **S** are excited (Fig. 11.7b). Excitation waves propagate inside and outside of the domain **S** of initial excitation; we can ignore behaviour of outward waves. Waves travelling inside the contour **S** trigger precipitation at the sites of the waves' interaction (Fig. 11.7c–e). Distribution of precipitate (Fig. 11.7e) inside **S** approximates skeleton of **S** (Fig. 11.6c).

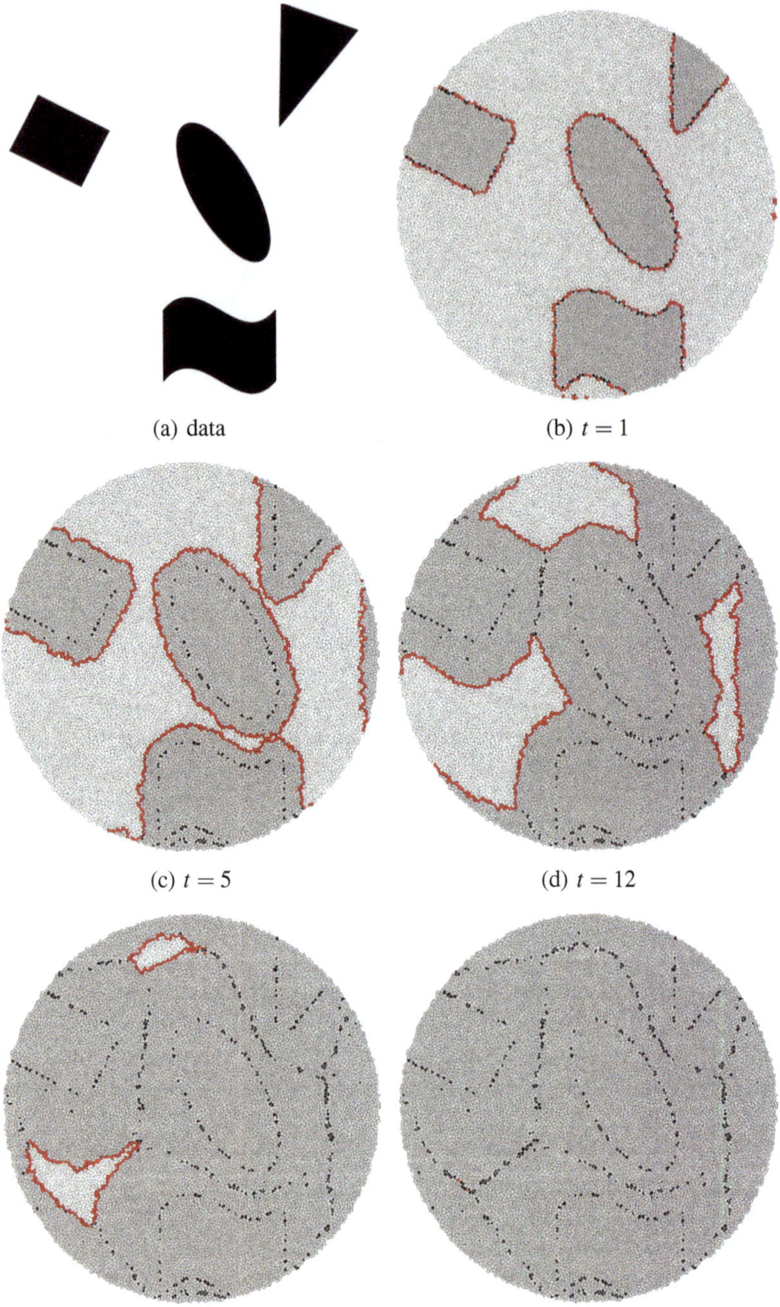

(a) data (b) $t = 1$

(c) $t = 5$ (d) $t = 12$

(e) $t = 20$ (f) $t = 28$

Fig. 11.5 Approximation of Voronoi diagram $\mathfrak{V}(\mathbf{O})$ of planar shapes $\mathbf{O}$ in Voronoi automaton $\mathscr{V}$. Voronoi cell of $\mathscr{V}$ in precipitate state are black, cells in refractory states are blank. Excited cells are red (gray). (a) representation of $\mathbf{O}$, (b)–(f) development of $\mathscr{V}$. $\mathfrak{V}(\mathbf{O})$ is represented by black pixels in (f).

(a) data (b) $t = 1$ (c) $t = 5$

Fig. 11.6 Schematic representation of data objects and their Voronoi diagram or skeleton for precipitation-based approximation shown in (a) Fig. 11.2, (b) Fig. 11.5, and (c) Fig. 11.7

11.4 Summary

We solve some problems of computational geometry on a discrete model of irregularly automaton networks, which imitate reaction-diffusion and excitable system. Approximation of a Voronoi diagram on Voronoi automata is the perfect example. A threshold of relative local excitation density of the Voronoi automata can be used to parameterise excitation and precipitation dynamics, enhance results of the computation and reduce noise-to-signal ratio. The same optimal threshold of precipitation works well not only for approximating Voronoi diagram of planar sets but also for arbitrary shapes and skeletonisation of planar contours. The parameterisation developed could be used in the designing experimental laboratory prototypes of fine-grained compartmentalised excitable chemical processors, e.g. using microemulsion approach [154], developing nano-scale massively parallel computers [47] and computing architectures based on ensembles of non-uniform vesicles filled with excitable and precipitating chemical mixture [26, 27, 140].

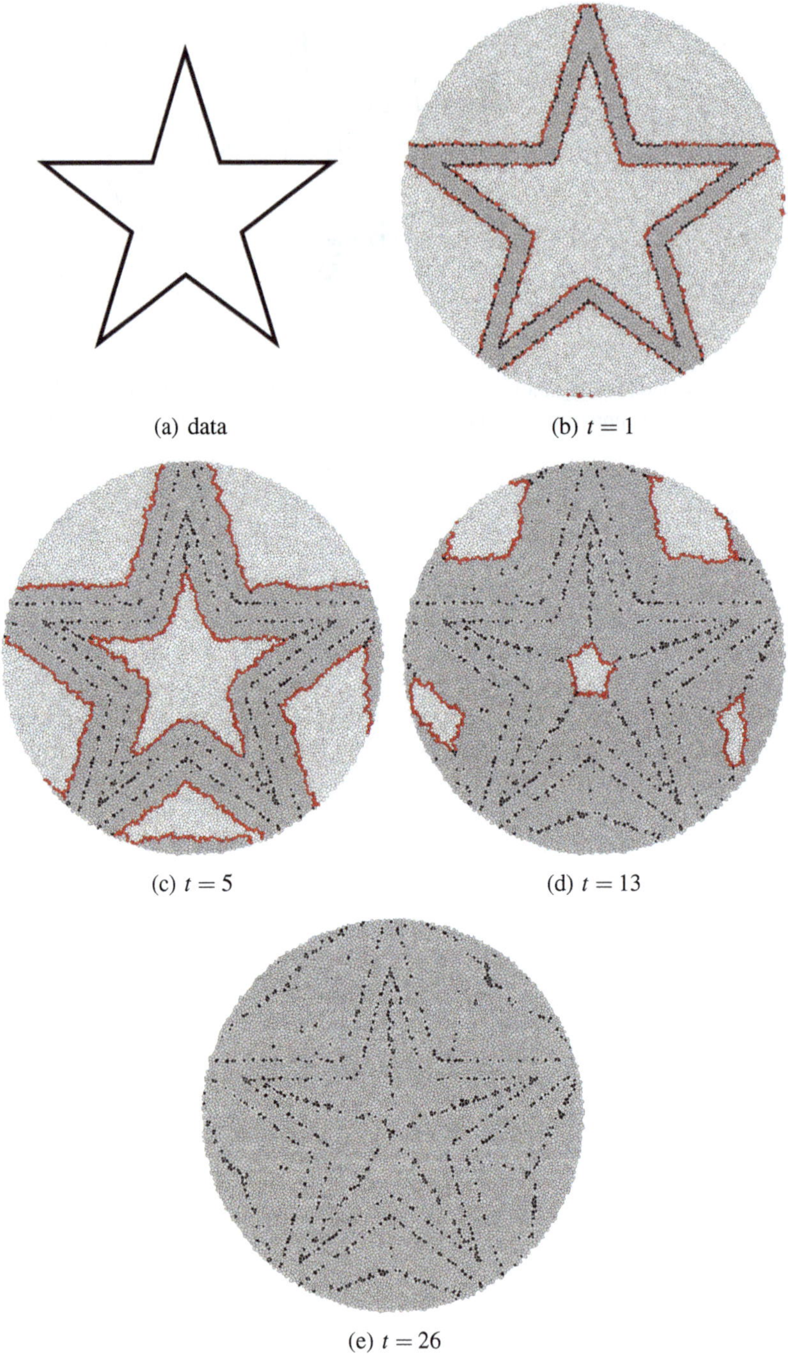

Fig. 11.7 Approximation of a skeleton of a star-contour **S** by $\mathcal{V}$. Voronoi cell of $\mathcal{V}$ in precipitate state are black, cells in refractory states are blank. Excited cells are red (grey). (a) representation of a star-contour, (b)–(e) development of $\mathcal{V}$.

Chapter 12
Adders and Multipliers in Sub-excitable Automata

To implement binary arithmetics we employ a cellular-automaton model of an excitable medium — the 2^+-medium, [2, 3, 8]. The 2^+-medium is a two-dimensional three-state — resting ($\cdot$), excited ($+$) and refractory ($-$) — cellular automaton where a resting cell becomes excited only exactly two neighbours are excited. The transitions from excited state to refractory state, and from refractory state to resting state are unconditional.

The automaton is an imitation of sub-excitable Belousov-Zhabotinsky medium, where a singular excitation does not lead to propagation of a target or spiral wave but rather formation of a travelling localisation, or a wave-fragment similar to dissipative soliton, which conserves its shape and velocity vector [14].

Most common localizations observed in 2^+-medium, are particles traveling along horizontal and vertical directions of the cellular lattice (e.g. Fig. 12.1a) and along diagonals of the lattice (e.g. Fig. 12.1b). A 2^+-particle consists of two excited cell-states and two refractory cell-states (Fig. 12.1a). The particle can travel in four possible directions: East, West, North and South, and the particle conserves its configuration. A 3^+-particle consists of three cells in excited states and three cells

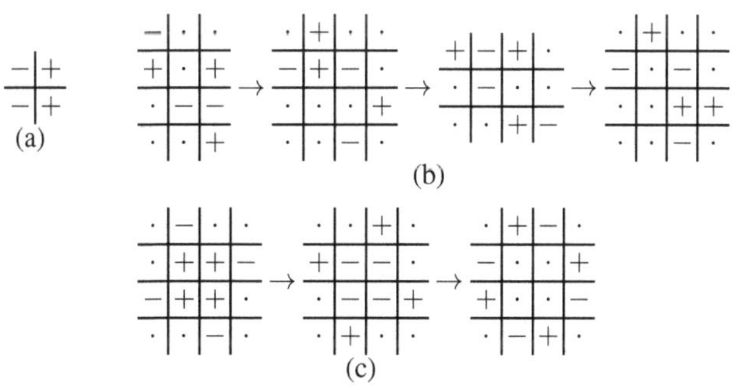

Fig. 12.1 Species of traveling localizations used to implement arithmetic circuits in 2^+-medium [3]: (a) 2^+-particle traveling East, (b) 3^+-particle traveling North-East, (c) stationary oscillator. Excited states are shown by '$+$', refractory '$-$' and resting '$\cdot$'.

A. Adamatzky: Reaction-Diffusion Automata, ECC 1, pp. 209–228.
springerlink.com

in refractory states (Fig. 12.1b). The 3^+-particle travels in the direction of North-East, South-East, North-West and South-West. Its topology changes in a cycle of four configurations.

There is the only one stationary localisation in 2^+-medium. The stationary oscillator (Fig. 12.1c) consists of four excited states and four refractory states. The oscillator's configuration changes in a cycle of three configurations and it remains at its original position all the time.

We are employing these two species of traveling localizations and the oscillator in further designs. Presence of a particle represents value 1, Boolean TRUTH, and absence of the particle value 0, Boolean FALSE. When the particles, produced by certain gates, are not used in further operations, we can annihilate them by directly changing their excited states to refractory states.

12.1 Adders

One-Bit Half Adder

There are many versions of particular implementations of a one-bit adder. The most common one takes outputs as sum of signal $S = (A \oplus B) \oplus C_{in}$ and carry out value $C_{out} = A \cdot B + C_{in}(A \oplus B)$, where A and B are input bits. A one-bit half adder is a simplified logical circuit: $S = A \oplus B$ and $C = A \cdot B$[1].

The initial configuration of implementing the half adder in the 2^+-medium is shown in Fig. 12.2a. Two aforementioned inputs, A and B, are represented by a 2^+-particle traveling South at the top of the graph and a 3^+-particle traveling South-East in the middle. The presence (or the absence) of these particles denotes the assignment of logical value '1' (or '0') to the input values A and B, therefore in this example, $A = 1$ and $B = 1$. Moreover, there are two additional control particles, a 2^+-particle traveling East on the far left of the graph and a 3^+-particle traveling North-East at the bottom. These two additional particles are always present regardless of the values of A and B, which means they always hold the value of logical '1' or TRUTH.

The one-bit half adder in the 2^+-medium occupies 18×45 cells. All particles are injected in the system, i.e. start their travel in predetermined places at the same time (Fig. 12.2a). The outputs are read at the right side of the lattice: the upper output represents CARRY value and the bottom output represents SUM value. Trajectories of the particles constituting the one-bit half adder are shown in Fig. 12.2b to Fig. 12.2e: when both input particles A and B hold logical value '0' (Fig. 12.2b), the control particles collide with and annihilate each other so that no particles reach the right side of the lattice and both outputs equal logical '0'; when only one of the input particles has logical value '1' (Fig. 12.2c and 12.2d), this input particle collide with and annihilate the control particle at the bottom, giving the 2^+-particle on the far left a chance to travel all along to the right side of the lattice; while if both input

[1] Substantial part of circuits presented are designed by Liang Zhang [281, 282].

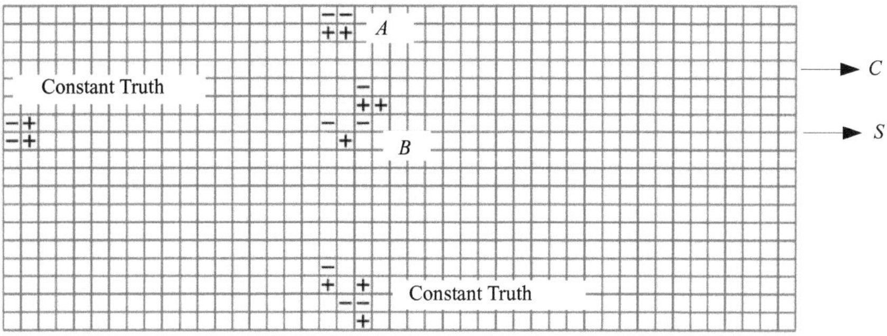

(a)

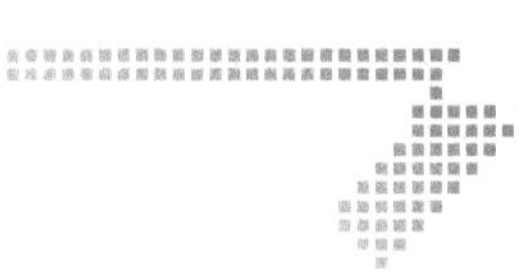

(b) $A=0, B=0; C=0, S=0$

(c) $A=0, B=1; C=0, S=1$

(d) $A=1, B=0; C=0, S=1$

(e) $A=1, B=1; C=1, S=0$

Fig. 12.2 Space-time configurations of excitable cellular automaton implementing one-bit half adder: (a) the initial configuration of input and control particles, both inputs A and B have value TRUTH in this example and represented by 2^+-particle at the top, and 3^+-particle in the middle of the lattice, while the control particles always hold TRUTH value; (b)–(e) trajectories of particles depending on different input values, produced as time lapsed superpositions of several consecutive configurations of the cellular automata, time-related positions of the particles are encoded by values of grey.

particles are present (Fig. 12.2e), they produce a new 2^+-particle traveling East after their collision, making the CARRY output having logical value ‘1’, and again, the control particles annihilate each other.

One-Bit Full Adder

A one-bit full adder has three binary inputs, addends A and B together with a CARRY-IN value C_{in}. Also the adder has two binary outputs, a SUM value S and a CARRY-OUT value C_{out}.

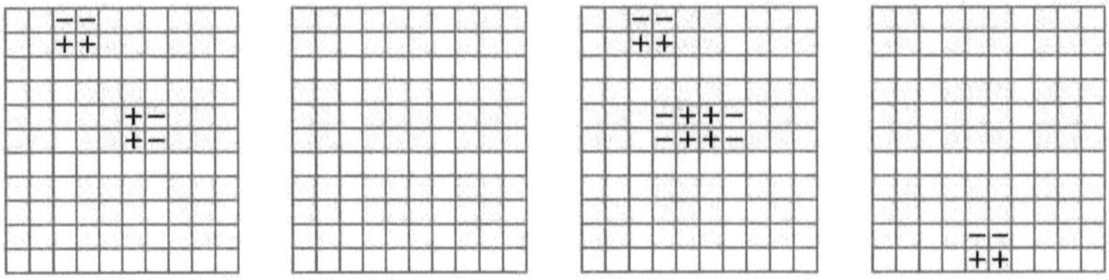

(a) (b)

Fig. 12.3 The redirection gate: (a) the redirection gate itself (at time step $t = 1$) vanishes ($t = 6$) when there is no particle to redirect, (b) 2^+-particle traveling East ($t = 1$) is redirected and heads South ($t = 6$)

$$\begin{aligned} S &= A \oplus B \oplus C_{in} \\ &= (C_{in} \oplus A) \oplus B \\ C_{out} &= A \cdot B + A \cdot C_{in} + B \cdot C_{in} \\ &= C_{in} \cdot A + ((C_{in} \oplus A) \cdot B) \end{aligned}$$

A one-bit full adder can be realised by using two one-bit half adders in the following way:

1. two of the three inputs are input values of the first half adder (C_{in} and A, for example);
2. the third input (B) and the SUM value of the first half adder ($C_{in} \oplus A$) are input values of the second half adder;
3. the SUM value of the full adder is equivalent to that of the second half adder ($(C_{in} \oplus A) \oplus B$);
4. the CARRY-OUT value of the full adder is equivalent to either the CARRY value of the first half adder ($C_{in} \cdot A$) or that of the second half adder ($(C_{in} \oplus A) \cdot B$).

The one-bit full adder in the 2^+-medium is implemented exactly the same way described above. In addition to the two half adders, there are some control elements used to complete the implementation. The redirection gate (Fig. 12.3), consisting of two 2^+-particles traveling in directions orthogonal to each other, functions as a gate that changes a 2^+-particle's traveling direction by 90 degrees. There are eight different instances of redirection gates in total depending on the traveling directions and relative positions of the consisting 2^+-particles, and the redirection gate shown in Fig. 12.3 is one of them. Fig. 12.3a shows the redirection gate *per se*, which simply disappears when there is no 2^+-particle to be redirected; whereas when collide with a 2^+-particle traveling East (Fig. 12.3b), a 2^+-particle traveling South emerges after 5 time steps.

A one-bit full adder in the 2^+-medium occupies 28×59 cells (Fig. 12.4). It consists of two one-bit half adders on the left, a redirection gate ($RG1$) in the middle and a block of redirection gates on the right of the lattice. The block of redirection gates encompasses a stationary oscillator (OSC), five redirection gates ($RG2 \ldots RG6$) and

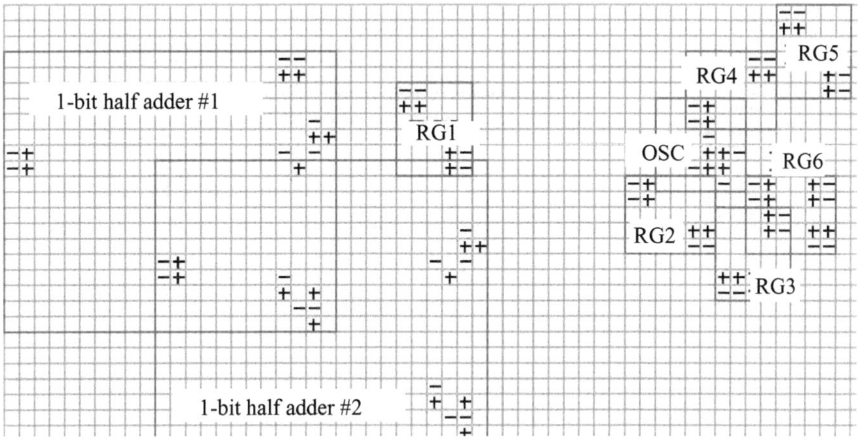

Fig. 12.4 The configuration layout of a one-bit full adder, composed of two one-bit half adders and control elements. Different elements are injected in the system at different time steps.

an additional 2^+-particle traveling West within the range of *RG*6. The gate *RG*1 is used to redirect the SUM value of the first half adder to become an input value of the second half adder. The block of redirection gates is used to delay the CARRY value of the first half adder by exactly 13 time steps, synchronising it with the CARRY value of the second half adder. Therefore there is eventually only one 2^+-particle to represent the CARRY-OUT value of the full adder. The stationary oscillator can perform other operations, we only use it for redirection purpose here. We use a stationary oscillator instead of another redirection gate because we want the 2^+-particle to be redirected to travel one time-step faster. The block of redirection gates disappears if there is no particle to redirect and/or delay.

Note that different elements are injected in the system at different time steps when implementing the one-bit full adder, contrary to the implementation of the one-bit half adder, where all particles are injected at once. Here the first one-bit half adder is added at time step $t = 1$, followed by the redirection gate *RG*1 and the stationary oscillator *OSC* at time step $t = 28$, and the second one-bit half adder at time step $t = 29$. The rest of the redirection gates *RG*2 to *RG*6 are added at time steps $t = 34, 37, 45, 48$ and 56. The additional 2^+-particle traveling West is injected in the system at time step $t = 57$. The one-bit full adder completes its computation and produces a result at time step $t = 60$.

The input values are represented by particles as follows:

- CARRY-IN: 2^+-particle traveling South in the first one-bit half adder;
- *A*: 3^+-particle traveling South-East in the first one-bit half adder;
- *B*: 3^+-particle traveling South-East in the second one-bit half adder.

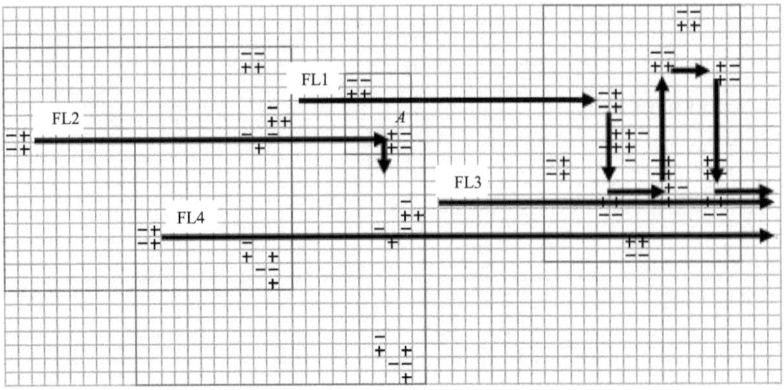

Fig. 12.5 Flows of information during operating of one-bit full adder

If we only implement a single isolate one-bit full adder, the input values do not necessarily have to be assigned to the particles exactly the same way described above. The CARRY-IN value can also be represented by the 3^+-particle traveling South-East in either of the half adder. However, when we use the one-bit full adder as a module to implement multiple-bit adders, the above assignment is highly recommended, so that it is much easier to propagate the signal of CARRY-OUT value of one full adder to become the CARRY-IN input value of the next full adder.

The possible flows of information when the one-bit full adder computes are shown in Fig. 12.5. Four flows of information, namely $FL1$ to $FL4$, are depicted. The $FL1$ appears only when both CARRY-IN=1 and A=1, thus the 2^+-particle traveling South and the 3^+-particle traveling South-East in the first one-bit half adder collide and produce a 2^+-particle traveling East, which is then redirected by the block of redirection gates six times (first the *OSC*, followed by the *RG*2, *RG*3, *RG*4, *RG*5 and *RG*6) and reach the right side of the adder. The $FL2$ appears when either CARRY-IN=1 or A=1, thus their sum equals 1 and the 2^+-particle traveling East in the first one-bit half adder would remain, be redirected by the redirection gate RG1 and become the 2^+-particle traveling South in the second one-bit half adder. The $FL3$ appears when the $FL2$ is appearing and B=1, thus the 2^+-particle traveling South and the 3^+-particle traveling South-East in the second one-bit half adder collide and produce a 2^+-particle traveling East, reaching the right side in the end. The $FL4$ appears when either the $FL2$ is appearing or B=1, thus the 2^+-particle traveling East in the second one-bit half adder would remain and reach the right side of the adder. Note that the $FL1$ and $FL3$ can not appear at the same time. However, if appears, both the $FL1$ and $FL3$ would reach the right side of the adder at the same time step $t = 60$.

Trajectories of the particles constituting the adder are shown in Fig. 12.6. Results of the computation are read at the right side of the adder: upper output represents the CARRY-OUT value and the bottom output represents the SUM value.

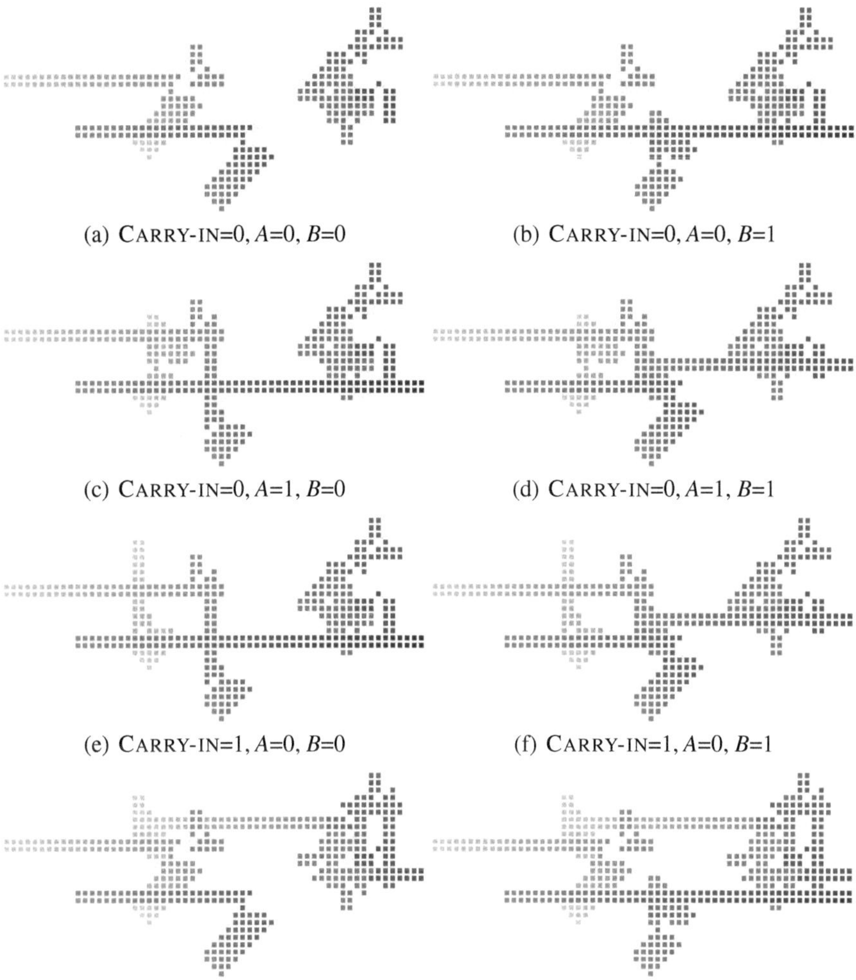

(a) CARRY-IN=0, A=0, B=0 (b) CARRY-IN=0, A=0, B=1

(c) CARRY-IN=0, A=1, B=0 (d) CARRY-IN=0, A=1, B=1

(e) CARRY-IN=1, A=0, B=0 (f) CARRY-IN=1, A=0, B=1

(g) CARRY-IN=1, A=1, B=0 (h) CARRY-IN=1, A=1, B=1

Fig. 12.6 Trajectories of particles implementing one-bit full adder

Multiple-bit Adders

The one-bit adders modular, so that we can use them to produce multiple-bit adders in the 2^+-medium as long as we deal with the carry bit, i.e. to propagate the carry information from one bit to its next higher significant bit. Here we show the realisation of two-bit adders and demonstrate one of possible ways to propagate the carry information with a redirection gate. Adders of arbitrary-bit length can be implemented in the same way, provided all input values are correctly assigned to the input particles within each one-bit adders.

The two-bit half adder occupies 38×97 cells, consisting of an one-bit half adder added at time step $t = 1$, a redirection gate at time step $t = 28$ and a one-bit full adder at time step $t = 37$. The redirection gate is used to propagate the CARRY value from the less significant bit to the next higher significant bit.

The input values are represented as follows:

- A_0: 2^+-particle traveling South in the one-bit half adder;
- B_0: 3^+-particle traveling South-East in the one-bit half adder;
- A_1: 3^+-particle traveling South-East in the first one-bit half adder of the one-bit full adder;
- B_1: 3^+-particle traveling South-East in the second one-bit half adder of the one-bit full adder.

The two-bit full adder occupies 45×109 cells, consisting of two one-bit full adders activated in the system at time step $t = 1$ and $t = 67$, as well as a redirection gate at time step $t = 60$. The redirection gate is again used to propagate the carry information.

The input values are represented as follows:

- CARRY-IN: 2^+-particle traveling South in the first one-bit half adder of the first one-bit full adder;
- A_0: 3^+-particle traveling South-East in the first one-bit half adder of the first one-bit full adder;
- B_0: 3^+-particle traveling South-East in the second one-bit half adder of the first one-bit full adder;
- A_1: 3^+-particle traveling South-East in the first one-bit half adder of the second one-bit full adder;
- B_1: 3^+-particle traveling South-East in the second one-bit half adder of the second one-bit full adder.

Selected trajectories of the particles constituting the two-bit adders are shown in Fig. 12.7. Results of the computation are read at the right side of the adder, with the outputs from top to bottom representing the SUM value of the less significant bit, the CARRY-OUT value and the SUM value of the higher significant bit respectively for both two-bit adders.

12.2 Multipliers

The initial configuration of the AND gate or the one-bit multiplier at time step $t = 1$ consists of at most three 2^+-particles (Fig. 12.8a). Two of the particles traveling South and North (grey regions) represent two operands and one particle traveling East represents constant TRUTH. When both operands are TRUE, collision of these three 2^+-particles produce a new 2^+-particle traveling East. This newly born particle indicates that output of the conjunction is TRUE. When there is only one operand

(a) a two-bit half adder, A_1=1, B_1=1, A_0=1, B_0=0

(b) a two-bit half adder, A_1=1, B_1=1, A_0=1, B_0=1

(c) a two-bit full adder, A_1=1, B_1=1, CARRY-IN=1, A_0=1, B_0=1

Fig. 12.7 Trajectories of particles implementing two-bit adders

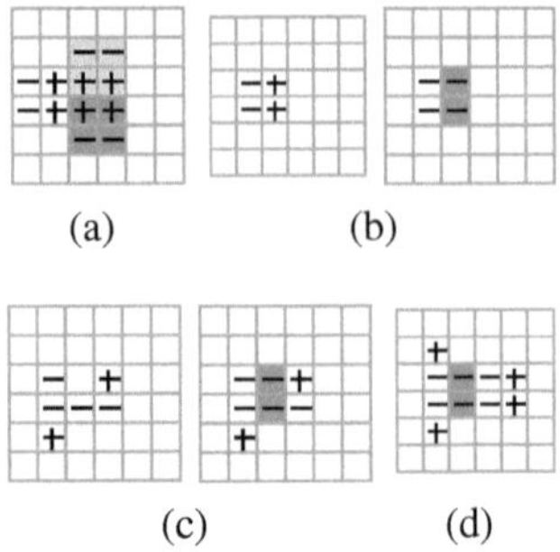

Fig. 12.8 Collision-based implementation of AND gate: (a) initial configuration (within grey regions) with both inputs TRUE, (b) 2^+-particle produced by the gate when both operands are FALSE is terminated at time step $t = 2$, (c) the termination operation affects the configuration produced by the gate when there is only one operand with value TRUE, this does not influence the overall result of the computation, (d) the termination operation does not change the configuration produced by the gate when both operands are TRUE.

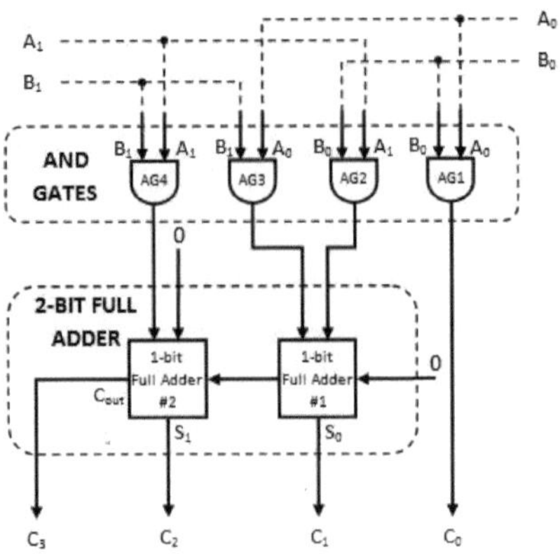

Fig. 12.9 Design of the two-bit multiplier

TRUE, no excitation is generated in the result of collision, thus representing the output FALSE.

When both operands are FALSE, the 2^+-particle traveling East is unaffected, this garbage-particle is annihilated at time step $t = 2$ (Fig. 12.8b). Note that the cells affected by the termination of particles (garbage collection) are always changed to refractory states at time step $t = 2$ no matter what initial values of the operands are. Fig. 12.8c shows that the configuration produced by the gate when there is only one operand TRUE at time step $t = 2$ is affected by the garbage-collection operation, whereas the overall result is not affected. The configuration produced by the gate when both operands are TRUE is not affected by the garbage-collection operation at all (Fig. 12.8d). Thus construction of collision-based AND gate is completed.

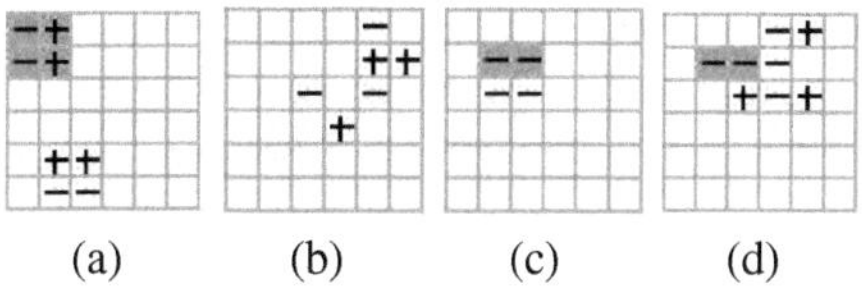

Fig. 12.10 Converting a 2^+-particle traveling East to a 3^+-particle traveling South-East. (a) Configuration at time step $t = 1$. The particle in the grey region is the one to be converted and the other one is the control particle. (b) A 3^+-particle traveling South-East is produced at time step $t = 5$. (c) If there is no particle to be converted, the control particle will be annihilated at time step $t = 4$. (d) The termination operation affects the configuration at time step $t = 4$ when there is a particle to be converted, but does not affect the final result.

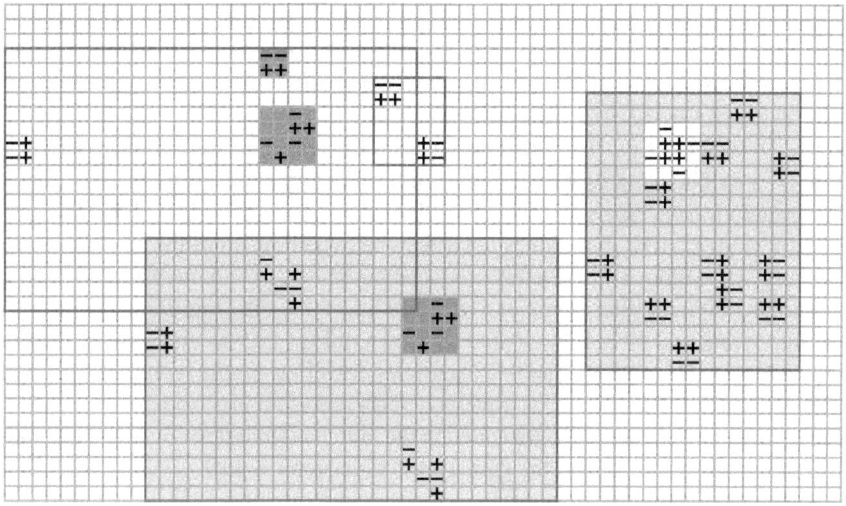

Fig. 12.11 A one-bit collision-based full adder used in designing of multipliers. Particles within the light grey regions are injected in the medium, as local perturbations, at different time steps and positions comparing to those constructed in Sect. 12.1. Particles within the dark grey regions are inputs of the adder, where the 2^+-particle traveling South represents the CARRY-IN value and two 3^+-particles traveling South-East represent addends. In this exemplar configuration, all three inputs take value TRUTH.

A two-bit multiplier is implemented using four AND gates and a two-bit full adder, as shown in Fig. 12.9. Boolean values of multiplicands A_1A_0 and B_1B_0 are distributed and fed to AND gates. Output of the AND gate $AG1$: A_0 AND B_0, is the lowest bit of the product C_0. Outputs of other AND gates ($AG2 \ldots AG4$) become addends of the two-bit full adder, and output of the adder is comprised of the bits of the product C_1, C_2 and C_3.

The two-bit full adder is constructed similarly to that done in Sect. 12.1, with a difference that we adopt a modified construction of one-bit full adder (Fig. 12.11), where the second one-bit half adder (the left-bottom grey region) as well as all particles except the stationary oscillator in the block of redirection gates (the right grey region) are added to the array six time steps later and six cells lower than those in the original construction. This leaves for newly introduced AND gates enough time and space to produce addends for the one-bit adder (0 or 1, depending on the absence or presence of a 3^+-particle traveling South-East within the dark grey regions after conjunction operations take place). Since the AND gates have beed already described the only design task left to complete is to convert the output of the AND gate, i.e. 2^+-particle traveling East, into the addend of the adder, i.e. 3^+-particle traveling South-East. This is done by the gate shown in Fig. 12.10a, where the 2^+-particle within the grey region is the one to be converted and the particle traveling North is the control particle being constant truth. Using this gate at time step $t = 1$, we produce a 3^+-particle traveling South-East at time step $t = 5$ (Fig. 12.10b). However, if the output of the AND gate is FALSE, i.e. there is no 2^+-particle in the grey region of Fig. 12.10a, we would be left with an unwanted 2^+-particle traveling North. To avoid this particle interfering with other particles, we terminate it at time step $t = 4$ by changing its excited states into refractory states (Fig. 12.10c). Again, this termination operation affects the configuration when the output of the AND gate is TRUE (Fig. 12.10d), but it does not affect the overall result of the gate, because a 3^+-particle traveling South-East is produced at time step $t = 5$ anyway.

The two-bit multiplier occupies a lattice of 82×149 cells (Fig. 12.12). Four input bits A_0, A_1, B_0 and B_1 are distributed as shown in Fig. 12.9 as operands of the four AND gates ($AG1 \ldots AG4$). There 2^+-particles traveling South represent bits from multiplicand A and the particles traveling North represent bits from multiplicand B. The AND gates $AG1$ and $AG2$ are added to the lattice at time step $t = 1$, and the gates $AG3$ at time step $t = 35$, and the gate $AG4$ at time step $t = 110$.

The control particles, responsible for converting outputs of the AND gates $AG2$, …, $AG4$ into addends of the adders, are 'injected' in the lattice one time step after the gates are introduced. The two one-bit full adders are added at time steps $t = 6$ and 81, respectively, and the redirection gate joining up the one-bit adders is added at time step $t = 71$. In addition, two more redirection gates are added after the to-bit adder at time steps $t = 146$ and 174, ensuring output bits of the multiplier in order of the lowest bit (C_0) to the highest bit (C_3) when they are read at the right side of the lattice from the top to the bottom.

Exemplar trajectories of the particles constituting the two-bit multiplier are shown in Fig. 12.13. Figure 12.13a demonstrate the case when both A_0 and b_0 equal 1, the output of AND gate $AG1$ which is also the lowest bit of the product C_0 equals 1. In Fig. 12.13b we see trajectories of particles for the situation when SUM values of the two-bit adder, which are also two bits from the product C_1 and C_2 being TRUE, and the connection between two one-bit full adders as well as the

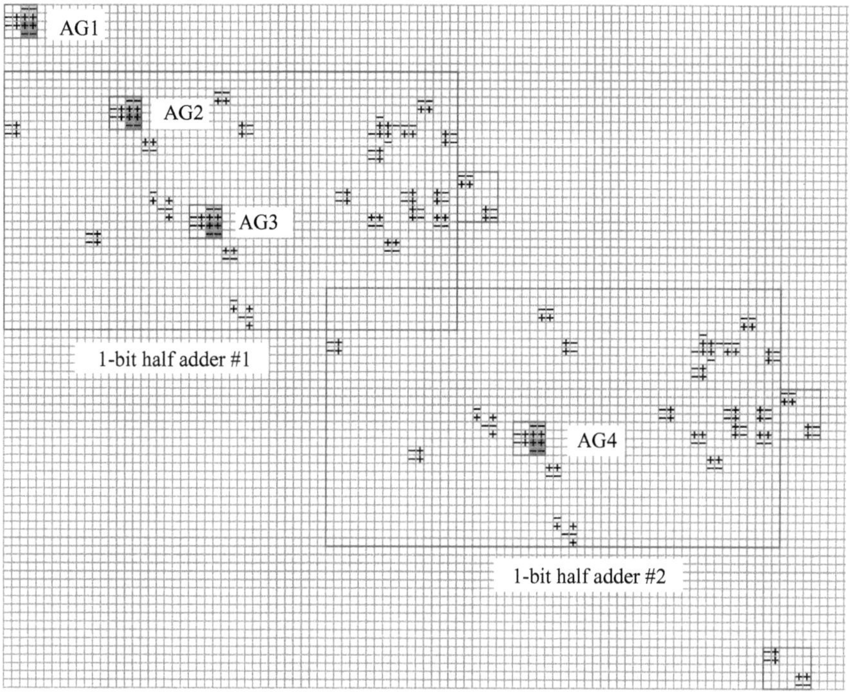

Fig. 12.12 Configuration layout of the two-bit multiplier, formed by four AND gates ($AG1 \ldots AG4$) and a two-bit full adder. Two additional redirection gates are used to make outputs of the multiplier, which will be read at the right side of the lattice (four grey regions), in order of the lowest bit (C_0) to the highest bit (C_3) from the top to the bottom.

CARRY-OUT value of the two-bit adder being redirected to become the highest bit of the product C_3 can be seen in Fig. 12.13c.

The three-bit multiplier is implemented using nine AND gates and two four-bit full adders, as shown in Fig. 12.14. Binary bits of the multiplicands $A_2A_1A_0$ and $B_2B_1B_0$ are distributed and are fed to the AND gates. Output of the AND gate $AG1$ — A_0 AND B_0 — is the lowest bit of the product C_0. Outputs of other AND gates ($AG2 - AG9$) become addends of the four-bit full adders. The lowest bit of the SUM value from the first four-bit adder is another bit of the product C_1, the rest of bits of the SUM value from this four-bit adder become input addends of the second four-bit adder, whose SUM value constitutes the rest bits of the product, C_2 to C_6.

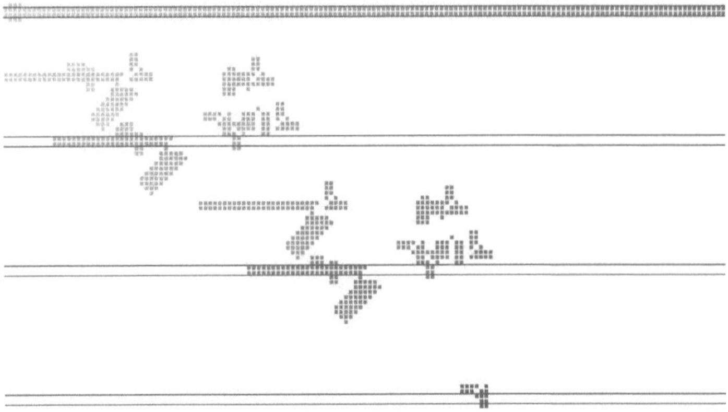

(a) A=01, B=01; C=0001

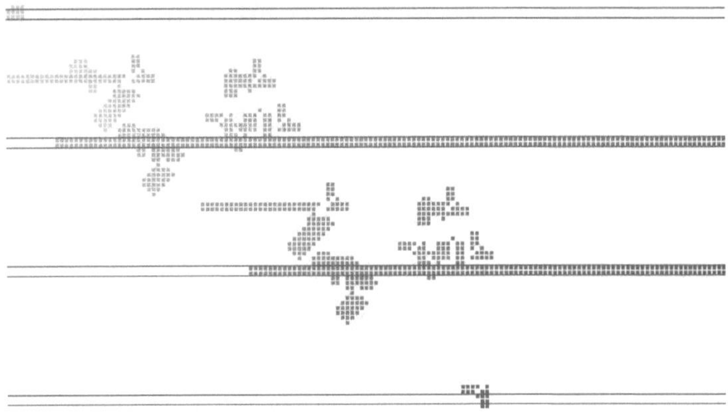

(b) A=11, B=10; C=0110

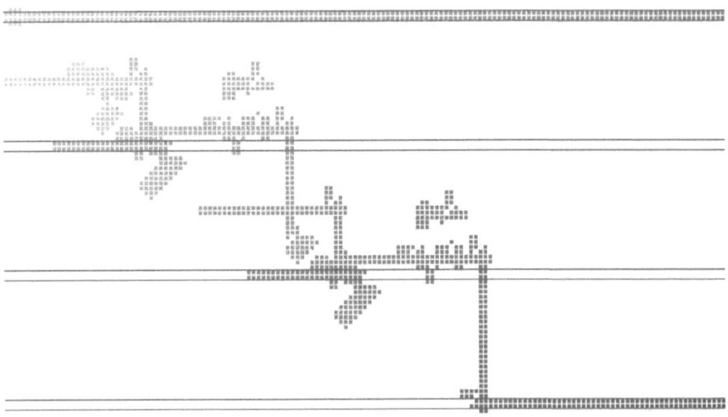

(c) A=11, B=11; C=1001

Fig. 12.13 Trajectories of particles implementing two-bit multiplier. The configuration of trajectories is produced as time-lapsed superpositions of several consecutive configurations of the cellular automaton implementing the multiplier. Level of grey encodes time elapsed from the beginning of computation.

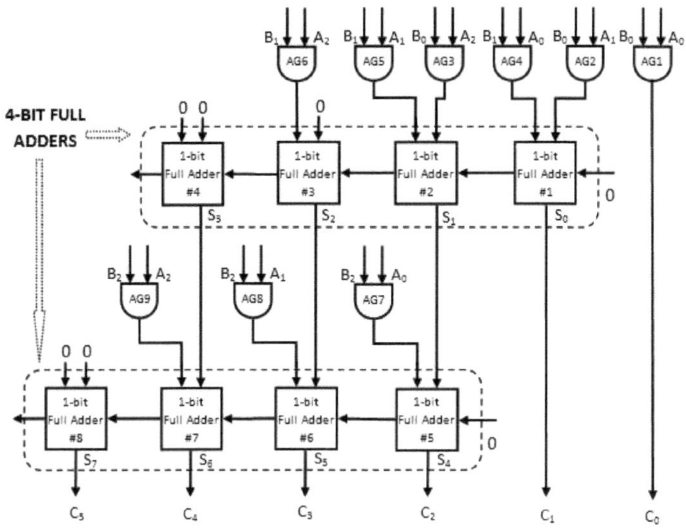

Fig. 12.14 Design of the three-bit multiplier

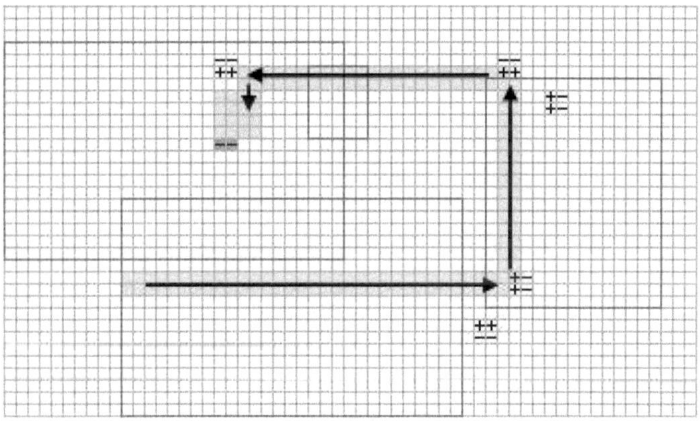

Fig. 12.15 Trajectories of particles which establish connections between full adders. The 2^+-particle traveling East, which represents SUM value of a first one-bit full adder, is converted to 3^+-particle traveling South-East, which represents an addend of a second one-bit full adder added to the lattice at the same position but 107 time steps later than the first one-bit full adder.

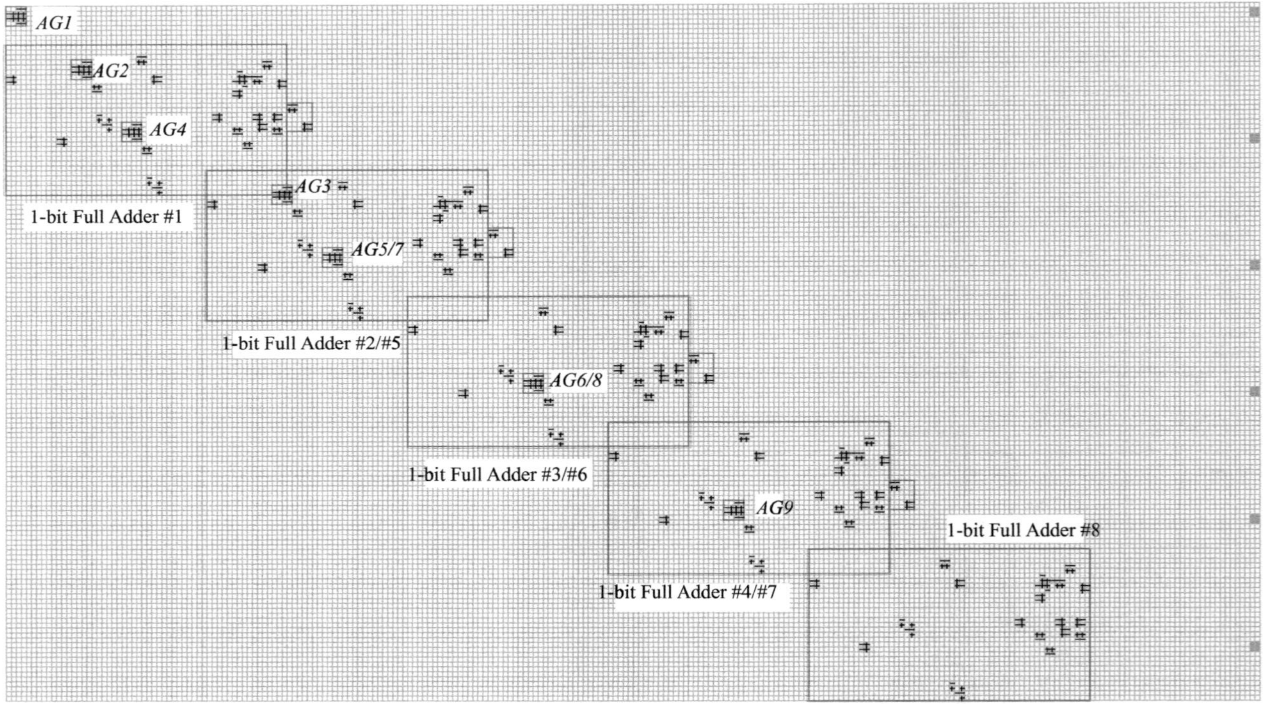

Fig. 12.16 Configuration layout of the three-bit multiplier, formed by nine AND gates (*AG*1 - *AG*9) and two four-bit full adders or eight one-bit full adders. Some of AND gates and one-bit full adders are added at the same position in the lattice but with a time difference of 107 time steps. All input values to the AND gates are set to 1 in the example provided. Outputs of the multiplier are read at the right side of the lattice (six grey regions) in order of the lowest bit (C_0) to the highest bit (C_5) from top to bottom. Controlling particles responsible for linking four-bit full adders are not shown in the picture because otherwise they would be overlapping with the block of the redirection gates.

The AND gate and conversion from output of the AND gate to addend of the adder are implemented in previous sections and the four-bit full adder is implemented using the way described in Sect. 12.1. Therefore the only thing we need to complete the construction of the three-bit multiplier is to establish a connection between two 4-bit full adders. This is done by redirecting bits from SUM value of the first four-bit adder (S_1, S_2 and S_3 in Fig. 12.14), which are all represented by 2^+-particle traveling East, to addends of the second foru-bit adder, which are 3^+-particles traveling South-East. This task is demonstrated in Fig. 12.15.

Let a first one-bit full adder be added to the lattice at time step $t = 1$. The adder's SUM value equals 1 and represented by 2^+-particle traveling East. This particle is deflected North by a redirection gate added at time step $t = 66$, and then to the West by another redirection gate added at time step $t = 82$. After the deflections the particle collides with 2^+-particle traveling South at time step $t = 102$. A 3^+-particle is produced in the result of collision, $t = 108$, when a second one-bit full adder is added to the lattice at the same position as the previous one. If the SUM value of the first adder is 0, two redirection gates would disappear themselves and the 2^+-particle traveling South would be terminated at time step $t = 108$. The termination operation does not affect the formation of the 3^+-particle if the SUM value of the first adder equals 1.

The three-bit multiplier occupies a lattice of 143×250 cells, see Fig. 12.16. Six input bits A_0, A_1, A_2, B_0, B_1 and B_2 are distributed as shown in Fig. 12.14 as operands of the nine AND gates ($AG1 - AG9$), which are added to the lattice at time steps $t = 1, 1, 76, 35, 110, 185, 217, 292$ and 367, respectively. The control particles responsible for converting outputs of the AND gates ($AG2 - AG9$) to addends of the adders are added one time step after introduction of the gates. The eight one-bit full adders are initiated on the lattice at time steps $t = 6, 81, 156, 231, 188, 263, 338$ and 421, respectively. The six redirection gates joining up the one-bit adders are added at time step $t = 71$, 146, 221, 253, 328 and 403. Control particles responsible for converting SUM values of the first four-bit full adder to addends of the second one are added at the time steps 65, 81 and 101 later than the first adder, as described above. Product of the three-bit multiplier is read at the right side of the lattice in order of the lowest bit (C_0) to the highest bit (C_5) from top to bottom. Sample trajectories of the particles constituting the three-bit multiplier are shown in Fig. 12.17, where connections of one-bit full adders within and between the two four-bit full adders can be seen.

Using the constructs described above, we can implement multipliers of an arbitrary length. For any n-bit multiplier ($n \geq 3$), we will need n^2 AND gates and $n - 1$ $(n+1)$-bit full adders to complete the implementation, with the same mechanism used to construct the three-bit multiplier.

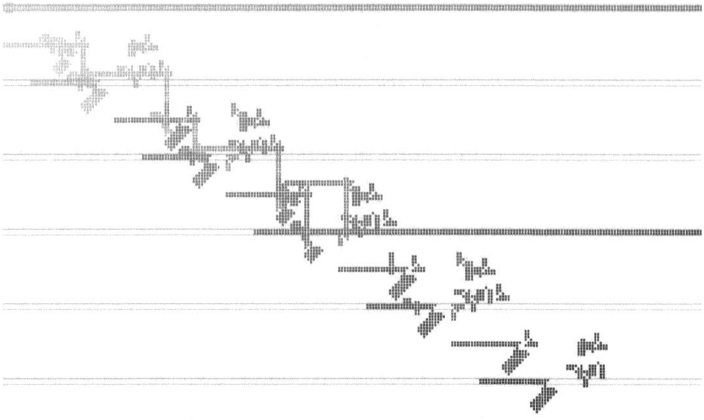

(a) A=011, B=011; C=001001

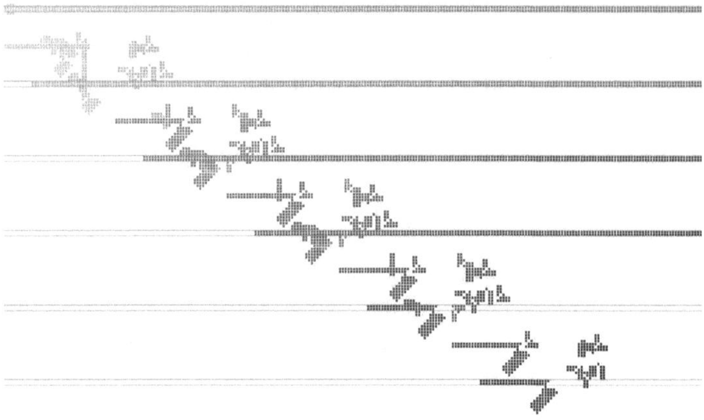

(b) A=011, B=101; C=001111

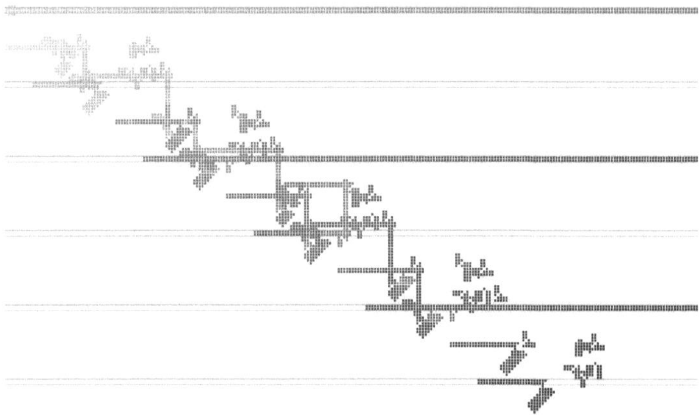

(c) A=011, B=111; C=010101

Fig. 12.17 Exemplar trajectories of particles implementing three-bit multiplier

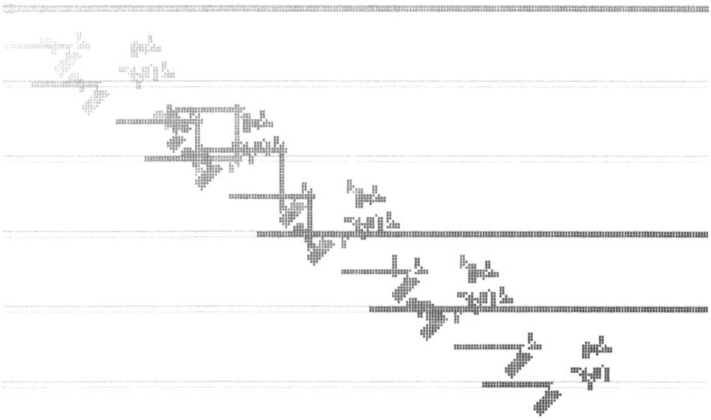

(d) A=101, B=101; C=011001

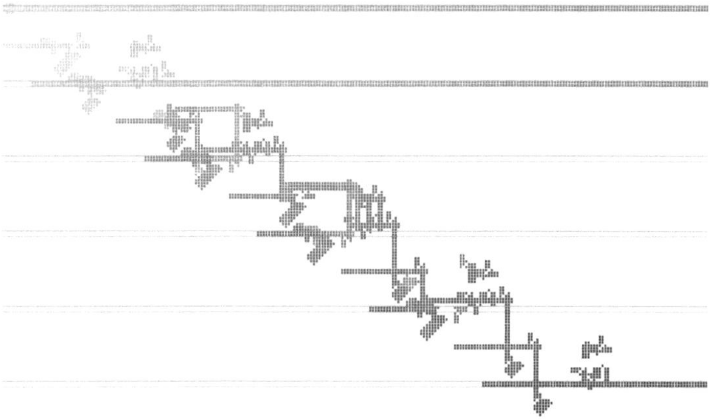

(e) A=101, B=111; C=100011

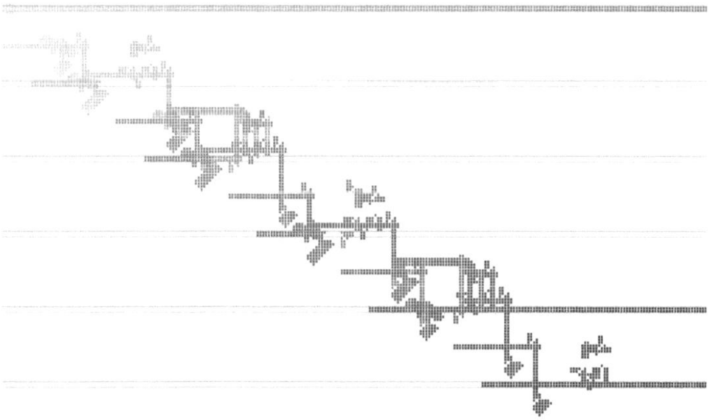

(f) A=111, B=111; C=110001

Fig. 12.17 (*continued*)

12.3 Summary

A two-dimensional three-state cellular automaton imitating a sub-excitable reaction-diffusion medium capable for implementing binary arithmetics circuits in space-time dynamics of excitation. The computation is realised by colliding travelling localisations (gliders, analogs of wave-fragments in sub-excitable Belousov-Zhabotinsky system). Circuits presented may be not optimal because they are rather illustrational implementations aimed to demonstrate how — in principle — binary arithmetics can be executed using travelling localisatons in excitable medium. The designs may also suffer from accumulation of 'garbage localisation' produced in a collision between signal particles and not used in further cascading of signals. The only way to remove such localisation is to directly extinguish them by direct external inhibition of their excited states. The collision-based circuits presented played a very key role in development of experimental laboratory prototypes of collision-based computers in Belousov-Zhabotinsky excitable chemical system [18, 26, 89, 246]. Principles of logical operation and cascading of Boolean gates into circuits, developed using 2^+-medium, were employed in design of collision-based half-adder in a model of propagating finger in a liquid-crystal [31].

Chapter 13
Computing in Hexagonal Reaction-Diffusion Automaton

We design a totalistic cellular automaton, where a cell updates its state depending on just the numbers of different cell-states in its neighbourhoods. Consider a ternary state automaton, where every cell takes one of the following cell-states: substrate S, activator A and inhibitor I. The update rule can be written as follows:

$$x^{t+1} = f(\sigma_I(x)^t, \sigma_A(x)^t, \sigma_S(x)^t),$$

where $\sigma_p(x)^t$ is the number of cell x's neighbours with cell-state $p \in \{I, A, S\}$ at time step t. Our automaton is based on a two-dimensional lattice with hexagonal tiling. The neighbourhood size is seven — the central cell and its six closest neighbours.

To give a compact representation of the cell-state transition rule, we adopt the formalism in [15], and represent the cell-state transition rule as a matrix $\mathbf{M} = (m_{ij})$, where $0 \leq i \leq j \leq 7$, $0 \leq i + j \leq 7$, and $m_{ij} \in \{I, A, S\}$. The output state of each neighbourhood is given by the row-index i (the number of neighbours in cell-state I) and column-index j (the number of neighbours in cell-state A). We do not have to count the number of neighbours in cell-state S, because it is given by $7 - (i + j)$. A cell with a neighbourhood represented by indices i and j will update to cell-state M_{ij} which can be read off the matrix. In terms of the cell-state transition function this can be presented as follows: $x^{t+1} = M_{\sigma_I(x)^t \sigma_A(x)^t}$.

The exact matrix structure, which corresponds to matrix M_3 in [271], is as follows:

$$M = \begin{Bmatrix} S\,A\,I\,A\,I\,I\,I\,I \\ S\,I\,I\,A\,I\,I\,I \\ S\,S\,I\,A\,I\,I \\ S\,I\,I\,A\,I \\ S\,S\,I\,A \\ S\,S\,I \\ S\,S \\ S \\ \end{Bmatrix}$$

A. Adamatzky: Reaction-Diffusion Automata, ECC 1, pp. 229–262.
springerlink.com

Do these matrix entries correspond to phenomena in reaction-diffusion chemical systems? They do. Thus, $M_{01} = A$ symbolises the diffusion of activator A, $M_{11} = I$ represents the suppression of activator A by the inhibitor I, $M_{z2} = I$ $(z = 0, \cdots, 5)$ can be interpreted as self-inhibition of the activator in particular concentrations. $M_{z3} = A$ $(z = 0, \cdots, 4)$ means a sustained excitation under particular concentrations of the activator. $M_{z0} = S$ $(z = 1, \cdots, 7)$ means that the inhibitor is dissociated in absence of the activator, and that the activator does not diffuse in sub-threshold concentrations. And, finally, $M_{zp} = I$, $p \geq 4$ is an upper-threshold self-inhibition.

The cell-state transition rule reflects the nonlinearity of activator-inhibitor interactions for sub-threshold concentrations of the activator. Namely, for small concentration of the inhibitor and for threshold concentrations (values 1 and 3), the activator is suppressed by the inhibitor, while for critical concentrations of the inhibitor (value 2) both inhibitor and activator dissociate producing the substrate, as symbolised in the following set of quasi-chemical reactions:

$$\begin{array}{ccc} A + 6S \to A & A + I \to I & A + 3I \to I \\ & A + 2I \to S & 2A \to I \\ & 3A \to A & \beta A \to I \\ & & I \to S. \end{array}$$

Starting in a random initial configuration the automaton will evolve towards a quasi-stationary configuration, with typically two types of stationary localisations, and a spiral generator of mobile localisations (Fig. 13.1). By analogy with Conway's Game-of-Life we call mobile localisations gliders, the generators of mobile localisations — glider-guns, and stationary localisations — (glider) eaters. Eaters usually annihilate gliders that collide into their central body, but they can also modify gliders that brush past interacting with the outer edge. The core of a glider-gun is a discrete analog of a 'classical' spiral wave (commonly found in excitable chemical systems like the Belousov-Zhabotinsky reaction) (Fig. 13.2). However, at some distance from the spiral wave tip the wave front becomes unstable and splits into localised wave-fragments. The wave-fragments continue traveling along their originally determined trajectories and keep their shape and velocity vector unchanged unless disturbed by other localisations. So, the wave-fragments behave as in sub-excitable Belousov-Zhabotinsky systems [14].

Basic gliders, those with one (activator) head, are found in five types (Fig. 13.3), which vary by the number of trailing inhibitors. Three types (G_{34}, G_{24}, G_{43}) alternate between two forms. Two types (G_4, G_5) have just one form. The spiral glider-gun in Figs. 13.1 and 13.2 release G_{34} gliders. An alternative, low frequency, spiral glider-gun (not shown) releases G_4 gliders. These basic gliders, and also a variety of more complicated gliders including mobile glider-guns, are also generated by many other interactions.

The existence of stationary localisations, or eaters, (Fig. 13.4) is yet one more important feature of the cellular automata. Eater E_3 (Fig. 13.4a) consists of three activator states surrounded by nine inhibitor states. Eater E_6 (Fig. 13.4b) has a core

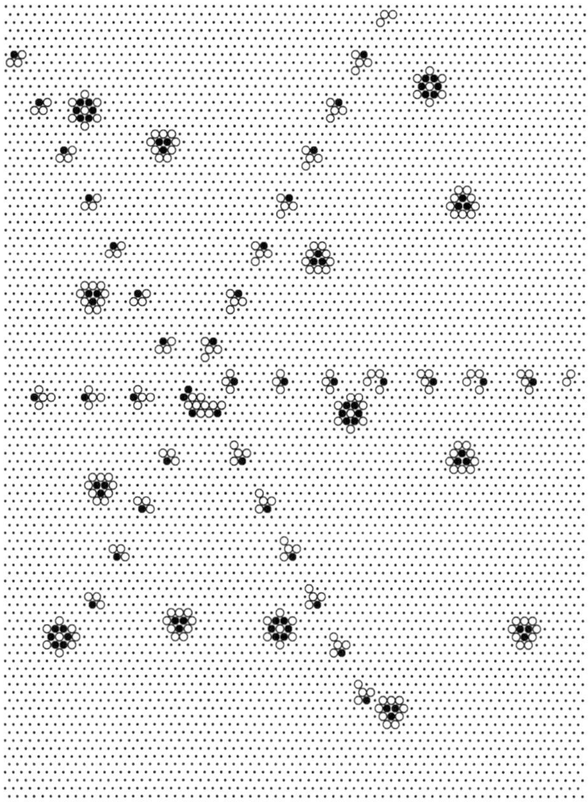

Fig. 13.1 A typical quasi-stable configuration of the cellular automaton which started its development in a random initial configuration (with $1/3$ probability of each cell-state). Cell-state I (inhibitor) is shown by a black disk, cell-state A (activator) by a circle, and cell-state S (substrate) by a dot. We can see there are two types of stationary localisations (glider eaters) and a spiral glider-gun, which emits six streams of gliders, with frequency a glider per six time steps in each glider stream.

Fig. 13.2 The principle scheme of the glider-gun in 'spiral rule' hexagonal cellular automaton: the core of the spiral wave rotates clockwise, wave-fragments break off from the tail of the spiral wave and travel in six directions: East, South-East, South-West, West, and North-West; the wave-fragment that will travel North-East has not been is generated yet.

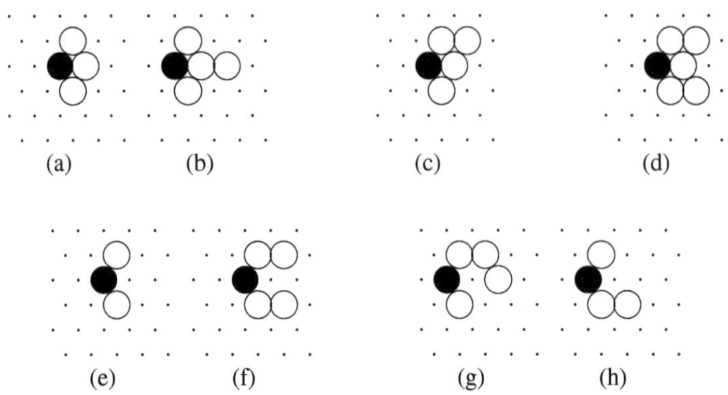

Fig. 13.3 The basic gliders come in five types, shown here traveling West, in the direction of their activator head (cell-state A), with a tail of trailing inhibitors made up of several cell-states I. The glider designator G_{ab} refers to the numbers of trailing inhibitors. (a) and (b) two forms of glider G_{34}. (c) glider G_4. (d) glider G_5. (e) and (f) two forms of glider G_{24}, (g) and (h) two forms of glider G_{43}.

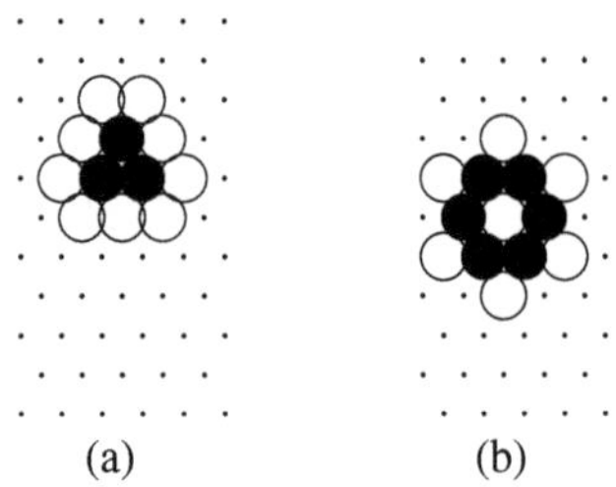

Fig. 13.4 Stationary localisations (eaters). (a) eater E_3, (b) eater E_6.

of one inhibitor-state surrounded by six activator-states, which in turn are encircled by six (in its minimal symmetric form) inhibitor-states.

We can speculate that our cellular automaton is analogous to a combination of two types of chemical system in one physical space: excitable systems where classical spiral waves are formed, and sub-excitable systems where no spiral waves are formed, but only traveling localised wave-fragments (assuming space is uniform and homogeneous). Such 'hybrid-functionality' systems were never observed experimentally, however there is evidence of complete spiral breakup and a subsequent transition to spatio-temporal chaotic states, e.g. reported in [204]. Also, in a modified Barkley model of an excitable reaction-diffusion system, a break-up of a spiral wave far away from the rotating tip was reported in [48]. However, this was achieved in somewhat "artificial" conditions, in which the ratio of time-scales, of the local dynamics of the activator and inhibitor variables, were dynamically changing, increasing during simulation.

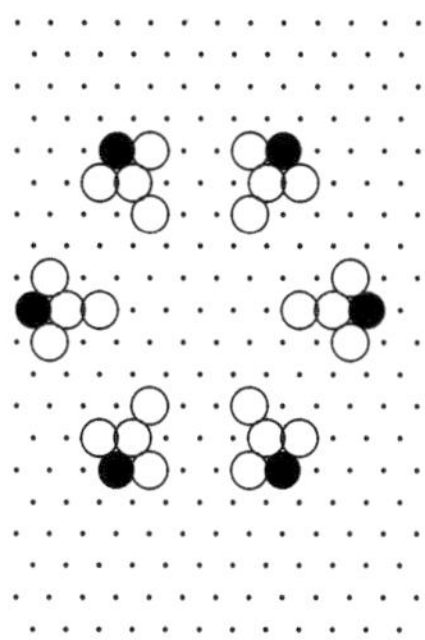

Fig. 13.5 Activation of one site of the lattice leads to formation of six G_{34} gliders

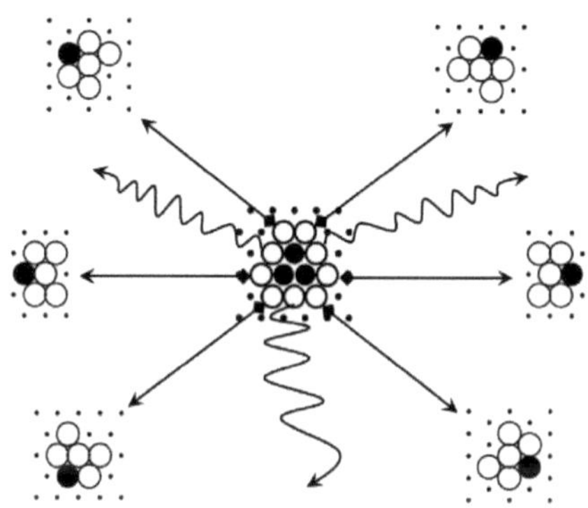

Fig. 13.6 Outcomes of the activation of the inhibitor-sites of the stationary localisation (eater) E_3. When an inhibitor-site (marked by the rhomboid end of an arrow) is switched externally to the activator-state, E_3 is transformed into a G_5 glider. The activation of sites marked my the zig-zag arrows leads to the annihilation of E_3.

13.1 Input Interface

How can we input information into the hexagonal cellular automata computing device? One sensible way to input a quantum of information might be to activate (or inhibit) just one site of the lattice, however such an action can lead to the generation of several gliders (Fig. 13.5) [270], and thus potentially 'pollute' the computational space.

What if we try to 'stimulate' localisations E_3 and E_6, so they can play the role of stationary sensors, or elements of an input interface?

Let us take the eater E_3 and stimulate — switch to the activator-state — one of its inhibitor-sites (Fig. 13.6). The activation of six of the nine inhibitor-sites leads to the transformation of E_3 into a G_5 glider traveling in one of six directions, as shown by the straight arrows in Fig. 13.6. The activation of the other inhibitor-sites (zig-zag arrows in Fig. 13.6) cause the annihilation of E_3.

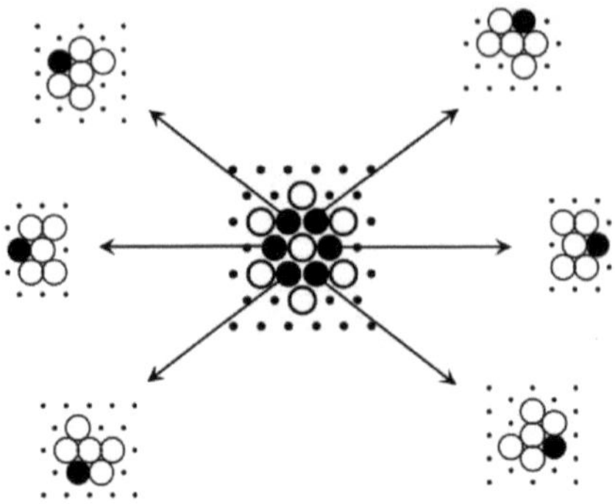

Fig. 13.7 Outcomes of external switching of an activator-site in the stationary localisation (eater) E_6 to a resting or inhibitor state. In both cases E_6 is transformed to a G_5 glider.

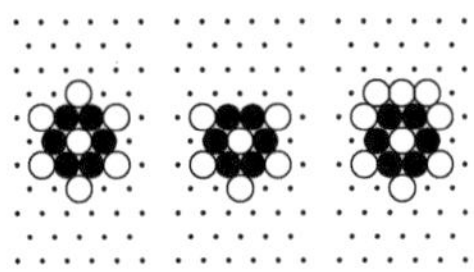

(a) (b) (c)

Fig. 13.8 Switching the northern inhibitor-site of eater to E_6 to the substrate-state leads to the formation of two more inhibitor-sites in the 'inhibitor-shell' of E_6. (a) E_6 in its "normal" form, (b) the northern inhibitor-site is forced to a substrate-state, (c) the configuration of inhibitor-sites is updated.

The inhibition of any of three activator-sites in E_3 will destroy it. E_3 is also stable to the external switching of an inhibitor-state to a resting substrate-state: an inhibitor-site is restored in one time-step. However, when one of activator-sites in E_3 is switched to a substrate-state, the localisation E_3 is destroyed.

Eater E_6 is more sensitive to external inputs comparing to E_3. Thus when we switch one of the activator-sites in E_6 to either a substrate or inhibitor state, E_6 is transformed to an G_5 glider (Fig. 13.7).

The external switching of one of the distal inhibitor-sites in E_6, the northern inhibitor-site in Fig 13.8a, to the substrate-state (Fig 13.8b), leads to the recovery of the site and the switching of two neighbouring sites to the inhibitor state (Fig 13.8c). The updated configuration (Fig 13.8c) can be detected by gliders, as shown in later sections.

The activation one of the outer inhibitor-sites of E_6 transforms the localisation to two G_4 gliders, as shown in Fig. 13.9.

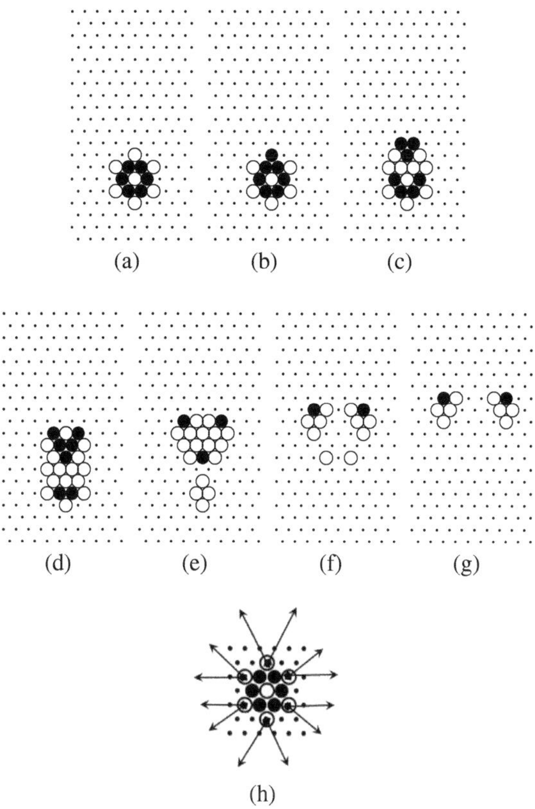

Fig. 13.9 The external switching of one of distal inhibitor-sites of E_6 to the activator-state, transforms E_6 into two G_4 gliders (a) the "normal" form E_6, (b) the northern inhibitor-site is switched to the activator-state, (c)–(f) two G_4 gliders are formed and (g) travel outward. The velocity vectors of the gliders formed by activating the inhibitor-sites are shown in (h).

13.2 Memory Device

The eater E_6 can play the role of a six-bit flip-flop memory device. The substrate-sites (bit-down) between inhibitor-sites (Fig. 13.4) can be switched to an inhibitor-state (bit-up) by a colliding glider.

An example of writing one bit of information in E_6 is shown in Fig. 13.10. Initially E_6 stores no information. We aim to write one bit in the substrate-site between the northern and north-western inhibitor-sites (Fig. 13.10a). We generate a glider G_{34} (Fig. 13.10bc) traveling West. G_{34} collides with (or brushes past) the North edge of E_6 resulting in G_{34} being transformed to a different type of glider, G_4 (Fig. 13.10gh). There is now a record of the collision — evidence that

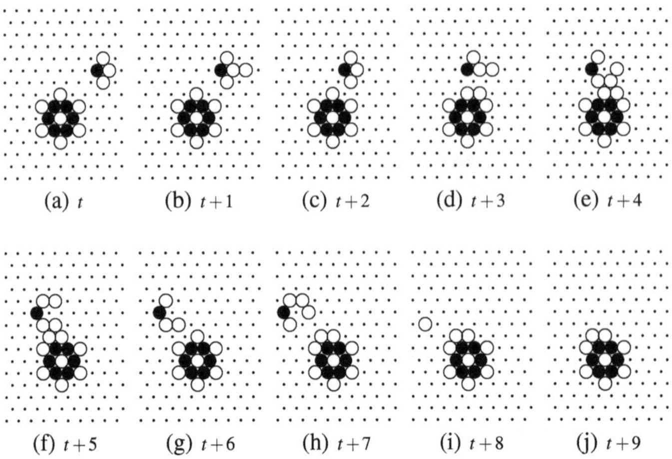

Fig. 13.10 Write bit

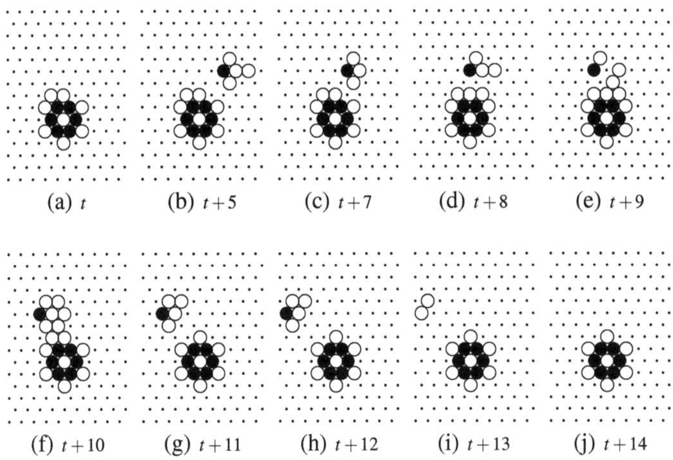

Fig. 13.11 Read and erase bit

writing was successful. The structure of E_6 now has one site (between the northern and north-western inhibitor-sites) changed to an inhibitor-state (Fig. 13.10j) — a bit was saved.

To read a bit from the E_6 memory device with one bit-up (Fig. 13.11a), we collide (or brush past) with glider G_{34} (Fig. 13.11b). Following the collision, the glider G_{34} is transformed into a different type of basic glider, G_{34} (Fig. 13.11g), and the bit is erased (Fig. 13.11j).

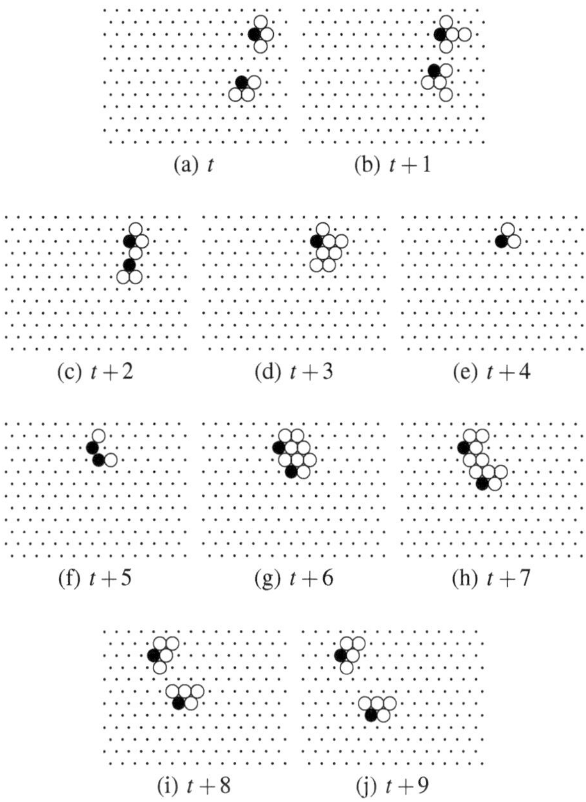

Fig. 13.12 Glider reflection

13.3 Routing and Tuning Signals

To route signals we can potentially employ either stationary localizations (to act as reflectors) or use other gliders to act as mobile reflectors. In practice, we were unable to find a stationary (eater) reflector of gliders; in all cases studied, gliders where either transformed to a different type, or where annihilated, but never changed their trajectory when colliding with an eater. However, mobile reflectors do exist.

Figure 6.8 shows how a glider traveling North-West collides with a glider traveling West, and is reflected South-West as a result of the collision. However both gliders are transformed to different types of gliders. This is acceptable on condition that both types of glider represent the same signal, or signal modality.

There are two more gates which though not essential in demonstrating computational universality, are nevertheless useful in designing practical collision-based computational schemes. They are the FANOUT gate and the ERASE gate.

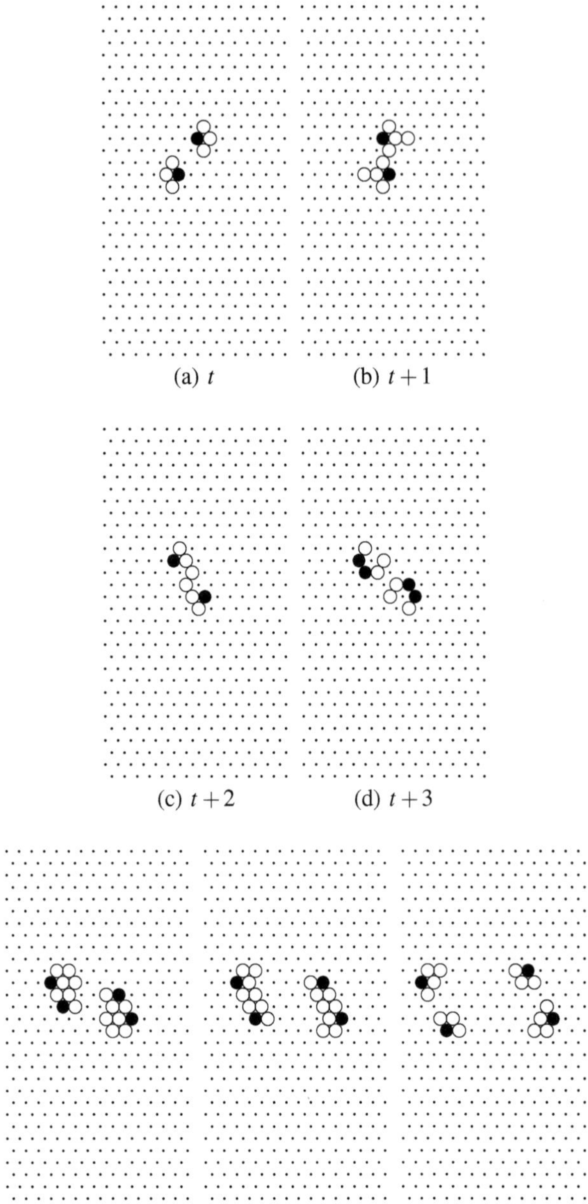

(a) t (b) $t+1$ (c) $t+2$ (d) $t+3$ (e) $t+4$ (f) $t+5$ (g) $t+6$

Fig. 13.13 Signal one to three multiplication, FANOUT gate

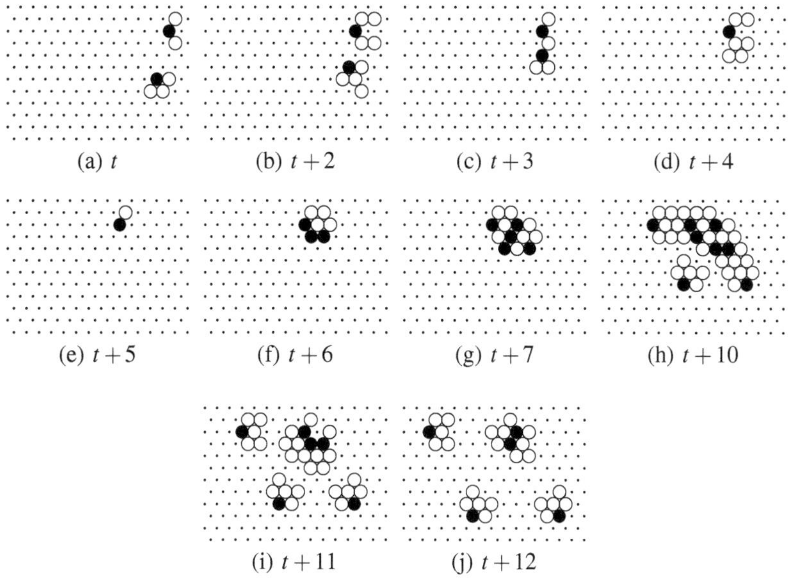

Fig. 13.14 Signal one to two multiplication, FANOUT gate

The FANOUT gate is based on glider multiplication. There are a few scenarios where one glider can be multiplied by another glider (for details see the original beehive rule [270], though this does not feature a spiral glider-gun). In Fig. 13.13 we see how a glider moving East collides with another moving West (Fig. 13.13ab), four new gliders are formed as a result of the collision (Fig. 13.13g), traveling East, West, North-East and South-West. This is an example of a one-to-three FANOUT gate.

We can make a FANOUT gate by colliding glider G_{34} with glider G_{24}, as shown in Fig. 13.14. Glider G_{34} traveling North-West collides with glider G_{24} traveling West (Fig. 13.14ab). The gliders almost annihilate as a result of the collision — just a tiny fragment, two sites made up of one activator and one inhibitor state remain (Fig. 13.14e). The activator-inhibitor pair grows into a more complicated pattern (Fig. 13.14fg), which finally splits into three G_5 gliders. One glider continues traveling West along the original trajectory of glider G_{24}. Ignoring the fact that the glider types change in the collision, we can assume that both G_{24} and G_5 gliders represent a *control* signal traveling West. Two other gliders, the result of multiplication, travel South-West and South-East (Fig. 13.14h–j), while gliders initially involved in the collision continue along their original trajectories.

To annihilate a glider we can collide it with the central bogy of an eater, as demonstrated in Fig. 13.15, or with another glider (head-on collisions usually lead to annihilation).

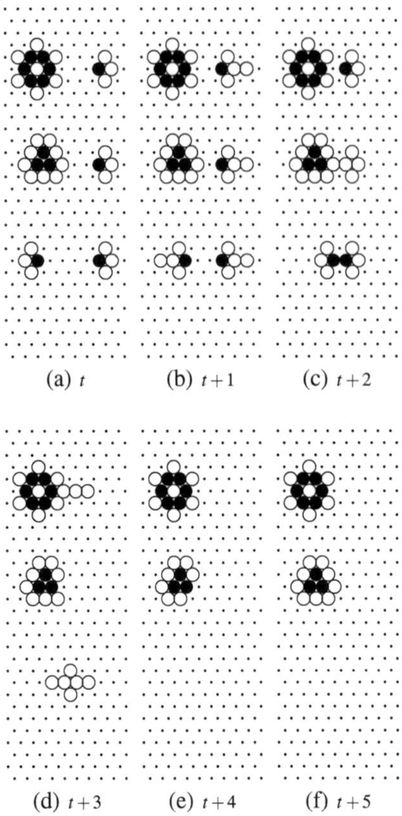

Fig. 13.15 Glider annihilation, ERASE gate, with eaters (top and middle collisions in each subfigure), and another glider (bottom collision)

13.4 Binary Operations

The Boolean logical universality of the spiral rule cellular automaton can be proved using the collision-based computing paradigm [10], where a glider represents the value TRUE, and the absence of a glider represents the value FALSE. When two gliders collide their trajectories may change or new gliders may be generated — a glider emerging on a new trajectory stands for conjunction, the gate AND. So Boolean variables can be represented by colliding gliders.

The computing medium represented by the spiral rule automaton is fully programmable because not only can we generate stationary localizations (eaters) in collisions between gliders, but we can also transform stationary localisation to make generators of mobile localisations (glider-guns), and destroy glider-guns when required. Conjunction and negation are sufficient to demonstrate the logical functional completeness. In this section we go a bit further and discuss the implementation of an asynchronous XOR gate and a five-valued binary operation.

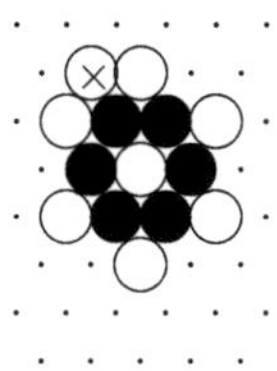

(a)

x_1	x_2	y
0	0	S
G_{34}	0	I
0	G_{34}	I
G_{34}	G_{34}	S

(b)

Fig. 13.16 Asynchronous XOR gate. (a) position of output cell is shown by ⊗. (b) operation implemented by the gate, input state G_{34} is logical TRUE, output state S is FALSE, output state I is TRUE.

The *asynchronous* XOR gate can be constructed from the memory device in Figs. 13.10 and 13.11, employing the eater E_6 and the glider G_{34}. The incoming trajectory of gliders is an input $x = \langle x, y \rangle$ of the gate, and the state of the cell which is switched to the inhibitor state by gliders, is an output z of the gate (this cell is shown by ⊗ in Fig. 13.16a). As seen in Fig. 13.10, when glider G_{34} brushes by the eater E_6 it 'adds' one inhibitor state to the eater configuration (Fig. 13.10, $t+7$), and transforms itself in glider G_{43}. If glider G_{34} brushes by E_6 with an additional inhibitor state (Fig. 13.11, t) it 'removes' this additional state and transforms itself into glider G_4 (Fig. 13.11, $t+11$).

Assume that the presence of glider G_{34} symbolises input logical TRUE and its absence — input FALSE, inhibitor state I in cell ⊗ — output TRUE and substrate state S — output FALSE. The result of this logical operation can be read directly from the configuration of E_6 or by sending a control glider to brush by E_6 to detect how the glider is transformed (see details of glider transformations in Sect. 13.5). Then the structure implements exclusive disjunction (Fig. 13.16b). The gate constructed is asynchronous because the output of the operation does not depend on the time interval between the signals but only on the value of signals: when inhibitor state is added or removed from E_6 the configuration of E_6 remains stable and does not change till another glider collides into it.

Interpreting different gliders as states of a multiple-valued logic variable could bring a new dimension to the study of collision-based computing. Multiple-valued gates will be invaluable in designing cellular-automaton representations of fuzzy reasoning, emotions and consciousness.

Let us look at just one example of the interpretation, and consider pairwise collisions involving any two out of four types of glider: G_{34}, G_4, G_5, and G_{24}. One of the pair moves West, the other North-West, positioned before the collision as follows. In this particular example of binary collisions we assume the activator-head (state A) of the glider traveling West is positioned at cell (i, j). Then $(0,0)$ is a northwest-most corner of the lattice and the activator-head of the glider traveling North-West occupies the cell with coordinates $(i-1, j+2)$ (see the example of the initial position at Fig. 6.8a). We assume that the glider traveling West represents the value of

z_1	0	a	b	c	d
0	0	0	0	0	0
a	$\underline{a}$	b	b	a	c
b	$\underline{b}$	a	a	a	b
c	c	b	a	a	c
d	d	b	a	a	$\underline{b}$

(a)

z_2	0	a	b	c	d
0	0	0	0	0	0
a	$\underline{0}$	b	b	a	c
b	$\underline{0}$	a	a	a	b
c	c	b	a	a	c
d	d	b	a	a	$\underline{a}$

(b)

Fig. 13.17 Binary operations realised in a collision between a glider traveling West and a glider traveling North-West. Two different operations are represented by West (a), and South-West (b), output trajectories. Glider G_{34} is represented by a, G_4 by b, G_5 by c, G_{24} by d, and the absence of glider is 0.

variable x, and glider traveling North-West represents the value of variable y. Following the collision, one new glider continues traveling West (let it represent the value of variable z_1, the result of operation $\odot_1$), another is "reflected" South-West (let it represent the value of variable z_2, the result of operation $\odot_2$). We encode glider G_{34} by the symbol a, G_4 by b, G_5 by c and G_{24} by d, and the absence of a glider by 0.

Operations realised by this gate are shown in Fig. 13.17. Let us briefly discuss the algebraic systems $A_1 = \langle \odot_1, \{0,a,b,c,d\} \rangle$ and $A_2 = \langle \odot_2, \{0,a,b,c,d\} \rangle$ implemented in the glider collision. Both systems have neither identities nor zeros. The element 0 is the only idempotent ($0 \odot_1 0 = 0$ and $0 \odot_2 0 = 0$).

However 0 is right zero in A_1 and A_2 (for any $x \in \{0,a,b,c,d\}$ we have $x \odot_1 0 = 0$ and $x \odot_2 0 = 0$), and left identity in A_1 (for any $x \in \{0,a,b,c,d\}$ we have $0 \odot_1 x = x$). Operations $\odot_1$ and $\odot_2$ are not associative and not commutative. Singleton $\{d\}$ is the only minimal generator in A_1 and A_2: $d \odot_1 d = b$, $d \odot_1 b = a$, $d \odot_1 a = c$; and $d \odot_2 d = a$, $a \odot_2 a = b$, $a \odot_2 d = c$.

13.5 Implementation of the Finite State Machine

While developing the exact construction of a memory device, described in Sect. 7.16, we discovered that eater E_6 can take four different configurations resulting from the interactions of gliders brushing past, and there are seven types of glider produced in collisions with the eater (including some basic types flipped). We therefore envisaged that a finite state machine can be implemented in the eater-glider system. The internal state of such a machine is represented by the configuration of the eater, the type of incoming glider symbolises the input symbol of the machine, and the type of outgoing glider represents the output state of the machine.

To construct the full state transition table of the eater-glider machine we collided seven types of glider into four configurations of the eater and recorded the results of the collisions. For the sake of compact representation we encoded the configurations of the eater as shown in Fig. 13.18. We denote the gliders as follows: G_{34} as a, G_{43} as

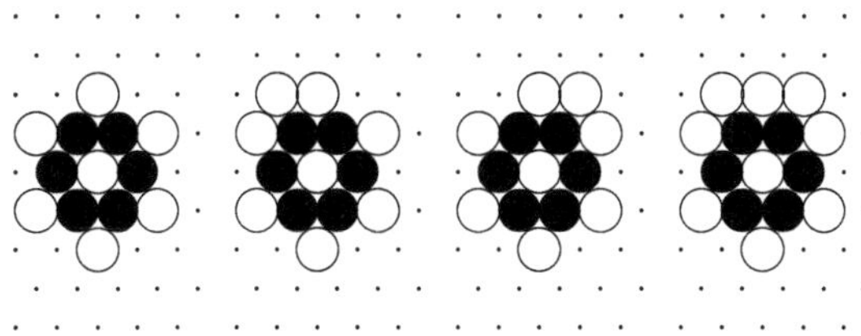

(a) α (b) β (c) χ (d) δ

Fig. 13.18 Encoding the internal states of the glider-eater machine in the configuration of eater E_6

	a	b	c	d	e	f	g
α	βb	δc	αb	αe	δd	αe	δc
β	αd	δe	βc	βc	χg	αa	χe
χ	χd	βe	δf	χa	βb	χa	βe
δ	δb	βc	χg	χe	αf	δe	αa

Fig. 13.19 The state transition table of the eater-glider machine. Tuple xy, a pair made up of an eater state x and glider state y, at the intersection of the ith row and jth column, signifies that being in state i while receiving input j the machine takes state x and generates output y.

operator	limit set
a	$\{\alpha,\beta\},\{\delta\}$
b	$\{\beta,\delta\}$
c	$\{\alpha\},\{\beta\},\{\delta,\chi\}$
d	$\{\alpha\},\{\beta\},\{\chi\}$
e	$\{\alpha,\delta\},\{\beta,\chi\}$
f	$\{\alpha\},\{\chi\},\{\delta\}$
g	$\{\alpha,\delta\},\{\beta,\chi\}$

Fig. 13.20 Limit sets of unary operators $a, \cdots, g$

b, G_5 as c, G_4 as d, G_{24} as e, G^4 (glider G_4 flipped horizontally) is f, and G^{43} (glider G_{43} flipped horizontally) is g. The state transition table is shown in Fig. 13.19.

Consider the internal states of the eater-glider machine as unary operators on the set $\{a,b,c,d,e,f,g\}$, i.e. the machine's state is reset to its initial state after the collision with the glider. For example, the unary operator α implements the following transformation: $a \to b$, $b \to c$, $c \to a$, $d \to a$, $e \to d$, $f \to e$, $g \to e$. The operators have the following limit sets: operator α has the limit set $\{a,b,c\}$, β — set $\{c\}$, χ has two limit sets $\{a,d\}$ and $\{b,c\}$, and operator δ — two limit sets $\{a,b,c,d\}$ and $\{e,f\}$. Considering unary operators $a, \cdots, g$ operating on set $\{\alpha,\beta,\chi,\delta\}$ we obtain the limit sets shown in Fig. 13.20. Many of the operators have more then two

	a^*	b^*	c^*	d^*	e^*	f^*	g^*
α	$(bd)^*$	$c(ce)^*$	b^*	e^*	$(de)^*$	e^*	$(ca)^*$
β	$(db)^*$	$(ec)^*$	c^*	c^*	$(gb)^*$	ae^*	e^*
χ	d^*	$e(ec)^*$	$(fg)^*$	a^*	$b(gb)^*$	a^*	e^*
δ	b^*	$(ce)^*$	$(gf)^*$	ea^*	$(ed)^*$	e^*	$(ac)^*$

Fig. 13.21 Input string to output string transformations implemented by the eater-glider machine. String s, at the intersection of the ith row and jth column, tells us that being initially in state i and receiving a uniform string j, the machine generates string s.

limits sets, which may indicate significant computational potential of the eater-glider machine.

To characterise the eater-glider machine in more detail we studied what output strings are generated by the machine when the machine receives the uniform infinite string s^*, $s \in \{a, \cdots g\}$ on its input. These input string to output string transformations are shown in Fig. 13.21.

Input string $abcdefg$ evokes the following output strings when fed into the machine. The machine starting in state α generates string $begabac$, in state β — string $dcgabac$, in state χ — string $deccgae$, and in state δ — string $bcccgae$.

13.6 Transformation of Two- and Four-Bit Strings

The eater can play a role of a non-volatile memory device and also implement an asynchronous XOR gate. What if we consider the states of several substrate-sites at once? Then gliders brushing the eater can change these states recursively and thus modify the string content represented by the states.

Let us encode two substrate-states in the northern part of the eater as x_1 and x_2 (Fig. 13.22a). If the cell x_i takes the inhibitor state it represents a bit-up state x_i of the string (x_1x_2), if the cell x_i is in the substrate-state then it represents a bit-down state of the string (x_1x_2). We have four possible combination of cell-states x_1 and x_2: *SS*, *SI*, *IS*, *II*, which corresponds to binary string 00, 01, 10 and 11, respectively (Fig. 13.22b).

An example of a binary string transformation is shown in Fig. 13.23. Initially, the eater represents string 00, cells x_1 and x_2 are in the substrate state (Fig. 13.23a). We launch glider G_{34} traveling West (Fig. 13.23b). When the glider brushes along the outer edge of the eater (Fig. 13.23a)f–k) it causes a minor perturbation of the marginal cell-states. The result of the perturbation is that one substrate-cell x_1 becomes an inhibitor (Fig. 13.23a)k). The glider changes its state as well — $G_{34} \rightarrow G_{43}$ — because of its interaction with the eater (Fig. 13.23a)j–o).

All configurations, including those modified by gliders, are still configurations, they are not evolving by themselves but can only be changed again by another brushing glider. When we apply a sequence of transformation we do not have to worry about the precise timing of glider generation or a time interval between gliders (as

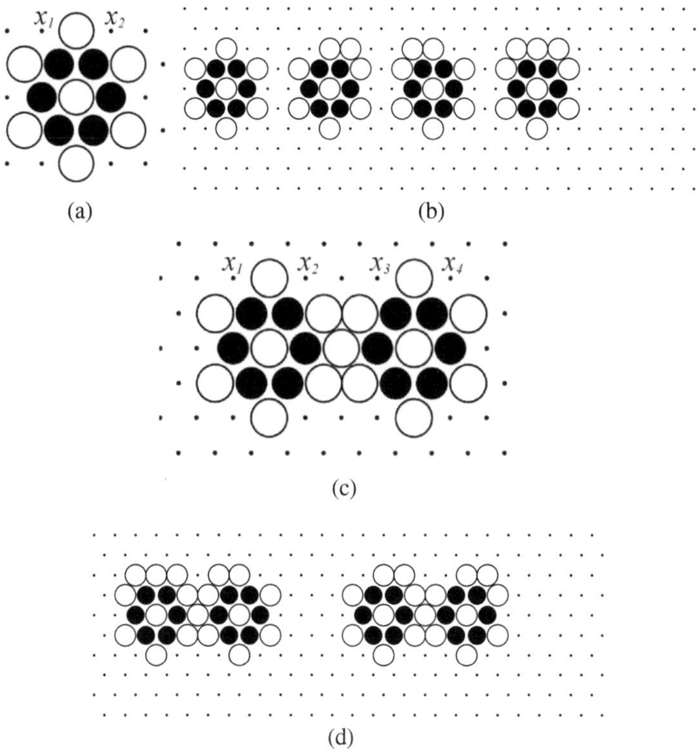

Fig. 13.22 Encoding two-bit strings in the states of eaters. The state of cells x_1 and x_4 for a single eater encoding (a) and cells $x_1 \dots x_4$ for the eater couple (c) encoding represent bits of strings (x_1x_2) and $(x_1x_2x_3x_4)$ respectively. Cell-state *I* (inhibitor) is shown by a circle, cell-state *A* (activator) by a solid disk, and cell-state *S* (substrate) by a dot. Cell x_i in state *I* represents '1', cell in state *S* represents '0'. Examples of encoding strings 00, 01, 10, 11 in a single eater are shown in (b). Examples of encoding strings 1110 and 0110 in the eater couple are shown in (d).

far as the glider phase is preserved). Therefore, eater-glider based transformations can be classified as asynchronous.

Finding 29. *The following transformations of two-bit strings, encoded below in digital numbers, are implemented by gliders and eaters in the spiral rule cellular automaton:*

$$
\begin{array}{lll}
L_1: & L_2: & \\
0 \to 0 & 0 \to 2 \to 0 & L_3: 0 \to 1 \to 2 \to 3 \to 0 \\
1 \to 3 \to 1 & 1 \to 1 & L_4: 0 \to 3 \to 2 \to 1 \to 0 \\
2 \to 2 & 3 \to 3 &
\end{array}
$$

Assuming states '0' and '1' are Boolean values, transformations $L_1 \cdots L_3$ can be expressed in the following manner:

$$L_1(x_i, x_{i-1}) = (x_i \vee y_i \wedge (x_i \oplus x_{i-1}), x_{i-1} \vee y_i \wedge x_{i-1}) \quad (13.1)$$

$$L_2(x_i, x_{i-1}) = (x_i \vee y_i \wedge (\overline{x_i \oplus x_{i-1}}), x_{i-1} \vee y_i \wedge x_{i-1}) \quad (13.2)$$

$$L_3(x_i, x_{i-1}) = (x_i \vee y_i \wedge (x_i \oplus x_{i-1}), x_{i-1} \vee y_i \wedge \overline{x_{i-1}}) \quad (13.3)$$

$$L_4(x_i, x_{i-1}) = (x_i \vee y_i \wedge (\overline{x_i \oplus x_{i-1}}), x_{i-1} \vee y_i \wedge \overline{x_{i-1}}). \quad (13.4)$$

Finding 30. *One can implement increment modulo 4 using transformation L_3 and decrement modulo 4 using transformation L_3*

This follows directly from Finding 29 and Boolean representation (13.3).

Similarly by positioning two eaters close to each other — the distance between central states of eaters is 4 — we can change four substrate-states just with one glider. The positions of substrate-states $x_1 \ldots x_4$ used for encoding are shown in Fig. 13.22c; examples of encoding four-bit binary sequences are demonstrated in Fig. 13.22d.

Example of transformation $0000 \to 1100$ is shown in Fig. 13.24, and transformation $1100 \to 0010$ in Fig. 13.25.

Finding 31. *The following transformations of four-bit strings, encoded below in digital numbers, are implemented by gliders and eaters in the spiral rule cellular automaton:*

T_1 :	T_2 :
$0 \to 12 \to 8 \to 4 \to 0$	$0 \to 6 \to 4 \to 10 \to 0$
$1 \to 3 \to 13 \to 7 \to 1$	$1 \to 9 \to 1$
$2 \to 2$	$2 \to 8 \to 14 \to 12 \to 2$
$5 \to 15 \to 9 \to 11 \to 5$	$3 \to 7 \to 11 \to 15 \to 3$
$6 \to 14 \to 6$	$5 \to 5$
$10 \to 10$	$13 \to 13$

$T3$:
$0 \to 9 \to 14 \to 3 \to 8 \to 1 \to 6 \to 11 \to 0$
$2 \to 7 \to 4 \to 5 \to 10 \to 15 \to 12 \to 13 \to 2$

$T4$:
$0 \to 3 \to 2 \to 13 \to 8 \to 11 \to 10 \to 5 \to 0$
$1 \to 12 \to 7 \to 14 \to 9 \to 4 \to 15 \to 6 \to 1$

The exact conditions $C(T_i)$ or $C(L_i)$ of transformations T_i and L_i are determined by two parameters (g,d), where a type g glider brushes the eater, and the initial distance d between the eater and the glider is even.

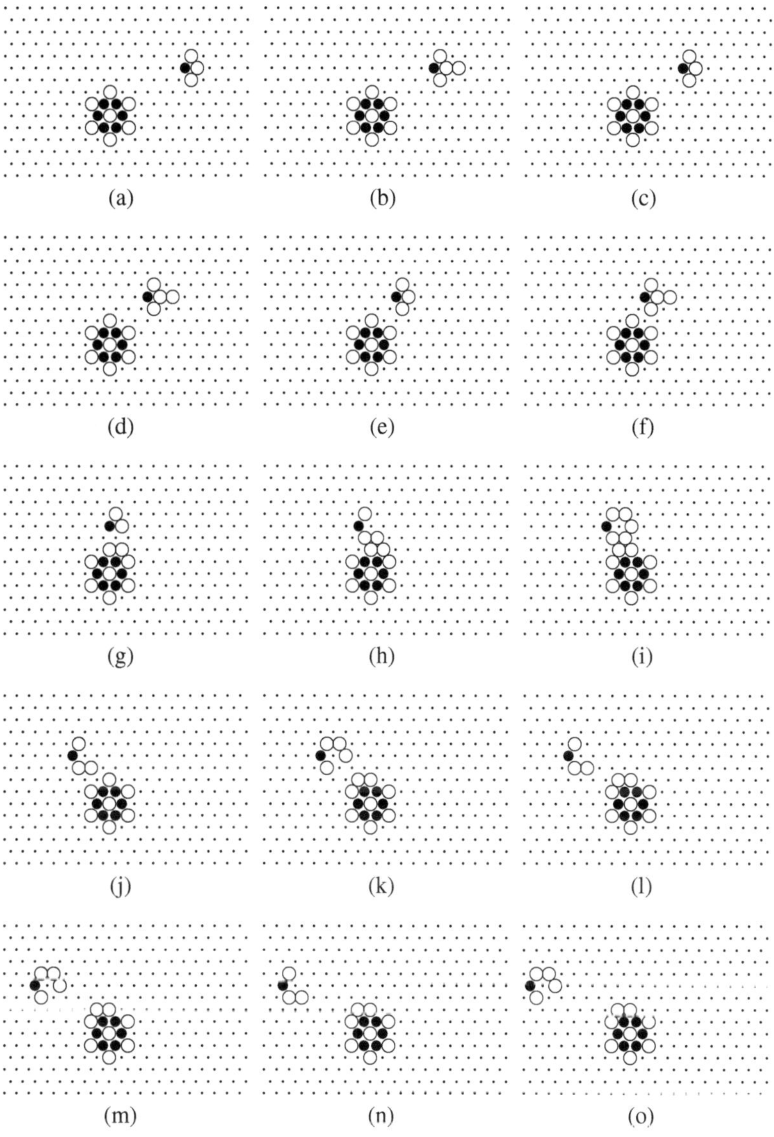

Fig. 13.23 Transformation 00 → 10 implemented by glider G_{34}

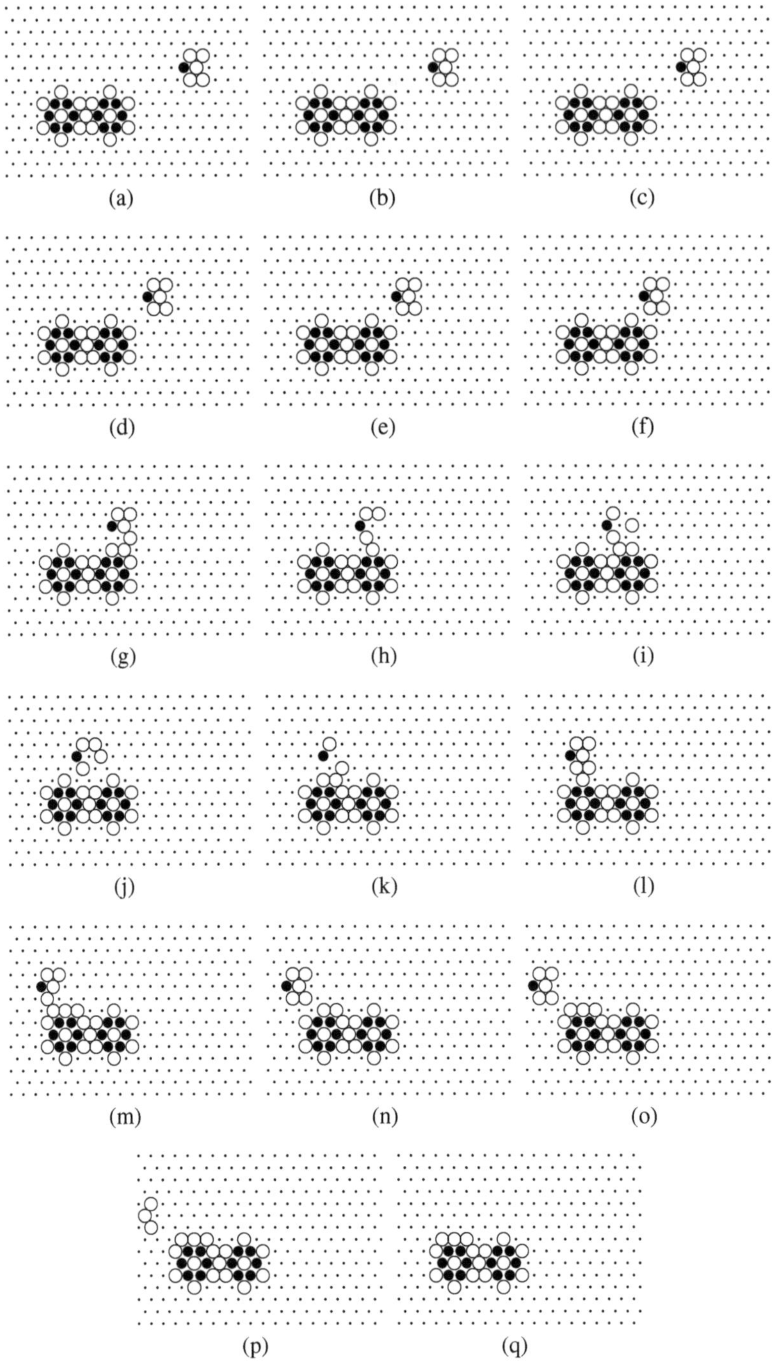

Fig. 13.24 Transformation 0000 → 1100 implemented by glider G_5

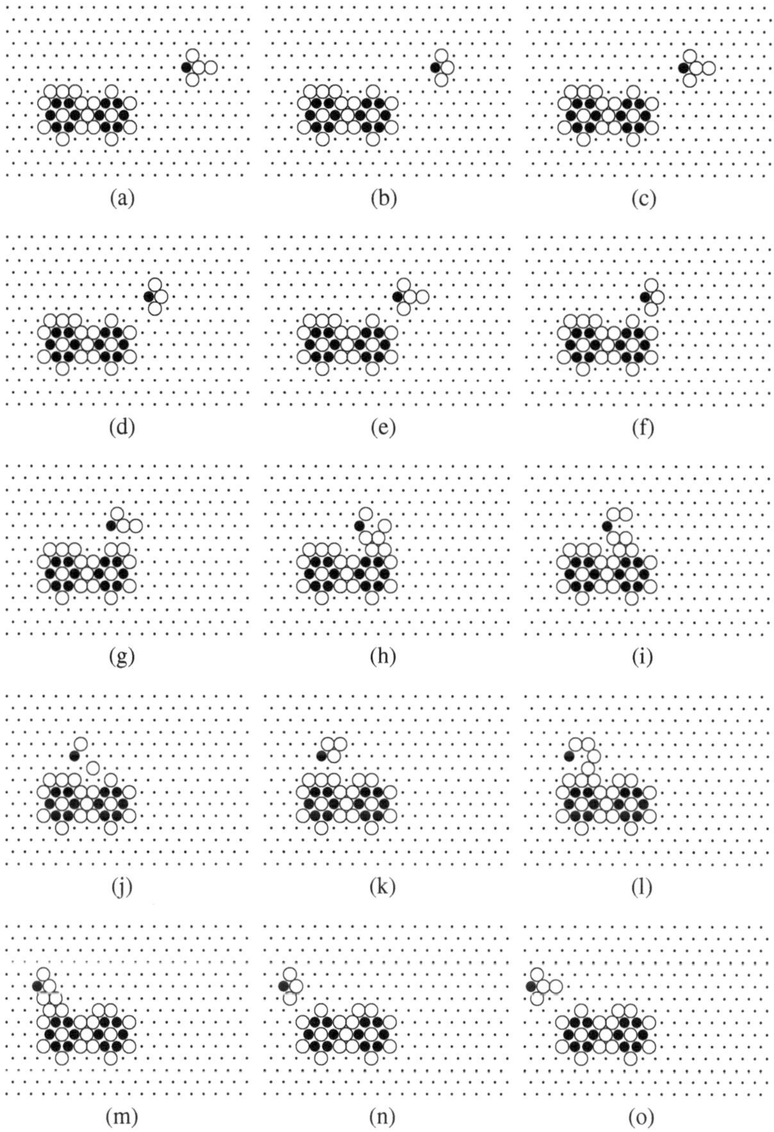

Fig. 13.25 Transformation $1100 \to 0010$ implemented by glider G_{34}

Finding 32. *The conditions of the transformations L_i and T_i, $i = 1 \ldots 4$ are as follows:*

- $C(L_1) = C(T_1) = (g', d')$, $g' = G_5, G_4$ *and* d'=*even, odd.*
- $C(L_2) = C(T_2) = (g', d')$, $g' = G_{34}$ *and* d'=*even, odd.*
- $C(L_3) = C(T_3) = \{G_{24}, even), (E_7, even), (G_{43}, odd)\}$
- $C(L_4) = C(T_4) = \{(G_{24}, odd), (E_7, odd), (G_{43}, even)\}$.

Finding 33. *Transformations T_1 and T_2 split the set $\{0, \cdots, 15\}$ into stable subsets of exclusively even numbers and subsets of exclusively odd numbers.*

This is an outcome of the experimental implementation of the transformations T_1 and T_2, and analysis of chains generated by the transformations in Finding 31.

Finding 34. *By using a series of transformations $T_1, \ldots, T_4$ one can transform a number from $\{0, \ldots, 15\}$ to any other number from $\{0, \ldots, 15\}$ in at most three steps.*

To prove this statement we compute a distance matrix $D = (d_{ij})_{0 \leq i,j \leq 15}$, where d_{ij} represents the minimal number of transformations $T_1, \ldots T_4$ that one needs to apply to get from number i to number j. The distance matrix has the following form:

$$D = \begin{pmatrix}
3&2&2&1&2&3&1&2&2&1&3&2&1&2&2&3\\
3&2&2&1&2&3&1&2&2&1&3&2&1&2&2&3\\
3&2&1&3&2&3&3&1&1&3&3&2&2&1&2&3\\
3&2&1&3&2&3&3&1&1&3&3&2&2&1&2&3\\
1&3&3&2&3&1&2&3&3&2&1&3&2&3&3&1\\
1&3&3&2&3&1&2&3&3&2&1&3&2&3&3&1\\
2&1&3&2&1&2&2&3&3&2&2&1&2&3&1&2\\
2&1&3&2&1&2&2&3&3&2&2&1&2&3&1&2\\
2&1&3&2&1&2&2&3&3&2&2&1&2&3&1&2\\
2&1&3&2&1&2&2&3&3&2&2&1&2&3&1&2\\
1&3&3&2&3&1&2&3&3&2&1&3&2&3&3&1\\
1&3&3&2&3&1&2&3&3&2&1&3&2&3&3&1\\
3&2&1&3&2&3&3&1&1&3&3&2&2&1&2&3\\
3&2&1&3&2&3&3&1&1&3&3&2&1&2&2&3\\
3&2&2&1&2&3&1&2&2&1&3&2&1&2&2&3\\
3&2&2&1&2&3&1&2&2&1&3&2&1&2&2&3
\end{pmatrix}$$

For example, to get 8 from 7 we must apply T_1 twice and then T_3 once: $T_3(T_1(T_1(7)))$ $= 8$. From D we can also tell that one needs 34 to generate a glider, which brushes the eater to generate a monotonous ordered sequence of numbers $(0, 1, \ldots, 15)$.

13.7 Six Bit Coding

Let us now consider the composition $\mathbf{x} \circ \mathbf{y}$ of two six bit binary strings $\mathbf{x} = (x_1 \cdots x_6)$ and $\mathbf{y} = (y_1 \cdots y_6)$. The string $\mathbf{x}$ is represented by states of six cells $x_1 \cdots x_6$ surrounding the eater, as shown in Fig. 13.26a. The string $\mathbf{y}$ is represented by gliders brushing the eater (as discussed in the previous sections). The trajectories of gliders are indicated in Fig. 13.26: $y_i = 1$ if there is a glider traveling along trajectory y_i (Fig. 13.26)a. Examples of encoding binary strings in the eater's states are shown in (Fig. 13.26)b.

Gliders representing string $\mathbf{y}$ are sent in the order $y_1, \cdots, y_6$. Values of Boolean variables y_i, $i = 1, \cdots, 6$, are represented by the presence or absence of a glider g, traveling along trajectory y_i. A minimum interval between two subsequent gliders is selected so the next glider is sent after the current glider has already passed the eater.

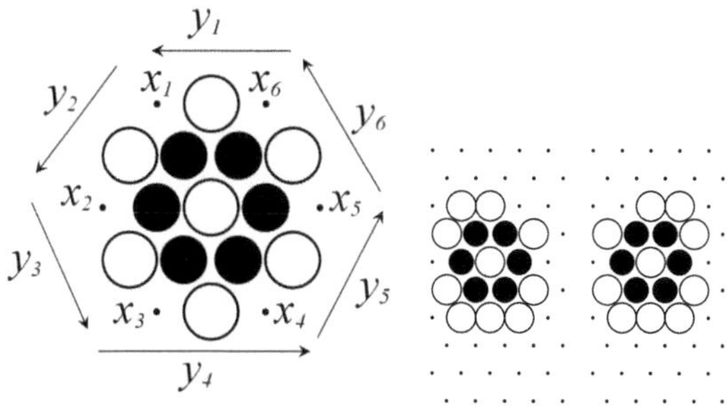

Fig. 13.26 Encoding six-bit strings in states of the eaters. (a) States of cells $x_1 \cdots x_6$ represent bits of strings $(x_1 \cdots x_6)$. Cell-state I (inhibitor) is shown by a black disk, cell-state A (activator) by a circle, and cell-state S (substrate) by a dot. Cell x_i in state I represents '1', cell in state S represents '0'. (b)–(c) Examples of encoding strings 101100 (a) and 001101 (b). Vectors $y_1 \cdots y_6$ in (a) show which direction gliders representing input string $y_1 \cdots y_6$ travel.

There is no maximal interval between the gliders. The exact structure of composition $\mathbf{x} \circ \mathbf{y}$ is determined by transformations $L_1 \cdots L_4$. For example, if the composition $\mathbf{x} \circ \mathbf{y}$ is based on L_4 then the result $\mathbf{x}' \leftarrow \mathbf{x} \circ \mathbf{y}$ can be calculated as follows:

$$x'_1 = x_1 \vee y_1 \wedge (\overline{x_1 \oplus x_6})$$
$$x'_6 = x_6 \vee y_1 \wedge \overline{x_6}$$

and for $i = 2, \ldots, 6$

$$x'_i = x_i \vee y_i \wedge (\overline{x_i \oplus x_{i-1}})$$
$$x'_i = x_{i-1} \vee y_i \wedge \overline{x_{i-1}}$$

We represent the composition $\mathbf{x} \circ \mathbf{y}$ implemented by the eater and glider g_z, $z = 1, \cdots, 6$, by matrix $M(L_i) = (m_{ij})_{0 \leq i,j \leq 63}$. Each entry m_{ij} of the matrix is an integer from $\{0, \cdots, 63\}$. The entry m_{ij} is a digital representation of the string $\mathbf{x} \leftarrow \mathbf{x} \circ \mathbf{y}$, and its indexes i and j are digital representations of strings $\mathbf{x}$ and $\mathbf{y}$, respectively.

Matrices of the compositions $\mathbf{x} \circ \mathbf{y}$ are shown in Appendix (Fig. 13.32–13.35). Structures of the matrices do not indicate any immediate implications for practical applications. There is however a range of properties which may be useful in further implementations of arithmetical operations with gliders and eaters.

Finding 35. *Let $\kappa(L_i)$, $i = 1, \ldots, 4$, be a ratio of commutative pairs in $\mathbf{x} \circ \mathbf{y}$ determined by transformation L_i, and $\alpha(L_i)$ a ratio of associative triples, then the following hierarchy of compositions can be drawn:*

$$\kappa(L_3) > \kappa(L_1) > \kappa(L_4) > \kappa(L_2)$$

$$\alpha(L_2) > \alpha(L_3) > \alpha(L_1) > \alpha(L_4)$$

This follows from straightforward analysis of the matrices $M(L_1), \ldots, M(L_4)$ in Tab. 13.1.

Table 13.1 Ratios of commutative pairs $\kappa(\cdot)$ and associative triples $\alpha(\cdot)$ in $M(L_1),\dots,M(L_4)$

L_i	$\kappa(L_i)$	$\alpha(L_i)$
L_1	0.33	0.79
L_2	0.11	0.85
L_3	0.35	0.81
L_4	0.15	0.63

Finding 36. *For compositions determined by transformations L_1 and L_2 we have $\mathbf{x} \circ \mathbf{x} = \mathbf{x}$ for any $\mathbf{x}$ from $\{0,\dots,63\}$.*

The following finding relates to identities and absorbing elements (nulls) of the compositions.

Finding 37. *Number 63 is the absorbing element of compositions determined by L_3 and L_4. The composition determined by L_1 has two left absorbing elements: 0 and 63. The composition determined by L_2 has eighteen left absorbing elements: 21, 23, 27, 29, 31, 42, 43, 45, 46, 47, 53, 54, 55, 58, 59, 61, 62, 63. None of the composition has left or right identities.*

The structure of the first two rows of $M(L_1)$ shows that stationary and traveling localizations in the spiral rule automaton can implement at least one sensible arithmetical operation: transformation of arbitrary stream of numbers to streams of even or odd numbers.

Finding 38. *Let number $a \in \{0,1\}$, encoded in a state of the eater, and number $b \in \{0,\dots,31\}$, encoded by a stream of gliders traveling along trajectories as as per Fig. 13.26, then the eater-gliders system executes operation $2b+a$; the result of the operation is encoded in state of the eater.*

The resultant state of the eater can be read and erased by additional stream of gliders.

13.8 Scylla and Charybdis: Outcomes of Passing between Two Eaters

There are endless possibilities of arrange interesting and useful interactions between gliders and eaters. Let us place two eaters so that they touch each other only by their extremal inhibitory states (Fig. 13.27a). We aim glider G_{24}, travelling west, to run between the eaters (Fig. 13.27abc), in such a phase that the glider crashes into the eaters while it is in one activator and four inhibitor states (Fig. 13.27d). All but activator states of the glider are destroyed in the crush (Fig. 13.27e) yet the crash causes emergence of excitation-inhibition perturbation (Fig. 13.27efg). A glider G_5 emerges as a result of the perturbation (Fig. 13.27h). This glider travels east (Fig. 13.27hi).

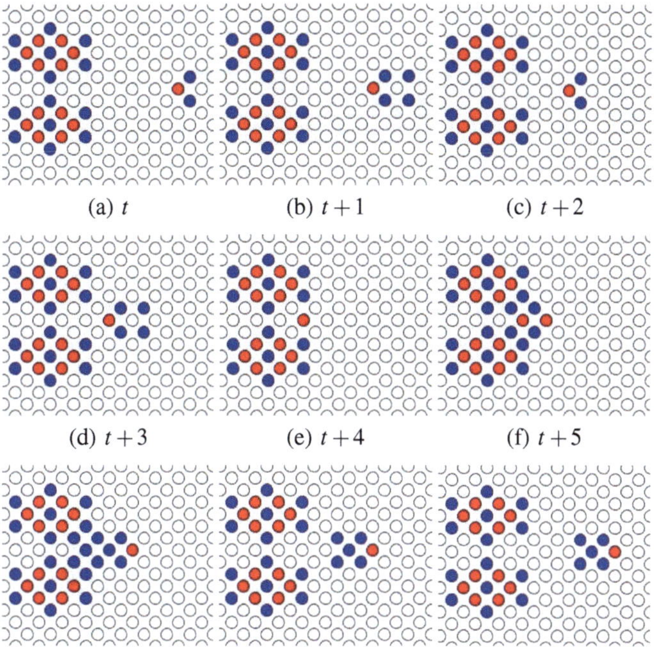

(a) t (b) $t+1$ (c) $t+2$

(d) $t+3$ (e) $t+4$ (f) $t+5$

(g) $t+6$ (h) $t+7$ (i) $t+8$

Fig. 13.27 Reflection and transformation of glider G_{24} to glider G_5 in collision with eater E_6

Finding 39. *Glider G_{24} is reflected and transformed into glider G_5 by two eaters E_6.*

Figure 13.28 shows what happens when glider G_{24}, travelling west (Fig. 13.28ab), collides to the eaters while in its two one activator and two inhibitor states phase (Fig. 13.28c). All but activator state of the glider are destroyed (Fig. 13.28e). Two inhibitor states are added to eaters: north eater gets the inhibitor state in position x_2 (see coding in Fig. 13.26a) and south eater in position x_6. Such local configuration causes the activator state, remainder of the glider, to reflect eastward (Fig. 13.28f) and a compact ensemble of activator states to emerge (Fig. 13.28g). The perturbation (Fig. 13.28hij) leads to formation of three gliders: G_5 travelling east (as reflection in Fig. 13.27) and two gliders G_4 travelling north-east and south-east (Fig. 13.28kl).

Finding 40. *Glider G_{24} is reflected and multiplied into gliders G_5 and G_4 by two eaters E_6.*

Setup in Figs. 13.29 – 13.31 shows 'capacitor'-like functioning of the two-eater complex. There is one-cell gap between the eaters. In the preparatory step (Figs. 13.29) we collide glider G_{34} into the gap between the eaters (Figs. 13.29af). The glider is transformed into glider G_5 but two inhibitor states are added to eaters in positions x_3 (north eater) and x_1 (south eater) (Figs. 13.29g). The 'hardware' is ready for operation.

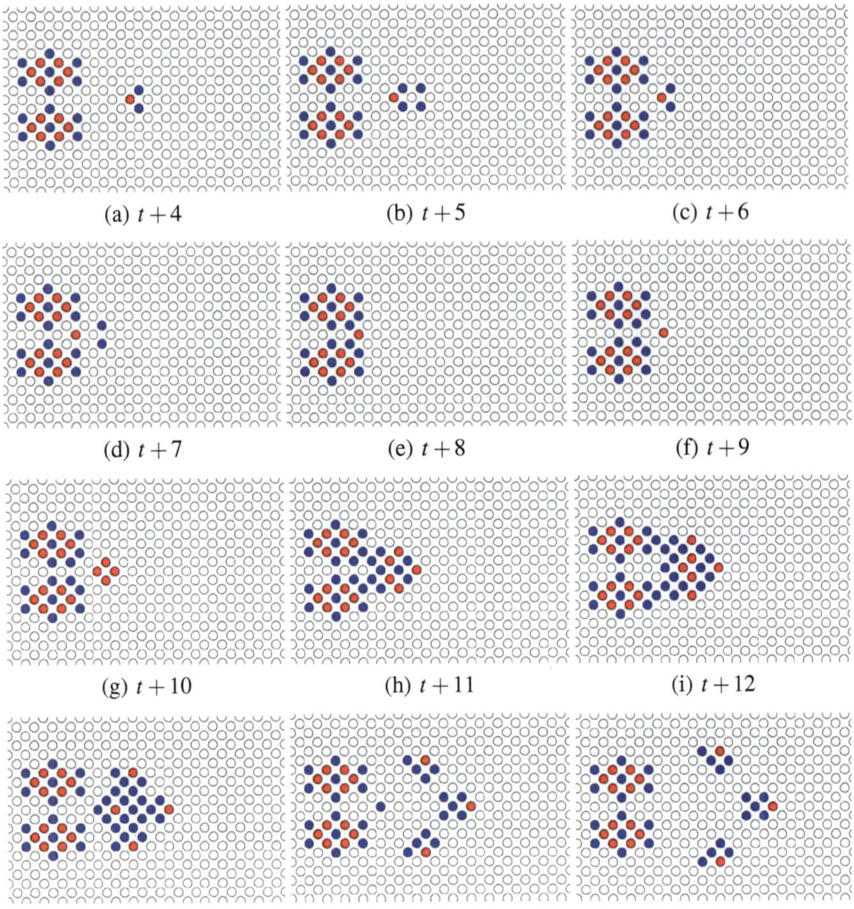

(a) $t+4$ (b) $t+5$ (c) $t+6$

(d) $t+7$ (e) $t+8$ (f) $t+9$

(g) $t+10$ (h) $t+11$ (i) $t+12$

(j) $t+13$ (k) $t+14$ (l) $t+15$

Fig. 13.28 Reflection, transformation and multiplication: glider G_{24} collides to eater E_6 and the glider reflected and transformed into two gliders G_4 and one glider G_5

Further we crash glider G_{24} into the 'capacitor'. In all examples Figs. 13.30 – 13.31 glider G_{24} crashes into the eaters in its phase with one activator and two inhibitor states.

The 'capacitor' is charged by glider G_{24} (Fig. 13.30a–h): inhibitor states are removed from positions x_3 of north eater and x_1 of south eater but the inhibitor states are added in positions x_2 of north eater and x_6 of south eater (Fig. 13.30a–h).

When glider G_{24}, travelling west, crushes into a charged 'capacitor' (Fig. 13.31abc). One activator state continues its propagation westward. Two additional activator states and two inhibitor states are formed, supported by inhibitor states in the charged positions x_2 (north eater) and x_6 (south eater) (Fig. 13.31d). This local configuration leads to a birth of glider G_5, which heads east (Fig. 13.31ef).

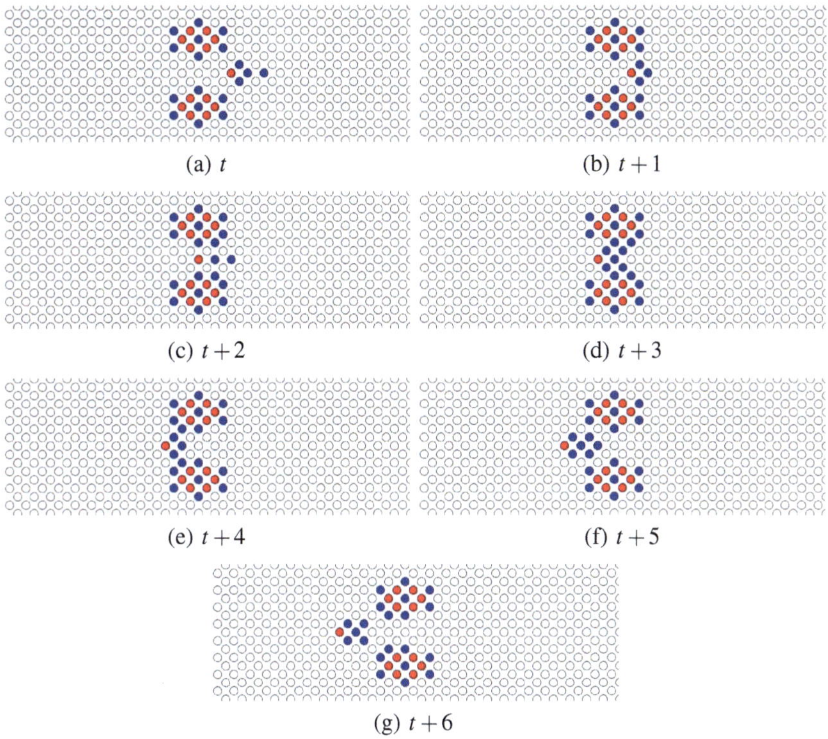

Fig. 13.29 'Capacitor': preparation of 'hardware'. Glider G_{34} transforms two eaters E_6 to their E_7 form, the glider is transformed into G_5 during the collision.

The activator state travelling west gets five inhibitor states (Fig. 13.31efg) and thus transformed to glider G_5 (Fig. 13.31h), it also fills sites x_3 (north eater) and x_1 (south eater) with inhibitor states. The 'capacitor' returns to its discharged state. The of cycle of charging and discharging can continue indefinitely.

Finding 41. *Two eaters E_6 can implement operations of counting and reflection.*

13.9 Summary

A hexagonal cellular automaton imitating an abstract spatially-extended three-species chemical system with non-trivial interactions between activator, inhibitor and substrate hosts a rich 'zoo' of travelling and stationary localisations, or gliders. All the basic components — necessary for constructing a general purpose computing device — are implemented in the space-time dynamics of the automaton. They include signal reflectors, multipliers, erasers, memory devices, binary and multiple-valued gates, and a finite state machine.

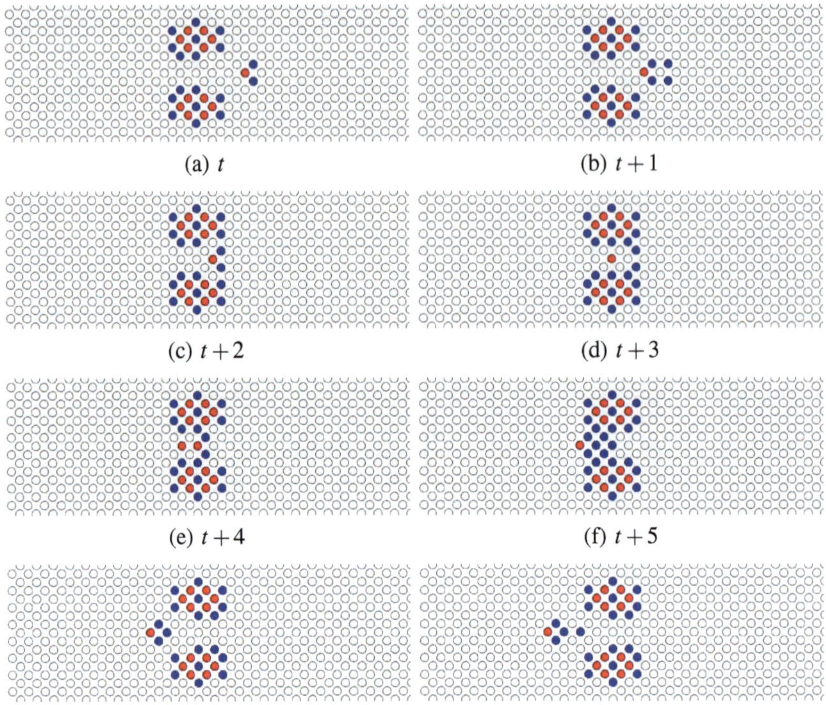

(a) t (b) $t+1$

(c) $t+2$ (d) $t+3$

(e) $t+4$ (f) $t+5$

(g) $t+6$ (h) $t+7$

Fig. 13.30 Charging the 'capacitor' by glider G_{24}

Amongst the problems that remain to be unravelled we can mention a few that are the most important:

- To build a configuration of reusable sensors, i.e. restorable eaters. Currently, to input a piece of information to the computing medium, we switch the state of one site of a stationary localisation, thus transforming the localisation to a mobile localisation. The sensor-localisation is destroyed as a result of 'sensing', which may be inconvenient for certain applications.
- To find ways of using eaters to change the trajectories of signal-gliders. So far, when a glider collides (or brushes past) an eater, the glider either changes its type or is annihilated, but the glider never starts moving along a new trajectory.
- To invent techniques for reading a bit from a memory device (see Sect. 7.16) without destroying the bit, at present, reading is associated with erasing.

On the experimental front, we are eager to see real-life chemical systems, which exhibit behaviour similar to that discovered in the present chapter, particularly concerning localisation dynamics. Some promising results have been obtained already by Vanag and Espstein [250], who demonstrated experimentally the existence of spiral waves emitting localised wave-fragments. However, this was not done in a 'conventional' setup of a liquid-phase chemical system, but in a Belousov-Zhabotinsky

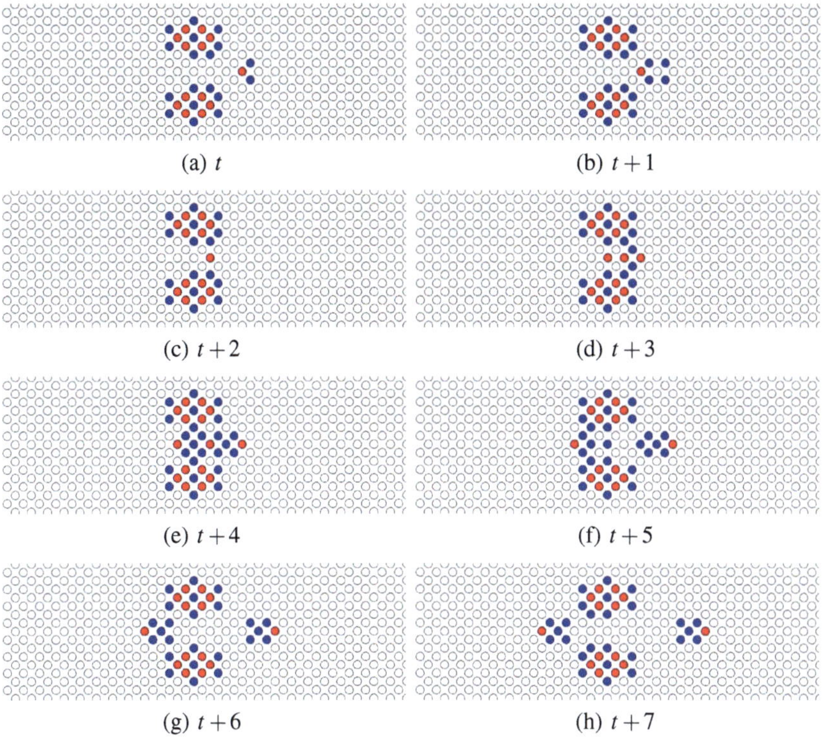

Fig. 13.31 Discharging the 'capacitor' by glider G_{24}

reaction dispersed in water nano-droplets of a water-in-oil micro-emulsion. We are not aware if any standing localizations existing in the same physical domain of the medium with travelling localizations. We hope to clarify these matters in future experiments. We uncovered ways of implementing asynchronous operations over sets of binary strings. We employ a sequence of mobile localizations (gliders, excitation wave-fragments) to represent one binary string, and a stationary localisation (eater, breather, standing wave) to represent another string in the composition. The operations are asynchronous in the sense that there is no precise time for a glider to collide with the eater, and there is no fixed time interval between gliders in a glider train representing a binary string. The order in which gliders hit certain parts of the eater is nevertheless important and must be executed precisely during the computation.

By colliding gliders with eaters we can implement transformations of sets of integers, which demonstrates collision based computation in the spiral rule. The methods could find applications for the design arithmetical chips. Most impressive so far were increments and decrements modulo four and subdivision of the digital number set onto prime and even numbers. We found operations on a set of six-bit integers which are rich yet irregular and require detailed study. Nevertheless are we convinced that by combining operations (based on $L_1,\ldots,L_4$) one can implement typical arithmetical operations. This will be subject of further analysis.

What are the potential practical outcomes of our designs and findings? As we demonstrated in [247], the spiral rule automaton is an ideal discrete approximation of excitation dynamics in a light-sensitive sub-excitable chemical medium on a patterned environment. Gliders are discrete analogs of localised wave-fragments. This does not mean that our computing schemes can be implemented directly in an excitable chemical medium, not least because there are no excitation-wave analogues of eaters. More feasible implementations would be in other substrates, possibly at the nano-scales as e.g. excitation patterns in networks of single-electron oscillators [39], excitons in mono-molecular layers and breathers in polymers and crystals [10], and dynamically evolving molecular circuits [46].

Appendix

0	0	0	0	0	0	0	0	0	0	0	0	0	0	0	0	0	0	0	0	0	0	0	0	0	0	0	0	0	0	0	0	0	0	0	0	0	0	0	0	0	0	0	0	0	0	0	0	0	0	0	0	0	0	0	0	0	0	0	0	0	0	0	0
1	1	1	1	1	1	1	1	1	1	1	1	1	1	1	1	1	1	1	1	1	1	1	1	1	1	1	1	1	1	1	1	33	33	33	33	33	33	33	33	33	33	33	33	33	33	33	33	49	49	49	49	49	49	49	49	57	57	57	57	61	61	63	63
2	3	2	3	2	3	2	3	2	3	2	3	2	3	2	3	2	3	2	3	2	3	2	3	2	3	2	3	2	3	2	3	2	3	2	3	2	3	2	3	2	3	2	3	2	3	2	3	2	3	2	3	2	3	2	3	2	3	2	3	2	3	2	3
3	3	3	3	3	3	3	3	3	3	3	3	3	3	3	3	3	3	3	3	3	3	3	3	3	3	3	3	3	3	3	3	35	35	35	35	35	35	35	35	35	35	35	35	35	35	35	35	51	51	51	51	51	51	51	51	59	59	59	59	63	63	63	63
4	4	6	7	4	4	6	7	4	4	6	7	4	4	6	7	4	4	6	7	4	4	6	7	4	4	6	7	4	4	6	7	4	4	6	7	4	4	6	7	4	4	6	7	4	4	6	7	4	4	6	7	4	4	6	7	4	4	6	7	4	4	6	7
5	5	7	7	5	5	7	7	5	5	7	7	5	5	7	7	5	5	7	7	5	5	7	7	5	5	7	7	5	5	7	7	37	37	39	39	37	37	39	39	37	37	39	39	37	37	39	39	53	53	55	55	53	53	55	55	61	61	63	63	61	61	63	63
6	7	6	7	6	7	6	7	6	7	6	7	6	7	6	7	6	7	6	7	6	7	6	7	6	7	6	7	6	7	6	7	6	7	6	7	6	7	6	7	6	7	6	7	6	7	6	7	6	7	6	7	6	7	6	7	6	7	6	7	6	7	6	7
7	7	7	7	7	7	7	7	7	7	7	7	7	7	7	7	7	7	7	7	7	7	7	7	7	7	7	7	7	7	7	7	39	39	39	39	39	39	39	39	39	39	39	39	39	39	39	39	55	55	55	55	55	55	55	55	63	63	63	63	63	63	63	63
8	8	8	8	12	12	14	15	8	8	8	8	12	12	14	15	8	8	8	8	12	12	14	15	8	8	8	8	12	12	14	15	8	8	8	8	12	12	14	15	8	8	8	8	12	12	14	15	8	8	8	8	12	12	14	15	8	8	8	8	12	12	14	15
9	9	9	9	13	13	15	15	9	9	9	9	13	13	15	15	9	9	9	9	13	13	15	15	9	9	9	9	13	13	15	15	41	41	41	41	45	45	47	47	41	41	41	41	45	45	47	47	57	57	57	57	61	61	63	63	57	57	57	57	61	61	63	63
10	11	10	11	14	15	14	15	10	11	10	11	14	15	14	15	10	11	10	11	14	15	14	15	10	11	10	11	14	15	14	15	10	11	10	11	14	15	14	15	10	11	10	11	14	15	14	15	10	11	10	11	14	15	14	15	10	11	10	11	14	15	14	15
11	11	11	11	15	15	15	15	11	11	11	11	15	15	15	15	11	11	11	11	15	15	15	15	11	11	11	11	15	15	15	15	43	43	43	43	47	47	47	47	43	43	43	43	47	47	47	47	59	59	59	59	63	63	63	63	59	59	59	59	63	63	63	63
12	12	14	15	12	12	14	15	12	12	14	15	12	12	14	15	12	12	14	15	12	12	14	15	12	12	14	15	12	12	14	15	12	12	14	15	12	12	14	15	12	12	14	15	12	12	14	15	12	12	14	15	12	12	14	15	12	12	14	15	12	12	14	15
13	13	15	15	13	13	15	15	13	13	15	15	13	13	15	15	13	13	15	15	13	13	15	15	13	13	15	15	13	13	15	15	45	45	47	47	45	45	47	47	45	45	47	47	45	45	47	47	61	61	63	63	61	61	63	63	61	61	63	63	61	61	63	63
14	15	14	15	14	15	14	15	14	15	14	15	14	15	14	15	14	15	14	15	14	15	14	15	14	15	14	15	14	15	14	15	14	15	14	15	14	15	14	15	14	15	14	15	14	15	14	15	14	15	14	15	14	15	14	15	14	15	14	15	14	15	14	15
15	15	15	15	15	15	15	15	15	15	15	15	15	15	15	15	15	15	15	15	15	15	15	15	15	15	15	15	15	15	15	15	47	47	47	47	47	47	47	47	47	47	47	47	47	47	47	47	63	63	63	63	63	63	63	63	63	63	63	63	63	63	63	63
16	16	16	16	16	16	16	16	24	24	24	24	28	28	30	31	16	16	16	16	16	16	16	16	24	24	24	24	28	28	30	31	16	16	16	16	16	16	16	16	24	24	24	24	28	28	30	31	16	16	16	16	16	16	16	16	24	24	24	24	28	28	30	31
17	17	17	17	17	17	17	17	25	25	25	25	29	29	31	31	17	17	17	17	17	17	17	17	25	25	25	25	29	29	31	31	49	49	49	49	49	49	49	49	57	57	57	57	61	61	63	63	49	49	49	49	49	49	49	49	57	57	57	57	61	61	63	63
18	19	18	19	18	19	18	19	26	27	26	27	30	31	30	31	18	19	18	19	18	19	18	19	26	27	26	27	30	31	30	31	18	19	18	19	18	19	18	19	26	27	26	27	30	31	30	31	18	19	18	19	18	19	18	19	26	27	26	27	30	31	30	31
19	19	19	19	19	19	19	19	27	27	27	27	31	31	31	31	19	19	19	19	19	19	19	19	27	27	27	27	31	31	31	31	51	51	51	51	51	51	51	51	59	59	59	59	63	63	63	63	51	51	51	51	51	51	51	51	59	59	59	59	63	63	63	63
20	20	22	23	20	20	22	23	28	28	30	31	28	28	30	31	20	20	22	23	20	20	22	23	28	28	30	31	28	28	30	31	20	20	22	23	20	20	22	23	28	28	30	31	28	28	30	31	20	20	22	23	20	20	22	23	28	28	30	31	28	28	30	31
21	21	23	23	21	21	23	23	29	29	31	31	29	29	31	31	21	21	23	23	21	21	23	23	29	29	31	31	29	29	31	31	53	53	55	55	53	53	55	55	61	61	63	63	61	61	63	63	53	53	55	55	53	53	55	55	61	61	63	63	61	61	63	63
22	23	22	23	22	23	22	23	30	31	30	31	30	31	30	31	22	23	22	23	22	23	22	23	30	31	30	31	30	31	30	31	22	23	22	23	22	23	22	23	30	31	30	31	30	31	30	31	22	23	22	23	22	23	22	23	30	31	30	31	30	31	30	31
23	23	23	23	23	23	23	23	31	31	31	31	31	31	31	31	23	23	23	23	23	23	23	23	31	31	31	31	31	31	31	31	55	55	55	55	55	55	55	55	63	63	63	63	63	63	63	63	55	55	55	55	55	55	55	55	63	63	63	63	63	63	63	63
24	24	24	24	28	28	30	31	24	24	24	24	28	28	30	31	24	24	24	24	28	28	30	31	24	24	24	24	28	28	30	31	24	24	24	24	28	28	30	31	24	24	24	24	28	28	30	31	24	24	24	24	28	28	30	31	24	24	24	24	28	28	30	31
25	25	25	25	29	29	31	31	25	25	25	25	29	29	31	31	25	25	25	25	29	29	31	31	25	25	25	25	29	29	31	31	57	57	57	57	61	61	63	63	57	57	57	57	61	61	63	63	57	57	57	57	61	61	63	63	57	57	57	57	61	61	63	63
26	27	26	27	30	31	30	31	26	27	26	27	30	31	30	31	26	27	26	27	30	31	30	31	26	27	26	27	30	31	30	31	26	27	26	27	30	31	30	31	26	27	26	27	30	31	30	31	26	27	26	27	30	31	30	31	26	27	26	27	30	31	30	31
27	27	27	27	31	31	31	31	27	27	27	27	31	31	31	31	27	27	27	27	31	31	31	31	27	27	27	27	31	31	31	31	59	59	59	59	63	63	63	63	59	59	59	59	63	63	63	63	59	59	59	59	63	63	63	63	59	59	59	59	63	63	63	63
28	28	30	31	28	28	30	31	28	28	30	31	28	28	30	31	28	28	30	31	28	28	30	31	28	28	30	31	28	28	30	31	28	28	30	31	28	28	30	31	28	28	30	31	28	28	30	31	28	28	30	31	28	28	30	31	28	28	30	31	28	28	30	31
29	29	31	31	29	29	31	31	29	29	31	31	29	29	31	31	29	29	31	31	29	29	31	31	29	29	31	31	29	29	31	31	61	61	63	63	61	61	63	63	61	61	63	63	61	61	63	63	61	61	63	63	61	61	63	63	61	61	63	63	61	61	63	63
30	31	30	31	30	31	30	31	30	31	30	31	30	31	30	31	30	31	30	31	30	31	30	31	30	31	30	31	30	31	30	31	30	31	30	31	30	31	30	31	30	31	30	31	30	31	30	31	30	31	30	31	30	31	30	31	30	31	30	31	30	31	30	31
31	31	31	31	31	31	31	31	31	31	31	31	31	31	31	31	31	31	31	31	31	31	31	31	31	31	31	31	31	31	31	31	63	63	63	63	63	63	63	63	63	63	63	63	63	63	63	63	63	63	63	63	63	63	63	63	63	63	63	63	63	63	63	63
32	32	32	32	32	32	32	32	32	32	32	32	32	32	32	32	48	48	48	48	48	48	48	48	56	56	56	56	60	60	62	63	32	32	32	32	32	32	32	32	32	32	32	32	32	32	32	32	48	48	48	48	48	48	48	48	56	56	56	56	60	60	62	63
33	33	33	33	33	33	33	33	33	33	33	33	33	33	33	33	49	49	49	49	49	49	49	49	57	57	57	57	61	61	63	63	33	33	33	33	33	33	33	33	33	33	33	33	33	33	33	33	49	49	49	49	49	49	49	49	57	57	57	57	61	61	63	63
34	35	34	35	34	35	34	35	34	35	34	35	34	35	34	35	50	51	50	51	50	51	50	51	58	59	58	59	62	63	62	63	34	35	34	35	34	35	34	35	34	35	34	35	34	35	34	35	50	51	50	51	50	51	50	51	58	59	58	59	62	63	62	63
35	35	35	35	35	35	35	35	35	35	35	35	35	35	35	35	51	51	51	51	51	51	51	51	59	59	59	59	63	63	63	63	35	35	35	35	35	35	35	35	35	35	35	35	35	35	35	35	51	51	51	51	51	51	51	51	59	59	59	59	63	63	63	63
36	36	38	39	36	36	38	39	36	36	38	39	36	36	38	39	52	52	54	55	52	52	54	55	60	60	62	63	60	60	62	63	36	36	38	39	36	36	38	39	36	36	38	39	36	36	38	39	52	52	54	55	52	52	54	55	60	60	62	63	60	60	62	63
37	37	39	39	37	37	39	39	37	37	39	39	37	37	39	39	53	53	55	55	53	53	55	55	61	61	63	63	61	61	63	63	37	37	39	39	37	37	39	39	37	37	39	39	37	37	39	39	53	53	55	55	53	53	55	55	61	61	63	63	61	61	63	63
38	39	38	39	38	39	38	39	38	39	38	39	38	39	38	39	54	55	54	55	54	55	54	55	62	63	62	63	62	63	62	63	38	39	38	39	38	39	38	39	38	39	38	39	38	39	38	39	54	55	54	55	54	55	54	55	62	63	62	63	62	63	62	63
39	39	39	39	39	39	39	39	39	39	39	39	39	39	39	39	55	55	55	55	55	55	55	55	63	63	63	63	63	63	63	63	39	39	39	39	39	39	39	39	39	39	39	39	39	39	39	39	55	55	55	55	55	55	55	55	63	63	63	63	63	63	63	63
40	40	40	40	44	44	46	47	40	40	40	40	44	44	46	47	56	56	56	56	60	60	62	63	56	56	56	56	60	60	62	63	40	40	40	40	44	44	46	47	40	40	40	40	44	44	46	47	56	56	56	56	60	60	62	63	56	56	56	56	60	60	62	63
41	41	41	41	45	45	47	47	41	41	41	41	45	45	47	47	57	57	57	57	61	61	63	63	57	57	57	57	61	61	63	63	41	41	41	41	45	45	47	47	41	41	41	41	45	45	47	47	57	57	57	57	61	61	63	63	57	57	57	57	61	61	63	63
42	43	42	43	46	47	46	47	42	43	42	43	46	47	46	47	58	59	58	59	62	63	62	63	58	59	58	59	62	63	62	63	42	43	42	43	46	47	46	47	42	43	42	43	46	47	46	47	58	59	58	59	62	63	62	63	58	59	58	59	62	63	62	63
43	43	43	43	47	47	47	47	43	43	43	43	47	47	47	47	59	59	59	59	63	63	63	63	59	59	59	59	63	63	63	63	43	43	43	43	47	47	47	47	43	43	43	43	47	47	47	47	59	59	59	59	63	63	63	63	59	59	59	59	63	63	63	63
44	44	46	47	44	44	46	47	44	44	46	47	44	44	46	47	60	60	62	63	60	60	62	63	60	60	62	63	60	60	62	63	44	44	46	47	44	44	46	47	44	44	46	47	44	44	46	47	60	60	62	63	60	60	62	63	60	60	62	63	60	60	62	63
45	45	47	47	45	45	47	47	45	45	47	47	45	45	47	47	61	61	63	63	61	61	63	63	61	61	63	63	61	61	63	63	45	45	47	47	45	45	47	47	45	45	47	47	45	45	47	47	61	61	63	63	61	61	63	63	61	61	63	63	61	61	63	63
46	47	46	47	46	47	46	47	46	47	46	47	46	47	46	47	62	63	62	63	62	63	62	63	62	63	62	63	62	63	62	63	46	47	46	47	46	47	46	47	46	47	46	47	46	47	46	47	62	63	62	63	62	63	62	63	62	63	62	63	62	63	62	63
47	47	47	47	47	47	47	47	47	47	47	47	47	47	47	47	63	63	63	63	63	63	63	63	63	63	63	63	63	63	63	63	47	47	47	47	47	47	47	47	47	47	47	47	47	47	47	47	63	63	63	63	63	63	63	63	63	63	63	63	63	63	63	63
48	48	48	48	48	48	48	48	56	56	56	56	60	60	62	63	48	48	48	48	48	48	48	48	56	56	56	56	60	60	62	63	48	48	48	48	48	48	48	48	56	56	56	56	60	60	62	63	48	48	48	48	48	48	48	48	56	56	56	56	60	60	62	63
49	49	49	49	49	49	49	49	57	57	57	57	61	61	63	63	49	49	49	49	49	49	49	49	57	57	57	57	61	61	63	63	49	49	49	49	49	49	49	49	57	57	57	57	61	61	63	63	49	49	49	49	49	49	49	49	57	57	57	57	61	61	63	63
50	51	50	51	50	51	50	51	58	59	58	59	62	63	62	63	50	51	50	51	50	51	50	51	58	59	58	59	62	63	62	63	50	51	50	51	50	51	50	51	58	59	58	59	62	63	62	63	50	51	50	51	50	51	50	51	58	59	58	59	62	63	62	63
51	51	51	51	51	51	51	51	59	59	59	59	63	63	63	63	51	51	51	51	51	51	51	51	59	59	59	59	63	63	63	63	51	51	51	51	51	51	51	51	59	59	59	59	63	63	63	63	51	51	51	51	51	51	51	51	59	59	59	59	63	63	63	63
52	52	54	55	52	52	54	55	60	60	62	63	60	60	62	63	52	52	54	55	52	52	54	55	60	60	62	63	60	60	62	63	52	52	54	55	52	52	54	55	60	60	62	63	60	60	62	63	52	52	54	55	52	52	54	55	60	60	62	63	60	60	62	63
53	53	55	55	53	53	55	55	61	61	63	63	61	61	63	63	53	53	55	55	53	53	55	55	61	61	63	63	61	61	63	63	53	53	55	55	53	53	55	55	61	61	63	63	61	61	63	63	53	53	55	55	53	53	55	55	61	61	63	63	61	61	63	63
54	55	54	55	54	55	54	55	62	63	62	63	62	63	62	63	54	55	54	55	54	55	54	55	62	63	62	63	62	63	62	63	54	55	54	55	54	55	54	55	62	63	62	63	62	63	62	63	54	55	54	55	54	55	54	55	62	63	62	63	62	63	62	63
55	55	55	55	55	55	55	55	63	63	63	63	63	63	63	63	55	55	55	55	55	55	55	55	63	63	63	63	63	63	63	63	55	55	55	55	55	55	55	55	63	63	63	63	63	63	63	63	55	55	55	55	55	55	55	55	63	63	63	63	63	63	63	63
56	56	56	56	60	60	62	63	56	56	56	56	60	60	62	63	56	56	56	56	60	60	62	63	56	56	56	56	60	60	62	63	56	56	56	56	60	60	62	63	56	56	56	56	60	60	62	63	56	56	56	56	60	60	62	63	56	56	56	56	60	60	62	63
57	57	57	57	61	61	63	63	57	57	57	57	61	61	63	63	57	57	57	57	61	61	63	63	57	57	57	57	61	61	63	63	57	57	57	57	61	61	63	63	57	57	57	57	61	61	63	63	57	57	57	57	61	61	63	63	57	57	57	57	61	61	63	63
58	59	58	59	62	63	62	63	58	59	58	59	62	63	62	63	58	59	58	59	62	63	62	63	58	59	58	59	62	63	62	63	58	59	58	59	62	63	62	63	58	59	58	59	62	63	62	63	58	59	58	59	62	63	62	63	58	59	58	59	62	63	62	63
59	59	59	59	63	63	63	63	59	59	59	59	63	63	63	63	59	59	59	59	63	63	63	63	59	59	59	59	63	63	63	63	59	59	59	59	63	63	63	63	59	59	59	59	63	63	63	63	59	59	59	59	63	63	63	63	59	59	59	59	63	63	63	63
60	60	62	63	60	60	62	63	60	60	62	63	60	60	62	63	60	60	62	63	60	60	62	63	60	60	62	63	60	60	62	63	60	60	62	63	60	60	62	63	60	60	62	63	60	60	62	63	60	60	62	63	60	60	62	63	60	60	62	63	60	60	62	63
61	61	63	63	61	61	63	63	61	61	63	63	61	61	63	63	61	61	63	63	61	61	63	63	61	61	63	63	61	61	63	63	61	61	63	63	61	61	63	63	61	61	63	63	61	61	63	63	61	61	63	63	61	61	63	63	61	61	63	63	61	61	63	63
62	63	62	63	62	63	62	63	62	63	62	63	62	63	62	63	62	63	62	63	62	63	62	63	62	63	62	63	62	63	62	63	62	63	62	63	62	63	62	63	62	63	62	63	62	63	62	63	62	63	62	63	62	63	62	63	62	63	62	63	62	63	62	63
63	63	63	63	63	63	63	63	63	63	63	63	63	63	63	63	63	63	63	63	63	63	63	63	63	63	63	63	63	63	63	63	63	63	63	63	63	63	63	63	63	63	63	63	63	63	63	63	63	63	63	63	63	63	63	63	63	63	63	63	63	63	63	63

Fig. 13.32 Matrix M_{L_1} of composition $\mathbf{x} \circ \mathbf{y}$ implemented by L_1. Symmetric ($m_{ij} = m_{ji}$) entries are underlined.

0	1	2	2	4	5	4	5	8	9	10	10	8	9	10	10	16	17	18	18	20	21	20	21	16	17	18	18	20	21	20	21	32	33	34	34	36	37	36	37	40	41	42	42	40	41	42	42	32	33	34	34	36	37	36	37	40	41	42	42	40	41	42	42
1	1	3	3	5	5	5	5	9	9	11	11	9	9	11	11	17	17	19	19	21	21	21	21	17	17	19	19	21	21	21	21	1	1	3	3	5	5	5	5	9	9	11	11	9	9	11	11	17	17	19	19	21	21	21	21	17	17	19	19	21	21	21	21
2	2	2	2	6	6	6	6	10	10	10	10	10	10	10	10	18	18	18	18	22	22	22	22	18	18	18	18	22	22	22	22	34	34	34	34	38	38	38	38	42	42	42	42	42	42	42	42	34	34	34	34	38	38	38	38	42	42	42	42	42	42	42	42
3	3	3	3	7	7	7	7	11	11	11	11	11	11	11	11	19	19	19	19	23	23	23	23	19	19	19	19	23	23	23	23	3	3	3	3	7	7	7	7	11	11	11	11	11	11	11	11	19	19	19	19	23	23	23	23	19	19	19	19	23	23	23	23
4	5	4	5	4	5	4	5	12	13	12	13	12	13	12	13	20	21	20	21	20	21	20	21	20	21	20	21	20	21	20	21	36	37	36	37	36	37	36	37	44	45	44	45	44	45	44	45	36	37	36	37	36	37	36	37	44	45	44	45	44	45	44	45
5	5	5	5	5	5	5	5	13	13	13	13	13	13	13	13	21	21	21	21	21	21	21	21	21	21	21	21	21	21	21	21	5	5	5	5	5	5	5	5	13	13	13	13	13	13	13	13	21	21	21	21	21	21	21	21	21	21	21	21	21	21	21	21
6	6	6	6	6	6	6	6	14	14	14	14	14	14	14	14	22	22	22	22	22	22	22	22	22	22	22	22	22	22	22	22	38	38	38	38	38	38	38	38	46	46	46	46	46	46	46	46	38	38	38	38	38	38	38	38	46	46	46	46	46	46	46	46
7	7	7	7	7	7	7	7	15	15	15	15	15	15	15	15	23	23	23	23	23	23	23	23	23	23	23	23	23	23	23	23	7	7	7	7	7	7	7	7	15	15	15	15	15	15	15	15	23	23	23	23	23	23	23	23	23	23	23	23	23	23	23	23
8	9	10	10	8	9	10	10	8	9	10	10	8	9	10	10	24	25	26	26	24	25	26	26	24	25	26	26	24	25	26	26	40	41	42	42	40	41	42	42	40	41	42	42	40	41	42	42	40	41	42	42	40	41	42	42	40	41	42	42	40	41	42	42
9	9	11	11	9	9	11	11	9	9	11	11	9	9	11	11	25	25	27	27	25	25	27	27	25	25	27	27	25	25	27	27	9	9	11	11	9	9	11	11	9	9	11	11	9	9	11	11	25	25	27	27	25	25	27	27	25	25	27	27	25	25	27	27
10	10	10	10	10	10	10	10	10	10	10	10	10	10	10	10	26	26	26	26	26	26	26	26	26	26	26	26	26	26	26	26	42	42	42	42	42	42	42	42	42	42	42	42	42	42	42	42	42	42	42	42	42	42	42	42	42	42	42	42	42	42	42	42
11	11	11	11	11	11	11	11	11	11	11	11	11	11	11	11	27	27	27	27	27	27	27	27	27	27	27	27	27	27	27	27	11	11	11	11	11	11	11	11	11	11	11	11	11	11	11	11	27	27	27	27	27	27	27	27	27	27	27	27	27	27	27	27
12	13	12	13	12	13	12	13	12	13	12	13	12	13	12	13	28	29	28	29	28	29	28	29	28	29	28	29	28	29	28	29	44	45	44	45	44	45	44	45	44	45	44	45	44	45	44	45	44	45	44	45	44	45	44	45	44	45	44	45	44	45	44	45
13	13	13	13	13	13	13	13	13	13	13	13	13	13	13	13	29	29	29	29	29	29	29	29	29	29	29	29	29	29	29	29	13	13	13	13	13	13	13	13	13	13	13	13	13	13	13	13	29	29	29	29	29	29	29	29	29	29	29	29	29	29	29	29
14	14	14	14	14	14	14	14	14	14	14	14	14	14	14	14	30	30	30	30	30	30	30	30	30	30	30	30	30	30	30	30	46	46	46	46	46	46	46	46	46	46	46	46	46	46	46	46	46	46	46	46	46	46	46	46	46	46	46	46	46	46	46	46
15	15	15	15	15	15	15	15	15	15	15	15	15	15	15	15	31	31	31	31	31	31	31	31	31	31	31	31	31	31	31	31	15	15	15	15	15	15	15	15	15	15	15	15	15	15	15	15	31	31	31	31	31	31	31	31	31	31	31	31	31	31	31	31
16	17	18	18	20	21	20	21	16	17	18	18	20	21	20	21	16	17	18	18	20	21	20	21	16	17	18	18	20	21	20	21	48	49	50	50	52	53	52	53	48	49	50	50	52	53	52	53	48	49	50	50	52	53	52	53	48	49	50	50	52	53	52	53
17	17	19	19	21	21	21	21	17	17	19	19	21	21	21	21	17	17	19	19	21	21	21	21	17	17	19	19	21	21	21	21	17	17	19	19	21	21	21	21	17	17	19	19	21	21	21	21	17	17	19	19	21	21	21	21	17	17	19	19	21	21	21	21
18	18	18	18	22	22	22	22	18	18	18	18	22	22	22	22	18	18	18	18	22	22	22	22	18	18	18	18	22	22	22	22	50	50	50	50	54	54	54	54	50	50	50	50	54	54	54	54	50	50	50	50	54	54	54	54	50	50	50	50	54	54	54	54
19	19	19	19	23	23	23	23	19	19	19	19	23	23	23	23	19	19	19	19	23	23	23	23	19	19	19	19	23	23	23	23	19	19	19	19	23	23	23	23	19	19	19	19	23	23	23	23	19	19	19	19	23	23	23	23	19	19	19	19	23	23	23	23
20	21	20	21	20	21	20	21	20	21	20	21	20	21	20	21	20	21	20	21	20	21	20	21	20	21	20	21	20	21	20	21	52	53	52	53	52	53	52	53	52	53	52	53	52	53	52	53	52	53	52	53	52	53	52	53	52	53	52	53	52	53	52	53
21	21	21	21	21	21	21	21	21	21	21	21	21	21	21	21	21	21	21	21	21	21	21	21	21	21	21	21	21	21	21	21	21	21	21	21	21	21	21	21	21	21	21	21	21	21	21	21	21	21	21	21	21	21	21	21	21	21	21	21	21	21	21	21
22	22	22	22	22	22	22	22	22	22	22	22	22	22	22	22	22	22	22	22	22	22	22	22	22	22	22	22	22	22	22	22	54	54	54	54	54	54	54	54	54	54	54	54	54	54	54	54	54	54	54	54	54	54	54	54	54	54	54	54	54	54	54	54
23	23	23	23	23	23	23	23	23	23	23	23	23	23	23	23	23	23	23	23	23	23	23	23	23	23	23	23	23	23	23	23	23	23	23	23	23	23	23	23	23	23	23	23	23	23	23	23	23	23	23	23	23	23	23	23	23	23	23	23	23	23	23	23
24	25	26	26	24	25	26	26	24	25	26	26	24	25	26	26	24	25	26	26	24	25	26	26	24	25	26	26	24	25	26	26	56	57	58	58	56	57	58	58	56	57	58	58	56	57	58	58	56	57	58	58	56	57	58	58	56	57	58	58	56	57	58	58
25	25	27	27	25	25	27	27	25	25	27	27	25	25	27	27	25	25	27	27	25	25	27	27	25	25	27	27	25	25	27	27	25	25	27	27	25	25	27	27	25	25	27	27	25	25	27	27	25	25	27	27	25	25	27	27	25	25	27	27	25	25	27	27
26	26	26	26	26	26	26	26	26	26	26	26	26	26	26	26	26	26	26	26	26	26	26	26	26	26	26	26	26	26	26	26	58	58	58	58	58	58	58	58	58	58	58	58	58	58	58	58	58	58	58	58	58	58	58	58	58	58	58	58	58	58	58	58
27	27	27	27	27	27	27	27	27	27	27	27	27	27	27	27	27	27	27	27	27	27	27	27	27	27	27	27	27	27	27	27	27	27	27	27	27	27	27	27	27	27	27	27	27	27	27	27	27	27	27	27	27	27	27	27	27	27	27	27	27	27	27	27
28	29	28	29	28	29	28	29	28	29	28	29	28	29	28	29	28	29	28	29	28	29	28	29	28	29	28	29	28	29	28	29	60	61	60	61	60	61	60	61	60	61	60	61	60	61	60	61	60	61	60	61	60	61	60	61	60	61	60	61	60	61	60	61
29	29	29	29	29	29	29	29	29	29	29	29	29	29	29	29	29	29	29	29	29	29	29	29	29	29	29	29	29	29	29	29	29	29	29	29	29	29	29	29	29	29	29	29	29	29	29	29	29	29	29	29	29	29	29	29	29	29	29	29	29	29	29	29
30	30	30	30	30	30	30	30	30	30	30	30	30	30	30	30	30	30	30	30	30	30	30	30	30	30	30	30	30	30	30	30	62	62	62	62	62	62	62	62	62	62	62	62	62	62	62	62	62	62	62	62	62	62	62	62	62	62	62	62	62	62	62	62
31	31	31	31	31	31	31	31	31	31	31	31	31	31	31	31	31	31	31	31	31	31	31	31	31	31	31	31	31	31	31	31	31	31	31	31	31	31	31	31	31	31	31	31	31	31	31	31	31	31	31	31	31	31	31	31	31	31	31	31	31	31	31	31
32	33	34	34	36	37	36	37	40	41	42	42	40	41	42	42	32	33	34	34	36	37	36	37	40	41	42	42	40	41	42	42	32	33	34	34	36	37	36	37	40	41	42	42	40	41	42	42	32	33	34	34	36	37	36	37	40	41	42	42	40	41	42	42
33	33	35	35	37	37	37	37	41	41	43	43	41	41	43	43	33	33	35	35	37	37	37	37	41	41	43	43	41	41	43	43	33	33	35	35	37	37	37	37	41	41	43	43	41	41	43	43	33	33	35	35	37	37	37	37	41	41	43	43	41	41	43	43
34	34	34	34	38	38	38	38	42	42	42	42	42	42	42	42	34	34	34	34	38	38	38	38	42	42	42	42	42	42	42	42	34	34	34	34	38	38	38	38	42	42	42	42	42	42	42	42	34	34	34	34	38	38	38	38	42	42	42	42	42	42	42	42
35	35	35	35	39	39	39	39	43	43	43	43	43	43	43	43	35	35	35	35	39	39	39	39	43	43	43	43	43	43	43	43	35	35	35	35	39	39	39	39	43	43	43	43	43	43	43	43	35	35	35	35	39	39	39	39	43	43	43	43	43	43	43	43
36	37	36	37	36	37	36	37	44	45	44	45	44	45	44	45	36	37	36	37	36	37	36	37	44	45	44	45	44	45	44	45	36	37	36	37	36	37	36	37	44	45	44	45	44	45	44	45	36	37	36	37	36	37	36	37	44	45	44	45	44	45	44	45
37	37	37	37	37	37	37	37	45	45	45	45	45	45	45	45	37	37	37	37	37	37	37	37	45	45	45	45	45	45	45	45	37	37	37	37	37	37	37	37	45	45	45	45	45	45	45	45	37	37	37	37	37	37	37	37	45	45	45	45	45	45	45	45
38	38	38	38	38	38	38	38	46	46	46	46	46	46	46	46	38	38	38	38	38	38	38	38	46	46	46	46	46	46	46	46	38	38	38	38	38	38	38	38	46	46	46	46	46	46	46	46	38	38	38	38	38	38	38	38	46	46	46	46	46	46	46	46
39	39	39	39	39	39	39	39	47	47	47	47	47	47	47	47	39	39	39	39	39	39	39	39	47	47	47	47	47	47	47	47	39	39	39	39	39	39	39	39	47	47	47	47	47	47	47	47	39	39	39	39	39	39	39	39	47	47	47	47	47	47	47	47
40	41	42	42	40	41	42	42	40	41	42	42	40	41	42	42	40	41	42	42	40	41	42	42	40	41	42	42	40	41	42	42	40	41	42	42	40	41	42	42	40	41	42	42	40	41	42	42	40	41	42	42	40	41	42	42	40	41	42	42	40	41	42	42
41	41	43	43	41	41	43	43	41	41	43	43	41	41	43	43	41	41	43	43	41	41	43	43	41	41	43	43	41	41	43	43	41	41	43	43	41	41	43	43	41	41	43	43	41	41	43	43	41	41	43	43	41	41	43	43	41	41	43	43	41	41	43	43
42	42	42	42	42	42	42	42	42	42	42	42	42	42	42	42	42	42	42	42	42	42	42	42	42	42	42	42	42	42	42	42	42	42	42	42	42	42	42	42	42	42	42	42	42	42	42	42	42	42	42	42	42	42	42	42	42	42	42	42	42	42	42	42
43	43	43	43	43	43	43	43	43	43	43	43	43	43	43	43	43	43	43	43	43	43	43	43	43	43	43	43	43	43	43	43	43	43	43	43	43	43	43	43	43	43	43	43	43	43	43	43	43	43	43	43	43	43	43	43	43	43	43	43	43	43	43	43
44	45	44	45	44	45	44	45	44	45	44	45	44	45	44	45	44	45	44	45	44	45	44	45	44	45	44	45	44	45	44	45	44	45	44	45	44	45	44	45	44	45	44	45	44	45	44	45	44	45	44	45	44	45	44	45	44	45	44	45	44	45	44	45
45	45	45	45	45	45	45	45	45	45	45	45	45	45	45	45	45	45	45	45	45	45	45	45	45	45	45	45	45	45	45	45	45	45	45	45	45	45	45	45	45	45	45	45	45	45	45	45	45	45	45	45	45	45	45	45	45	45	45	45	45	45	45	45
46	46	46	46	46	46	46	46	46	46	46	46	46	46	46	46	46	46	46	46	46	46	46	46	46	46	46	46	46	46	46	46	46	46	46	46	46	46	46	46	46	46	46	46	46	46	46	46	46	46	46	46	46	46	46	46	46	46	46	46	46	46	46	46
47	47	47	47	47	47	47	47	47	47	47	47	47	47	47	47	47	47	47	47	47	47	47	47	47	47	47	47	47	47	47	47	47	47	47	47	47	47	47	47	47	47	47	47	47	47	47	47	47	47	47	47	47	47	47	47	47	47	47	47	47	47	47	47
48	49	50	50	52	53	52	53	48	49	50	50	52	53	52	53	48	49	50	50	52	53	52	53	48	49	50	50	52	53	52	53	48	49	50	50	52	53	52	53	48	49	50	50	52	53	52	53	48	49	50	50	52	53	52	53	48	49	50	50	52	53	52	53
49	49	51	51	53	53	53	53	49	49	51	51	53	53	53	53	49	49	51	51	53	53	53	53	49	49	51	51	53	53	53	53	49	49	51	51	53	53	53	53	49	49	51	51	53	53	53	53	49	49	51	51	53	53	53	53	49	49	51	51	53	53	53	53
50	50	50	50	54	54	54	54	50	50	50	50	54	54	54	54	50	50	50	50	54	54	54	54	50	50	50	50	54	54	54	54	50	50	50	50	54	54	54	54	50	50	50	50	54	54	54	54	50	50	50	50	54	54	54	54	50	50	50	50	54	54	54	54
51	51	51	51	55	55	55	55	51	51	51	51	55	55	55	55	51	51	51	51	55	55	55	55	51	51	51	51	55	55	55	55	51	51	51	51	55	55	55	55	51	51	51	51	55	55	55	55	51	51	51	51	55	55	55	55	51	51	51	51	55	55	55	55
52	53	52	53	52	53	52	53	52	53	52	53	52	53	52	53	52	53	52	53	52	53	52	53	52	53	52	53	52	53	52	53	52	53	52	53	52	53	52	53	52	53	52	53	52	53	52	53	52	53	52	53	52	53	52	53	52	53	52	53	52	53	52	53
53	53	53	53	53	53	53	53	53	53	53	53	53	53	53	53	53	53	53	53	53	53	53	53	53	53	53	53	53	53	53	53	53	53	53	53	53	53	53	53	53	53	53	53	53	53	53	53	53	53	53	53	53	53	53	53	53	53	53	53	53	53	53	53
54	54	54	54	54	54	54	54	54	54	54	54	54	54	54	54	54	54	54	54	54	54	54	54	54	54	54	54	54	54	54	54	54	54	54	54	54	54	54	54	54	54	54	54	54	54	54	54	54	54	54	54	54	54	54	54	54	54	54	54	54	54	54	54
55	55	55	55	55	55	55	55	55	55	55	55	55	55	55	55	55	55	55	55	55	55	55	55	55	55	55	55	55	55	55	55	55	55	55	55	55	55	55	55	55	55	55	55	55	55	55	55	55	55	55	55	55	55	55	55	55	55	55	55	55	55	55	55
56	57	58	58	56	57	58	58	56	57	58	58	56	57	58	58	56	57	58	58	56	57	58	58	56	57	58	58	56	57	58	58	56	57	58	58	56	57	58	58	56	57	58	58	56	57	58	58	56	57	58	58	56	57	58	58	56	57	58	58	56	57	58	58
57	57	59	59	57	57	59	59	57	57	59	59	57	57	59	59	57	57	59	59	57	57	59	59	57	57	59	59	57	57	59	59	57	57	59	59	57	57	59	59	57	57	59	59	57	57	59	59	57	57	59	59	57	57	59	59	57	57	59	59	57	57	59	59
58	58	58	58	58	58	58	58	58	58	58	58	58	58	58	58	58	58	58	58	58	58	58	58	58	58	58	58	58	58	58	58	58	58	58	58	58	58	58	58	58	58	58	58	58	58	58	58	58	58	58	58	58	58	58	58	58	58	58	58	58	58	58	58
59	59	59	59	59	59	59	59	59	59	59	59	59	59	59	59	59	59	59	59	59	59	59	59	59	59	59	59	59	59	59	59	59	59	59	59	59	59	59	59	59	59	59	59	59	59	59	59	59	59	59	59	59	59	59	59	59	59	59	59	59	59	59	59
60	61	60	61	60	61	60	61	60	61	60	61	60	61	60	61	60	61	60	61	60	61	60	61	60	61	60	61	60	61	60	61	60	61	60	61	60	61	60	61	60	61	60	61	60	61	60	61	60	61	60	61	60	61	60	61	60	61	60	61	60	61	60	61
61	61	61	61	61	61	61	61	61	61	61	61	61	61	61	61	61	61	61	61	61	61	61	61	61	61	61	61	61	61	61	61	61	61	61	61	61	61	61	61	61	61	61	61	61	61	61	61	61	61	61	61	61	61	61	61	61	61	61	61	61	61	61	61
62	62	62	62	62	62	62	62	62	62	62	62	62	62	62	62	62	62	62	62	62	62	62	62	62	62	62	62	62	62	62	62	62	62	62	62	62	62	62	62	62	62	62	62	62	62	62	62	62	62	62	62	62	62	62	62	62	62	62	62	62	62	62	62
63	63	63	63	63	63	63	63	63	63	63	63	63	63	63	63	63	63	63	63	63	63	63	63	63	63	63	63	63	63	63	63	63	63	63	63	63	63	63	63	63	63	63	63	63	63	63	63	63	63	63	63	63	63	63	63	63	63	63	63	63	63	63	63

Fig. 13.33 Matrix M_{L_2} of composition $\mathbf{x} \circ \mathbf{y}$ implemented by L_2. Symmetric entries are underlined.

0	2	4	6	8	10	12	14	16	18	20	22	24	26	28	30	32	34	36	38	40	42	44	46	48	50	52	54	56	58	60	62	1	3	5	7	9	11	13	15	17	19	21	23	25	27	29	31	33	35	37	39	41	43	45	47	49	51	53	55	57	59	61	63
1	3	5	7	9	11	13	15	17	19	21	23	25	27	29	31	33	35	37	39	41	43	45	47	49	51	53	55	57	59	61	63	33	35	37	39	41	43	45	47	49	51	53	55	57	59	61	63	49	51	53	55	57	59	61	63	57	59	61	63	61	63	63	63
2	3	6	7	10	11	14	15	18	19	22	23	26	27	30	31	34	35	38	39	42	43	46	47	50	51	54	55	58	59	62	63	3	3	7	7	11	11	15	15	19	19	23	23	27	27	31	31	35	35	39	39	43	43	47	47	51	51	55	55	59	59	63	63
3	3	7	7	11	11	15	15	19	19	23	23	27	27	31	31	35	35	39	39	43	43	47	47	51	51	55	55	59	59	63	63	35	35	39	39	43	43	47	47	51	51	55	55	59	59	63	63	51	51	55	55	59	59	63	63	59	59	63	63	63	63	63	63
4	6	6	7	12	14	14	15	20	22	22	23	28	30	30	31	36	38	38	39	44	46	46	47	52	54	54	55	60	62	62	63	5	7	7	7	13	15	15	15	21	23	23	23	29	31	31	31	37	39	39	39	45	47	47	47	53	55	55	55	61	63	63	63
5	7	7	7	13	15	15	15	21	23	23	23	29	31	31	31	37	39	39	39	45	47	47	47	53	55	55	55	61	63	63	63	37	39	39	39	45	47	47	47	53	55	55	55	61	63	63	63	53	55	55	55	61	63	63	63	61	63	63	63	61	63	63	63
6	7	6	7	14	15	14	15	22	23	22	23	30	31	30	31	38	39	38	39	46	47	46	47	54	55	54	55	62	63	62	63	7	7	7	7	15	15	15	15	23	23	23	23	31	31	31	31	39	39	39	39	47	47	47	47	55	55	55	55	63	63	63	63
7	7	7	7	15	15	15	15	23	23	23	23	31	31	31	31	39	39	39	39	47	47	47	47	55	55	55	55	63	63	63	63	39	39	39	39	47	47	47	47	55	55	55	55	63	63	63	63	55	55	55	55	63	63	63	63	63	63	63	63	63	63	63	63
8	10	12	14	12	14	14	15	24	26	28	30	28	30	30	31	40	42	44	46	44	46	46	47	56	58	60	62	60	62	62	63	9	11	13	15	13	15	15	15	25	27	29	31	29	31	31	31	41	43	45	47	45	47	47	47	57	59	61	63	61	63	63	63
9	11	13	15	13	15	15	15	25	27	29	31	29	31	31	31	41	43	45	47	45	47	47	47	57	59	61	63	61	63	63	63	41	43	45	47	45	47	47	47	57	59	61	63	61	63	63	63	57	59	61	63	61	63	63	63	57	59	61	63	61	63	63	63
10	11	14	15	14	15	14	15	26	27	30	31	30	31	30	31	42	43	46	47	46	47	46	47	58	59	62	63	62	63	62	63	11	11	15	15	15	15	15	15	27	27	31	31	31	31	31	31	43	43	47	47	47	47	47	47	59	59	63	63	63	63	63	63
11	11	15	15	15	15	15	15	27	27	31	31	31	31	31	31	43	43	47	47	47	47	47	47	59	59	63	63	63	63	63	63	43	43	47	47	47	47	47	47	59	59	63	63	63	63	63	63	59	59	63	63	63	63	63	63	59	59	63	63	63	63	63	63
12	14	14	15	12	14	14	15	28	30	30	31	28	30	30	31	44	46	46	47	44	46	46	47	60	62	62	63	60	62	62	63	13	15	15	15	13	15	15	15	29	31	31	31	29	31	31	31	45	47	47	47	45	47	47	47	61	63	63	63	61	63	63	63
13	15	15	15	13	15	15	15	29	31	31	31	29	31	31	31	45	47	47	47	45	47	47	47	61	63	63	63	61	63	63	63	45	47	47	47	45	47	47	47	61	63	63	63	61	63	63	63	61	63	63	63	61	63	63	63	61	63	63	63	61	63	63	63
14	15	14	15	14	15	14	15	30	31	30	31	30	31	30	31	46	47	46	47	46	47	46	47	62	63	62	63	62	63	62	63	15	15	15	15	15	15	15	15	31	31	31	31	31	31	31	31	47	47	47	47	47	47	47	47	63	63	63	63	63	63	63	63
15	15	15	15	15	15	15	15	31	31	31	31	31	31	31	31	47	47	47	47	47	47	47	47	63	63	63	63	63	63	63	63	47	47	47	47	47	47	47	47	63	63	63	63	63	63	63	63	63	63	63	63	63	63	63	63	63	63	63	63	63	63	63	63
16	18	20	22	24	26	28	30	24	26	28	30	28	30	30	31	48	50	52	54	56	58	60	62	56	58	60	62	60	62	62	63	17	19	21	23	25	27	29	31	25	27	29	31	29	31	31	31	49	51	53	55	57	59	61	63	57	59	61	63	61	63	63	63
17	19	21	23	25	27	29	31	25	27	29	31	29	31	31	31	49	51	53	55	57	59	61	63	57	59	61	63	61	63	63	63	49	51	53	55	57	59	61	63	57	59	61	63	61	63	63	63	49	51	53	55	57	59	61	63	57	59	61	63	61	63	63	63
18	19	22	23	26	27	30	31	26	27	30	31	30	31	30	31	50	51	54	55	58	59	62	63	58	59	62	63	62	63	62	63	19	19	23	23	27	27	31	31	27	27	31	31	31	31	31	31	51	51	55	55	59	59	63	63	59	59	63	63	63	63	63	63
19	19	23	23	27	27	31	31	27	27	31	31	31	31	31	31	51	51	55	55	59	59	63	63	59	59	63	63	63	63	63	63	51	51	55	55	59	59	63	63	59	59	63	63	63	63	63	63	51	51	55	55	59	59	63	63	59	59	63	63	63	63	63	63
20	22	22	23	28	30	30	31	28	30	30	31	28	30	30	31	52	54	54	55	60	62	62	63	60	62	62	63	60	62	62	63	21	23	23	23	29	31	31	31	29	31	31	31	29	31	31	31	53	55	55	55	61	63	63	63	61	63	63	63	61	63	63	63
21	23	23	23	29	31	31	31	29	31	31	31	29	31	31	31	53	55	55	55	61	63	63	63	61	63	63	63	61	63	63	63	53	55	55	55	61	63	63	63	61	63	63	63	61	63	63	63	53	55	55	55	61	63	63	63	61	63	63	63	61	63	63	63
22	23	22	23	30	31	30	31	30	31	30	31	30	31	30	31	54	55	54	55	62	63	62	63	62	63	62	63	62	63	62	63	23	23	23	23	31	31	31	31	31	31	31	31	31	31	31	31	55	55	55	55	63	63	63	63	63	63	63	63	63	63	63	63
23	23	23	23	31	31	31	31	31	31	31	31	31	31	31	31	55	55	55	55	63	63	63	63	63	63	63	63	63	63	63	63	55	55	55	55	63	63	63	63	63	63	63	63	63	63	63	63	55	55	55	55	63	63	63	63	63	63	63	63	63	63	63	63
24	26	28	30	28	30	30	31	24	26	28	30	28	30	30	31	56	58	60	62	60	62	62	63	56	58	60	62	60	62	62	63	25	27	29	31	29	31	31	31	25	27	29	31	29	31	31	31	57	59	61	63	61	63	63	63	57	59	61	63	61	63	63	63
25	27	29	31	29	31	31	31	25	27	29	31	29	31	31	31	57	59	61	63	61	63	63	63	57	59	61	63	61	63	63	63	57	59	61	63	61	63	63	63	57	59	61	63	61	63	63	63	57	59	61	63	61	63	63	63	57	59	61	63	61	63	63	63
26	27	30	31	30	31	30	31	26	27	30	31	30	31	30	31	58	59	62	63	62	63	62	63	58	59	62	63	62	63	62	63	27	27	31	31	31	31	31	31	27	27	31	31	31	31	31	31	59	59	63	63	63	63	63	63	59	59	63	63	63	63	63	63
27	27	31	31	31	31	31	31	27	27	31	31	31	31	31	31	59	59	63	63	63	63	63	63	59	59	63	63	63	63	63	63	59	59	63	63	63	63	63	63	59	59	63	63	63	63	63	63	59	59	63	63	63	63	63	63	59	59	63	63	63	63	63	63
28	30	30	31	28	30	30	31	28	30	30	31	28	30	30	31	60	62	62	63	60	62	62	63	60	62	62	63	60	62	62	63	29	31	31	31	29	31	31	31	29	31	31	31	29	31	31	31	61	63	63	63	61	63	63	63	61	63	63	63	61	63	63	63
29	31	31	31	29	31	31	31	29	31	31	31	29	31	31	31	61	63	63	63	61	63	63	63	61	63	63	63	61	63	63	63	61	63	63	63	61	63	63	63	61	63	63	63	61	63	63	63	61	63	63	63	61	63	63	63	61	63	63	63	61	63	63	63
30	31	30	31	30	31	30	31	30	31	30	31	30	31	30	31	62	63	62	63	62	63	62	63	62	63	62	63	62	63	62	63	31	31	31	31	31	31	31	31	31	31	31	31	31	31	31	31	63	63	63	63	63	63	63	63	63	63	63	63	63	63	63	63
31	31	31	31	31	31	31	31	31	31	31	31	31	31	31	31	63	63	63	63	63	63	63	63	63	63	63	63	63	63	63	63	63	63	63	63	63	63	63	63	63	63	63	63	63	63	63	63	63	63	63	63	63	63	63	63	63	63	63	63	63	63	63	63
32	34	36	38	40	42	44	46	48	50	52	54	56	58	60	62	48	50	52	54	56	58	60	62	56	58	60	62	60	62	62	63	33	35	37	39	41	43	45	47	49	51	53	55	57	59	61	63	49	51	53	55	57	59	61	63	57	59	61	63	61	63	63	63
33	35	37	39	41	43	45	47	49	51	53	55	57	59	61	63	49	51	53	55	57	59	61	63	57	59	61	63	61	63	63	63	33	35	37	39	41	43	45	47	49	51	53	55	57	59	61	63	49	51	53	55	57	59	61	63	57	59	61	63	61	63	63	63
34	35	38	39	42	43	46	47	50	51	54	55	58	59	62	63	50	51	54	55	58	59	62	63	58	59	62	63	62	63	62	63	35	35	39	39	43	43	47	47	51	51	55	55	59	59	63	63	51	51	55	55	59	59	63	63	59	59	63	63	63	63	63	63
35	35	39	39	43	43	47	47	51	51	55	55	59	59	63	63	51	51	55	55	59	59	63	63	59	59	63	63	63	63	63	63	35	35	39	39	43	43	47	47	51	51	55	55	59	59	63	63	51	51	55	55	59	59	63	63	59	59	63	63	63	63	63	63
36	38	38	39	44	46	46	47	52	54	54	55	60	62	62	63	52	54	54	55	60	62	62	63	60	62	62	63	60	62	62	63	37	39	39	39	45	47	47	47	53	55	55	55	61	63	63	63	53	55	55	55	61	63	63	63	61	63	63	63	61	63	63	63
37	39	39	39	45	47	47	47	53	55	55	55	61	63	63	63	53	55	55	55	61	63	63	63	61	63	63	63	61	63	63	63	37	39	39	39	45	47	47	47	53	55	55	55	61	63	63	63	53	55	55	55	61	63	63	63	61	63	63	63	61	63	63	63
38	39	38	39	46	47	46	47	54	55	54	55	62	63	62	63	54	55	54	55	62	63	62	63	62	63	62	63	62	63	62	63	39	39	39	39	47	47	47	47	55	55	55	55	63	63	63	63	55	55	55	55	63	63	63	63	63	63	63	63	63	63	63	63
39	39	39	39	47	47	47	47	55	55	55	55	63	63	63	63	55	55	55	55	63	63	63	63	63	63	63	63	63	63	63	63	39	39	39	39	47	47	47	47	55	55	55	55	63	63	63	63	55	55	55	55	63	63	63	63	63	63	63	63	63	63	63	63
40	42	44	46	44	46	46	47	56	58	60	62	60	62	62	63	56	58	60	62	60	62	62	63	56	58	60	62	60	62	62	63	41	43	45	47	45	47	47	47	57	59	61	63	61	63	63	63	57	59	61	63	61	63	63	63	57	59	61	63	61	63	63	63
41	43	45	47	45	47	47	47	57	59	61	63	61	63	63	63	57	59	61	63	61	63	63	63	57	59	61	63	61	63	63	63	41	43	45	47	45	47	47	47	57	59	61	63	61	63	63	63	57	59	61	63	61	63	63	63	57	59	61	63	61	63	63	63
42	43	46	47	46	47	46	47	58	59	62	63	62	63	62	63	58	59	62	63	62	63	62	63	58	59	62	63	62	63	62	63	43	43	47	47	47	47	47	47	59	59	63	63	63	63	63	63	59	59	63	63	63	63	63	63	59	59	63	63	63	63	63	63
43	43	47	47	47	47	47	47	59	59	63	63	63	63	63	63	59	59	63	63	63	63	63	63	59	59	63	63	63	63	63	63	43	43	47	47	47	47	47	47	59	59	63	63	63	63	63	63	59	59	63	63	63	63	63	63	59	59	63	63	63	63	63	63
44	46	46	47	44	46	46	47	60	62	62	63	60	62	62	63	60	62	62	63	60	62	62	63	60	62	62	63	60	62	62	63	45	47	47	47	45	47	47	47	61	63	63	63	61	63	63	63	61	63	63	63	61	63	63	63	61	63	63	63	61	63	63	63
45	47	47	47	45	47	47	47	61	63	63	63	61	63	63	63	61	63	63	63	61	63	63	63	61	63	63	63	61	63	63	63	45	47	47	47	45	47	47	47	61	63	63	63	61	63	63	63	61	63	63	63	61	63	63	63	61	63	63	63	61	63	63	63
46	47	46	47	46	47	46	47	62	63	62	63	62	63	62	63	62	63	62	63	62	63	62	63	62	63	62	63	62	63	62	63	47	47	47	47	47	47	47	47	63	63	63	63	63	63	63	63	63	63	63	63	63	63	63	63	63	63	63	63	63	63	63	63
47	47	47	47	47	47	47	47	63	63	63	63	63	63	63	63	63	63	63	63	63	63	63	63	63	63	63	63	63	63	63	63	47	47	47	47	47	47	47	47	63	63	63	63	63	63	63	63	63	63	63	63	63	63	63	63	63	63	63	63	63	63	63	63
48	50	52	54	56	58	60	62	56	58	60	62	60	62	62	63	48	50	52	54	56	58	60	62	56	58	60	62	60	62	62	63	49	51	53	55	57	59	61	63	57	59	61	63	61	63	63	63	49	51	53	55	57	59	61	63	57	59	61	63	61	63	63	63
49	51	53	55	57	59	61	63	57	59	61	63	61	63	63	63	49	51	53	55	57	59	61	63	57	59	61	63	61	63	63	63	49	51	53	55	57	59	61	63	57	59	61	63	61	63	63	63	49	51	53	55	57	59	61	63	57	59	61	63	61	63	63	63
50	51	54	55	58	59	62	63	58	59	62	63	62	63	62	63	50	51	54	55	58	59	62	63	58	59	62	63	62	63	62	63	51	51	55	55	59	59	63	63	59	59	63	63	63	63	63	63	51	51	55	55	59	59	63	63	59	59	63	63	63	63	63	63
51	51	55	55	59	59	63	63	59	59	63	63	63	63	63	63	51	51	55	55	59	59	63	63	59	59	63	63	63	63	63	63	51	51	55	55	59	59	63	63	59	59	63	63	63	63	63	63	51	51	55	55	59	59	63	63	59	59	63	63	63	63	63	63
52	54	54	55	60	62	62	63	60	62	62	63	60	62	62	63	52	54	54	55	60	62	62	63	60	62	62	63	60	62	62	63	53	55	55	55	61	63	63	63	61	63	63	63	61	63	63	63	53	55	55	55	61	63	63	63	61	63	63	63	61	63	63	63
53	55	55	55	61	63	63	63	61	63	63	63	61	63	63	63	53	55	55	55	61	63	63	63	61	63	63	63	61	63	63	63	53	55	55	55	61	63	63	63	61	63	63	63	61	63	63	63	53	55	55	55	61	63	63	63	61	63	63	63	61	63	63	63
54	55	54	55	62	63	62	63	62	63	62	63	62	63	62	63	54	55	54	55	62	63	62	63	62	63	62	63	62	63	62	63	55	55	55	55	63	63	63	63	63	63	63	63	63	63	63	63	55	55	55	55	63	63	63	63	63	63	63	63	63	63	63	63
55	55	55	55	63	63	63	63	63	63	63	63	63	63	63	63	55	55	55	55	63	63	63	63	63	63	63	63	63	63	63	63	55	55	55	55	63	63	63	63	63	63	63	63	63	63	63	63	55	55	55	55	63	63	63	63	63	63	63	63	63	63	63	63
56	58	60	62	60	62	62	63	56	58	60	62	60	62	62	63	56	58	60	62	60	62	62	63	56	58	60	62	60	62	62	63	57	59	61	63	61	63	63	63	57	59	61	63	61	63	63	63	57	59	61	63	61	63	63	63	57	59	61	63	61	63	63	63
57	59	61	63	61	63	63	63	57	59	61	63	61	63	63	63	57	59	61	63	61	63	63	63	57	59	61	63	61	63	63	63	57	59	61	63	61	63	63	63	57	59	61	63	61	63	63	63	57	59	61	63	61	63	63	63	57	59	61	63	61	63	63	63
58	59	62	63	62	63	62	63	58	59	62	63	62	63	62	63	58	59	62	63	62	63	62	63	58	59	62	63	62	63	62	63	59	59	63	63	63	63	63	63	59	59	63	63	63	63	63	63	59	59	63	63	63	63	63	63	59	59	63	63	63	63	63	63
59	59	63	63	63	63	63	63	59	59	63	63	63	63	63	63	59	59	63	63	63	63	63	63	59	59	63	63	63	63	63	63	59	59	63	63	63	63	63	63	59	59	63	63	63	63	63	63	59	59	63	63	63	63	63	63	59	59	63	63	63	63	63	63
60	62	62	63	60	62	62	63	60	62	62	63	60	62	62	63	60	62	62	63	60	62	62	63	60	62	62	63	60	62	62	63	61	63	63	63	61	63	63	63	61	63	63	63	61	63	63	63	61	63	63	63	61	63	63	63	61	63	63	63	61	63	63	63
61	63	63	63	61	63	63	63	61	63	63	63	61	63	63	63	61	63	63	63	61	63	63	63	61	63	63	63	61	63	63	63	61	63	63	63	61	63	63	63	61	63	63	63	61	63	63	63	61	63	63	63	61	63	63	63	61	63	63	63	61	63	63	63
62	63	62	63	62	63	62	63	62	63	62	63	62	63	62	63	62	63	62	63	62	63	62	63	62	63	62	63	62	63	62	63	63	63	63	63	63	63	63	63	63	63	63	63	63	63	63	63	63	63	63	63	63	63	63	63	63	63	63	63	63	63	63	63
63	63	63	63	63	63	63	63	63	63	63	63	63	63	63	63	63	63	63	63	63	63	63	63	63	63	63	63	63	63	63	63	63	63	63	63	63	63	63	63	63	63	63	63	63	63	63	63	63	63	63	63	63	63	63	63	63	63	63	63	63	63	63	63

Fig. 13.34 Matrix M_{L_3} of composition $\mathbf{x} \circ \mathbf{y}$ implemented by L_3. Symmetric entries are underlined.

0	3	6	6	12	15	12	15	24	27	30	30	24	27	30	30	48	51	54	54	60	63	60	63	48	51	54	54	60	63	60	63	33	35	39	39	45	47	45	47	57	59	63	63	57	59	63	63	33	35	39	39	45	47	45	47	57	59	63	63	57	59	63	63
1	3	7	7	13	15	13	15	25	27	31	31	25	27	31	31	49	51	55	55	61	63	61	63	49	51	55	55	61	63	61	63	1	3	7	7	13	15	13	15	25	27	31	31	25	27	31	31	49	51	55	55	61	63	61	63	49	51	55	55	61	63	61	63
2	2	6	6	14	14	14	14	26	26	30	30	26	26	30	30	50	50	54	54	62	62	62	62	50	50	54	54	62	62	62	62	35	35	39	39	47	47	47	47	59	59	63	63	59	59	63	63	35	35	39	39	47	47	47	47	59	59	63	63	59	59	63	63
3	3	7	7	15	15	15	15	27	27	31	31	27	27	31	31	51	51	55	55	63	63	63	63	51	51	55	55	63	63	63	63	3	3	7	7	15	15	15	15	27	27	31	31	27	27	31	31	51	51	55	55	63	63	63	63	51	51	55	55	63	63	63	63
4	7	4	7	12	15	12	15	28	31	28	31	28	31	28	31	52	55	52	55	60	63	60	63	52	55	52	55	60	63	60	63	37	39	37	39	45	47	45	47	61	63	61	63	61	63	61	63	37	39	37	39	45	47	45	47	61	63	61	63	61	63	61	63
5	7	5	7	13	15	13	15	29	31	29	31	29	31	29	31	53	55	53	55	61	63	61	63	53	55	53	55	61	63	61	63	5	7	5	7	13	15	13	15	29	31	29	31	29	31	29	31	53	55	53	55	61	63	61	63	53	55	53	55	61	63	61	63
6	6	6	6	14	14	14	14	30	30	30	30	30	30	30	30	54	54	54	54	62	62	62	62	54	54	54	54	62	62	62	62	39	39	39	39	47	47	47	47	63	63	63	63	63	63	63	63	39	39	39	39	47	47	47	47	63	63	63	63	63	63	63	63
7	7	7	7	15	15	15	15	31	31	31	31	31	31	31	31	55	55	55	55	63	63	63	63	55	55	55	55	63	63	63	63	7	7	7	7	15	15	15	15	31	31	31	31	31	31	31	31	55	55	55	55	63	63	63	63	55	55	55	55	63	63	63	63
8	11	14	14	8	11	14	14	24	27	30	30	24	27	30	30	56	59	62	62	56	59	62	62	56	59	62	62	56	59	62	62	41	43	47	47	41	43	47	47	57	59	63	63	57	59	63	63	41	43	47	47	41	43	47	47	57	59	63	63	57	59	63	63
9	11	15	15	9	11	15	15	25	27	31	31	25	27	31	31	57	59	63	63	57	59	63	63	57	59	63	63	57	59	63	63	9	11	15	15	9	11	15	15	25	27	31	31	25	27	31	31	57	59	63	63	57	59	63	63	57	59	63	63	57	59	63	63
10	10	14	14	10	10	14	14	26	26	30	30	26	26	30	30	58	58	62	62	58	58	62	62	58	58	62	62	58	58	62	62	43	43	47	47	43	43	47	47	59	59	63	63	59	59	63	63	43	43	47	47	43	43	47	47	59	59	63	63	59	59	63	63
11	11	15	15	11	11	15	15	27	27	31	31	27	27	31	31	59	59	63	63	59	59	63	63	59	59	63	63	59	59	63	63	11	11	15	15	11	11	15	15	27	27	31	31	27	27	31	31	59	59	63	63	59	59	63	63	59	59	63	63	59	59	63	63
12	15	12	15	12	15	12	15	28	31	28	31	28	31	28	31	60	63	60	63	60	63	60	63	60	63	60	63	60	63	60	63	45	47	45	47	45	47	45	47	61	63	61	63	61	63	61	63	45	47	45	47	45	47	45	47	61	63	61	63	61	63	61	63
13	15	13	15	13	15	13	15	29	31	29	31	29	31	29	31	61	63	61	63	61	63	61	63	61	63	61	63	61	63	61	63	13	15	13	15	13	15	13	15	29	31	29	31	29	31	29	31	61	63	61	63	61	63	61	63	61	63	61	63	61	63	61	63
14	14	14	14	14	14	14	14	30	30	30	30	30	30	30	30	62	62	62	62	62	62	62	62	62	62	62	62	62	62	62	62	47	47	47	47	47	47	47	47	63	63	63	63	63	63	63	63	47	47	47	47	47	47	47	47	63	63	63	63	63	63	63	63
15	15	15	15	15	15	15	15	31	31	31	31	31	31	31	31	63	63	63	63	63	63	63	63	63	63	63	63	63	63	63	63	15	15	15	15	15	15	15	15	31	31	31	31	31	31	31	31	63	63	63	63	63	63	63	63	63	63	63	63	63	63	63	63
16	19	22	22	28	31	28	31	16	19	22	22	28	31	28	31	48	51	54	54	60	63	60	63	48	51	54	54	60	63	60	63	49	51	55	55	61	63	61	63	49	51	55	55	61	63	61	63	49	51	55	55	61	63	61	63	49	51	55	55	61	63	61	63
17	19	23	23	29	31	29	31	17	19	23	23	29	31	29	31	49	51	55	55	61	63	61	63	49	51	55	55	61	63	61	63	17	19	23	23	29	31	29	31	17	19	23	23	29	31	29	31	49	51	55	55	61	63	61	63	49	51	55	55	61	63	61	63
18	18	22	22	30	30	30	30	18	18	22	22	30	30	30	30	50	50	54	54	62	62	62	62	50	50	54	54	62	62	62	62	51	51	55	55	63	63	63	63	51	51	55	55	63	63	63	63	51	51	55	55	63	63	63	63	51	51	55	55	63	63	63	63
19	19	23	23	31	31	31	31	19	19	23	23	31	31	31	31	51	51	55	55	63	63	63	63	51	51	55	55	63	63	63	63	19	19	23	23	31	31	31	31	19	19	23	23	31	31	31	31	51	51	55	55	63	63	63	63	51	51	55	55	63	63	63	63
20	23	20	23	28	31	28	31	20	23	20	23	28	31	28	31	52	55	52	55	60	63	60	63	52	55	52	55	60	63	60	63	53	55	53	55	61	63	61	63	53	55	53	55	61	63	61	63	53	55	53	55	61	63	61	63	53	55	53	55	61	63	61	63
21	23	21	23	29	31	29	31	21	23	21	23	29	31	29	31	53	55	53	55	61	63	61	63	53	55	53	55	61	63	61	63	21	23	21	23	29	31	29	31	21	23	21	23	29	31	29	31	53	55	53	55	61	63	61	63	53	55	53	55	61	63	61	63
22	22	22	22	30	30	30	30	22	22	22	22	30	30	30	30	54	54	54	54	62	62	62	62	54	54	54	54	62	62	62	62	55	55	55	55	63	63	63	63	55	55	55	55	63	63	63	63	55	55	55	55	63	63	63	63	55	55	55	55	63	63	63	63
23	23	23	23	31	31	31	31	23	23	23	23	31	31	31	31	55	55	55	55	63	63	63	63	55	55	55	55	63	63	63	63	23	23	23	23	31	31	31	31	23	23	23	23	31	31	31	31	55	55	55	55	63	63	63	63	55	55	55	55	63	63	63	63
24	27	30	30	24	27	30	30	24	27	30	30	24	27	30	30	56	59	62	62	56	59	62	62	56	59	62	62	56	59	62	62	57	59	63	63	57	59	63	63	57	59	63	63	57	59	63	63	57	59	63	63	57	59	63	63	57	59	63	63	57	59	63	63
25	27	31	31	25	27	31	31	25	27	31	31	25	27	31	31	57	59	63	63	57	59	63	63	57	59	63	63	57	59	63	63	25	27	31	31	25	27	31	31	25	27	31	31	25	27	31	31	57	59	63	63	57	59	63	63	57	59	63	63	57	59	63	63
26	26	30	30	26	26	30	30	26	26	30	30	26	26	30	30	58	58	62	62	58	58	62	62	58	58	62	62	58	58	62	62	59	59	63	63	59	59	63	63	59	59	63	63	59	59	63	63	59	59	63	63	59	59	63	63	59	59	63	63	59	59	63	63
27	27	31	31	27	27	31	31	27	27	31	31	27	27	31	31	59	59	63	63	59	59	63	63	59	59	63	63	59	59	63	63	27	27	31	31	27	27	31	31	27	27	31	31	27	27	31	31	59	59	63	63	59	59	63	63	59	59	63	63	59	59	63	63
28	31	28	31	28	31	28	31	28	31	28	31	28	31	28	31	60	63	60	63	60	63	60	63	60	63	60	63	60	63	60	63	61	63	61	63	61	63	61	63	61	63	61	63	61	63	61	63	61	63	61	63	61	63	61	63	61	63	61	63	61	63	61	63
29	31	29	31	29	31	29	31	29	31	29	31	29	31	29	31	61	63	61	63	61	63	61	63	61	63	61	63	61	63	61	63	29	31	29	31	29	31	29	31	29	31	29	31	29	31	29	31	61	63	61	63	61	63	61	63	61	63	61	63	61	63	61	63
30	30	30	30	30	30	30	30	30	30	30	30	30	30	30	30	62	62	62	62	62	62	62	62	62	62	62	62	62	62	62	62	63	63	63	63	63	63	63	63	63	63	63	63	63	63	63	63	63	63	63	63	63	63	63	63	63	63	63	63	63	63	63	63
31	31	31	31	31	31	31	31	31	31	31	31	31	31	31	31	63	63	63	63	63	63	63	63	63	63	63	63	63	63	63	63	31	31	31	31	31	31	31	31	31	31	31	31	31	31	31	31	63	63	63	63	63	63	63	63	63	63	63	63	63	63	63	63
32	35	38	38	44	47	44	47	56	59	62	62	56	59	62	62	32	35	38	38	44	47	44	47	56	59	62	62	56	59	62	62	33	35	39	39	45	47	45	47	57	59	63	63	57	59	63	63	33	35	39	39	45	47	45	47	57	59	63	63	57	59	63	63
33	35	39	39	45	47	45	47	57	59	63	63	57	59	63	63	33	35	39	39	45	47	45	47	57	59	63	63	57	59	63	63	33	35	39	39	45	47	45	47	57	59	63	63	57	59	63	63	33	35	39	39	45	47	45	47	57	59	63	63	57	59	63	63
34	34	38	38	46	46	46	46	58	58	62	62	58	58	62	62	34	34	38	38	46	46	46	46	58	58	62	62	58	58	62	62	35	35	39	39	47	47	47	47	59	59	63	63	59	59	63	63	35	35	39	39	47	47	47	47	59	59	63	63	59	59	63	63
35	35	39	39	47	47	47	47	59	59	63	63	59	59	63	63	35	35	39	39	47	47	47	47	59	59	63	63	59	59	63	63	35	35	39	39	47	47	47	47	59	59	63	63	59	59	63	63	35	35	39	39	47	47	47	47	59	59	63	63	59	59	63	63
36	39	36	39	44	47	44	47	60	63	60	63	60	63	60	63	36	39	36	39	44	47	44	47	60	63	60	63	60	63	60	63	37	39	37	39	45	47	45	47	61	63	61	63	61	63	61	63	37	39	37	39	45	47	45	47	61	63	61	63	61	63	61	63
37	39	37	39	45	47	45	47	61	63	61	63	61	63	61	63	37	39	37	39	45	47	45	47	61	63	61	63	61	63	61	63	37	39	37	39	45	47	45	47	61	63	61	63	61	63	61	63	37	39	37	39	45	47	45	47	61	63	61	63	61	63	61	63
38	38	38	38	46	46	46	46	62	62	62	62	62	62	62	62	38	38	38	38	46	46	46	46	62	62	62	62	62	62	62	62	39	39	39	39	47	47	47	47	63	63	63	63	63	63	63	63	39	39	39	39	47	47	47	47	63	63	63	63	63	63	63	63
39	39	39	39	47	47	47	47	63	63	63	63	63	63	63	63	39	39	39	39	47	47	47	47	63	63	63	63	63	63	63	63	39	39	39	39	47	47	47	47	63	63	63	63	63	63	63	63	39	39	39	39	47	47	47	47	63	63	63	63	63	63	63	63
40	43	46	46	40	43	46	46	56	59	62	62	56	59	62	62	40	43	46	46	40	43	46	46	56	59	62	62	56	59	62	62	41	43	47	47	41	43	47	47	57	59	63	63	57	59	63	63	41	43	47	47	41	43	47	47	57	59	63	63	57	59	63	63
41	43	47	47	41	43	47	47	57	59	63	63	57	59	63	63	41	43	47	47	41	43	47	47	57	59	63	63	57	59	63	63	41	43	47	47	41	43	47	47	57	59	63	63	57	59	63	63	41	43	47	47	41	43	47	47	57	59	63	63	57	59	63	63
42	42	46	46	42	42	46	46	58	58	62	62	58	58	62	62	42	42	46	46	42	42	46	46	58	58	62	62	58	58	62	62	43	43	47	47	43	43	47	47	59	59	63	63	59	59	63	63	43	43	47	47	43	43	47	47	59	59	63	63	59	59	63	63
43	43	47	47	43	43	47	47	59	59	63	63	59	59	63	63	43	43	47	47	43	43	47	47	59	59	63	63	59	59	63	63	43	43	47	47	43	43	47	47	59	59	63	63	59	59	63	63	43	43	47	47	43	43	47	47	59	59	63	63	59	59	63	63
44	47	44	47	44	47	44	47	60	63	60	63	60	63	60	63	44	47	44	47	44	47	44	47	60	63	60	63	60	63	60	63	45	47	45	47	45	47	45	47	61	63	61	63	61	63	61	63	45	47	45	47	45	47	45	47	61	63	61	63	61	63	61	63
45	47	45	47	45	47	45	47	61	63	61	63	61	63	61	63	45	47	45	47	45	47	45	47	61	63	61	63	61	63	61	63	45	47	45	47	45	47	45	47	61	63	61	63	61	63	61	63	45	47	45	47	45	47	45	47	61	63	61	63	61	63	61	63
46	46	46	46	46	46	46	46	62	62	62	62	62	62	62	62	46	46	46	46	46	46	46	46	62	62	62	62	62	62	62	62	47	47	47	47	47	47	47	47	63	63	63	63	63	63	63	63	47	47	47	47	47	47	47	47	63	63	63	63	63	63	63	63
47	47	47	47	47	47	47	47	63	63	63	63	63	63	63	63	47	47	47	47	47	47	47	47	63	63	63	63	63	63	63	63	47	47	47	47	47	47	47	47	63	63	63	63	63	63	63	63	47	47	47	47	47	47	47	47	63	63	63	63	63	63	63	63
48	51	54	54	60	63	60	63	48	51	54	54	60	63	60	63	48	51	54	54	60	63	60	63	48	51	54	54	60	63	60	63	49	51	55	55	61	63	61	63	49	51	55	55	61	63	61	63	49	51	55	55	61	63	61	63	49	51	55	55	61	63	61	63
49	51	55	55	61	63	61	63	49	51	55	55	61	63	61	63	49	51	55	55	61	63	61	63	49	51	55	55	61	63	61	63	49	51	55	55	61	63	61	63	49	51	55	55	61	63	61	63	49	51	55	55	61	63	61	63	49	51	55	55	61	63	61	63
50	50	54	54	62	62	62	62	50	50	54	54	62	62	62	62	50	50	54	54	62	62	62	62	50	50	54	54	62	62	62	62	51	51	55	55	63	63	63	63	51	51	55	55	63	63	63	63	51	51	55	55	63	63	63	63	51	51	55	55	63	63	63	63
51	51	55	55	63	63	63	63	51	51	55	55	63	63	63	63	51	51	55	55	63	63	63	63	51	51	55	55	63	63	63	63	51	51	55	55	63	63	63	63	51	51	55	55	63	63	63	63	51	51	55	55	63	63	63	63	51	51	55	55	63	63	63	63
52	55	52	55	60	63	60	63	52	55	52	55	60	63	60	63	52	55	52	55	60	63	60	63	52	55	52	55	60	63	60	63	53	55	53	55	61	63	61	63	53	55	53	55	61	63	61	63	53	55	53	55	61	63	61	63	53	55	53	55	61	63	61	63
53	55	53	55	61	63	61	63	53	55	53	55	61	63	61	63	53	55	53	55	61	63	61	63	53	55	53	55	61	63	61	63	53	55	53	55	61	63	61	63	53	55	53	55	61	63	61	63	53	55	53	55	61	63	61	63	53	55	53	55	61	63	61	63
54	54	54	54	62	62	62	62	54	54	54	54	62	62	62	62	54	54	54	54	62	62	62	62	54	54	54	54	62	62	62	62	55	55	55	55	63	63	63	63	55	55	55	55	63	63	63	63	55	55	55	55	63	63	63	63	55	55	55	55	63	63	63	63
55	55	55	55	63	63	63	63	55	55	55	55	63	63	63	63	55	55	55	55	63	63	63	63	55	55	55	55	63	63	63	63	55	55	55	55	63	63	63	63	55	55	55	55	63	63	63	63	55	55	55	55	63	63	63	63	55	55	55	55	63	63	63	63
56	59	62	62	56	59	62	62	56	59	62	62	56	59	62	62	56	59	62	62	56	59	62	62	56	59	62	62	56	59	62	62	57	59	63	63	57	59	63	63	57	59	63	63	57	59	63	63	57	59	63	63	57	59	63	63	57	59	63	63	57	59	63	63
57	59	63	63	57	59	63	63	57	59	63	63	57	59	63	63	57	59	63	63	57	59	63	63	57	59	63	63	57	59	63	63	57	59	63	63	57	59	63	63	57	59	63	63	57	59	63	63	57	59	63	63	57	59	63	63	57	59	63	63	57	59	63	63
58	58	62	62	58	58	62	62	58	58	62	62	58	58	62	62	58	58	62	62	58	58	62	62	58	58	62	62	58	58	62	62	59	59	63	63	59	59	63	63	59	59	63	63	59	59	63	63	59	59	63	63	59	59	63	63	59	59	63	63	59	59	63	63
59	59	63	63	59	59	63	63	59	59	63	63	59	59	63	63	59	59	63	63	59	59	63	63	59	59	63	63	59	59	63	63	59	59	63	63	59	59	63	63	59	59	63	63	59	59	63	63	59	59	63	63	59	59	63	63	59	59	63	63	59	59	63	63
60	63	60	63	60	63	60	63	60	63	60	63	60	63	60	63	60	63	60	63	60	63	60	63	60	63	60	63	60	63	60	63	61	63	61	63	61	63	61	63	61	63	61	63	61	63	61	63	61	63	61	63	61	63	61	63	61	63	61	63	61	63	61	63
61	63	61	63	61	63	61	63	61	63	61	63	61	63	61	63	61	63	61	63	61	63	61	63	61	63	61	63	61	63	61	63	61	63	61	63	61	63	61	63	61	63	61	63	61	63	61	63	61	63	61	63	61	63	61	63	61	63	61	63	61	63	61	63
62	62	62	62	62	62	62	62	62	62	62	62	62	62	62	62	62	62	62	62	62	62	62	62	62	62	62	62	62	62	62	62	63	63	63	63	63	63	63	63	63	63	63	63	63	63	63	63	63	63	63	63	63	63	63	63	63	63	63	63	63	63	63	63
63	63	63	63	63	63	63	63	63	63	63	63	63	63	63	63	63	63	63	63	63	63	63	63	63	63	63	63	63	63	63	63	63	63	63	63	63	63	63	63	63	63	63	63	63	63	63	63	63	63	63	63	63	63	63	63	63	63	63	63	63	63	63	63

Fig. 13.35 Matrix M_{L_4} of composition $\mathbf{x} \circ \mathbf{y}$ implemented by L_4. Symmetric entries are underlined.

Chapter 14
Semi-memristive Automata: Retained Refractoriness

A cellular automaton $\mathscr{A}$ is an orthogonal array of uniform finite-state machines, or cells. Each cell takes the finite number of states and updates its states in discrete time depending on states of its closest neighbours. All cells update their states simultaneously by the same rule. We consider eight-cell neighbourhood and three cell-states: resting $\circ$, excited $+$, and refractory $-$. Let $u(x) = \{y : |x-y|_{L\infty} = 1\}$ be a neighbourhood of cell x. Let $\sigma_x^t = \sum_{y \in u(x)} \chi(y^t, +)$ be the sum of excited neighbours of cell x at time step t, where $\chi(y^t, +) = 1$ if $y^t = +$ and $\chi(y^t, +) = 0$ otherwise. Let $\Theta = [\theta_1, \theta_2]$, $1 \leq \theta_1 \leq \theta_2 \leq 8$, be an excitation interval, and $\Phi = [\phi_1, \phi_2]$, $1 \leq \phi_1 \leq \phi_2 \leq 8$, be a recovery interval. A cell x updates its state by the following rule:

$$x^{t+1} = \begin{cases} +, \text{ if } x^t = \circ \text{ and } \sigma_x^t \in \Theta \\ -, \text{ if } x^t = + \\ \circ, \text{ if } (x^t = \circ \text{ and } \sigma_x^t \notin \Theta) \text{ or } (x^t = - \text{ and } \sigma_x^t \in \Phi) \end{cases}$$

A resting cell becomes excited if the number of its excited neighbours lies in the interval Θ. An excited cell takes refractory state unconditionally. A refractory cell returns to its resting state if number of its excited neighbours lie in the interval Φ.

In computer simulations we consider $\theta_1 = 1$ or 2 and $\theta_1 < \theta_2 \leq 8$; there is no point to consider $\theta_1 > 2$ because for such lower boundary of excitation interval most excitation patterns quickly extinct. We adopt the same boundaries of recovery interval Φ: $\phi_1 = 1, 2$ and $\phi_1 \leq \phi_2 \leq 8$. Further we refer to excitation and retention of refractoriness functions as $\mathscr{M}(\theta_1, \theta_2, \phi_1, \phi_2)$.

Why do we call the automaton described above a *semi-memristive* automaton? Because we assume that the refractory state symbolises high-resistivity, or non-conductivity. A refractory cell can not be excited whatever number of excited neighbours it has, therefore it can not 'conduct' excitation. In a fully memristive automaton both transitions from excited to refractory and from refractory to resting must be controlled by a local excitation density. In the semi-memristive automaton, the transition from excited to refractory is unconditional and happens independently of a local excitation density but the transition from a refractory state to resting state depends on a local excitation density.

A. Adamatzky: Reaction-Diffusion Automata, ECC 1, pp. 263–285.
springerlink.com

14.1 Methods: Experiments and Classificiation

We experiment with cellular automaton array of $n \times n$ cells, $n = 500$, and absorbing boundary conditions. While testing automata's response to external stimulation we use point-wise and D-stimulation (a spatially extended stimulation in a disc). By point-wise stimulation we mean excitation of one cell (for functions, where $\theta_1 = 1$) or two neighbouring cells (for functions, where $\theta_1 = 2$) of elsewhere resting cellular array. The D-stimulation is implemented as follows. Let D-disc be a set of cells which lie at distance not more than r from the centre $(n/2, n/2)$ of the cellular array. When undertaking D-stimulation we assign excited state to a cell of D with probability 0.1. In computer experiments we used $r = 100$.

To classify morphology of configurations generated by functions $\mathcal{M}(\theta_1, \theta_2, \phi_1, \phi_2)$, $\theta_1 = 1,2$, $\theta_1 \leq \theta_2 \leq 8$, $\phi_1 \leq \phi_2 \leq 8$, we D-stimulate cellular automaton lattices, and subdivided them into classes by visually detected likeness of excitation and refractory patterns generated and based on the following characteristics:

- *Space-filling ratio*. Let automaton of $n \times n$ cells being D-stimulated with disc radius r, then space-filling ratio is a ratio of non-resting cells in configuration recorded at time step $n - r$ (i.e. when front of propagating disturbance just reaches edge of the automaton array) to the size of array n^2.
- *Morphological diversity* is the number of proportion of different neighbourhood configurations. It is evaluated using Shannon entropy and Simpson's index. Let $W = \{\circ, +, -\}$ be a set of all possible configurations of a 9-cell neighbourhood $w(x) = u(x) \cup x$, $x \in \mathbf{L}$. Let c be a configuration of automaton, we calculate number of non-resting configurations as $\eta = \sum_{x \in \mathbf{L}} \varepsilon(x)$, where $\varepsilon(x) = 0$ if for every resting x all its neighbours are resting, and $\varepsilon(x) = 1$ otherwise. The Shannon entropy is calculated as $-\sum_{w \in W} (\nu(w)/\eta \cdot ln(\nu(w)/\eta))$, where $\nu(w)$ is a number of times the neighbourhood configuration w is found in automaton configuration c. Simpson's index is calculated as $1 - \sum_{w \in W} (\nu(w)/\eta)^2$.
- *Generative diversity* is a diversity of a configuration developed from a point-wise excitation.
- *Expressiveness* of a function is calculated as the Shannon entropy divided by space-filling ratio. Generative expressiveness is the expressiveness calculated after point-wise stimulation.
- *Power of a class* is the number of functions in this class.
- *Ratios of excited and refractory states* are calculated as $n^{-2} \sum_{x \in L} \chi(x^t, +)$ and $n^{-2} \sum_{x \in \mathbf{L}} \chi(x^t, -)$, where $\chi(a,b) = 1$ if $a = b$, and $\chi(a,b) = 0$ otherwise.

14.2 Classes

We separate functions $\mathcal{M}(\theta_1, \theta_2, \phi_1, \phi_2)$ into eleven classes, characterise morphology of configurations, generated by automata from each class, and provide details on membership of each class.

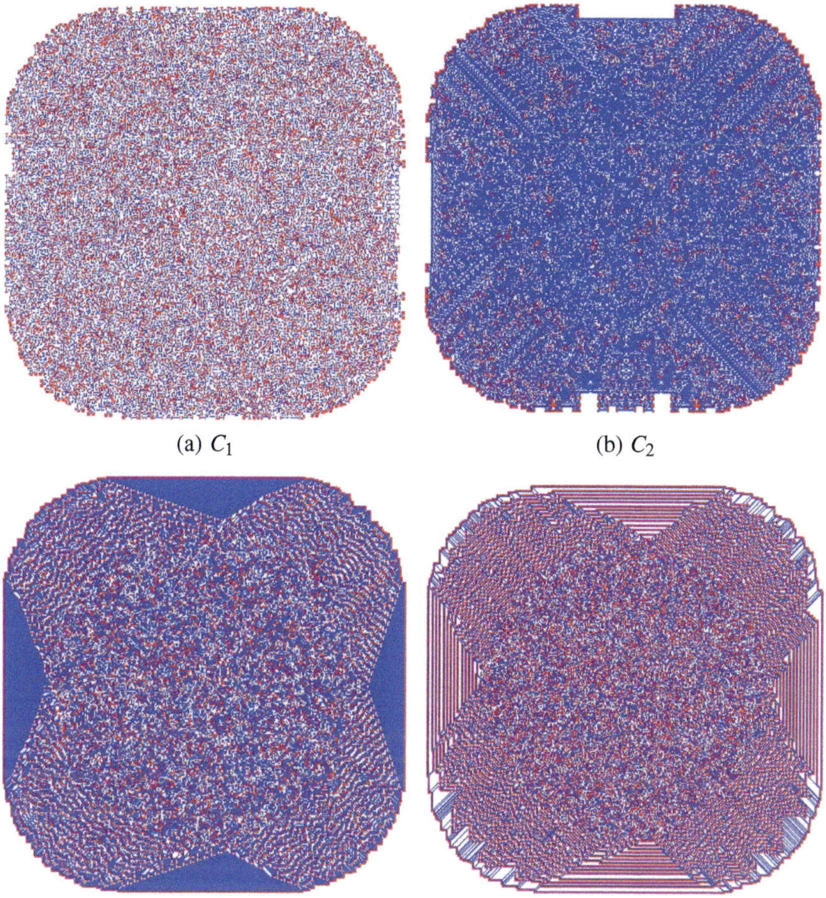

(a) C_1 (b) C_2

(c) C_3 (d) C_4

Fig. 14.1 Exemplars of configurations generated by functions from classes $C_1 \ldots C_4$ with D-stimulation. Excited cells are red (light grey), refractory cells are blue (dark grey) and resting cells are white.

Class C_1

D-stimulation gives birth to propagating quasi-chaotic pattern of excited and refractory states, the pattern fills the automaton array (Fig. 14.1a). A front of the pattern is comprised of branching localised excitations, which periodically merge in a single excitation wave-front; the pattern is almost isotropic with slight indentations along north-south and west-east axis. Excited and refractory states are exhibited equally in patterns generated by automata on class C_1 (Fig. 14.2, C_1). The class C_1 is comprised of 27 functions: $\mathscr{M}(111a)$, $a = 1,\ldots,8$; $\mathscr{M}(112a)$, $a = 3,\ldots,8$; $\mathscr{M}(121a)$, $a = 2,\ldots 8$; $\mathscr{M}(122a)$, $a = 3,\ldots 8$.

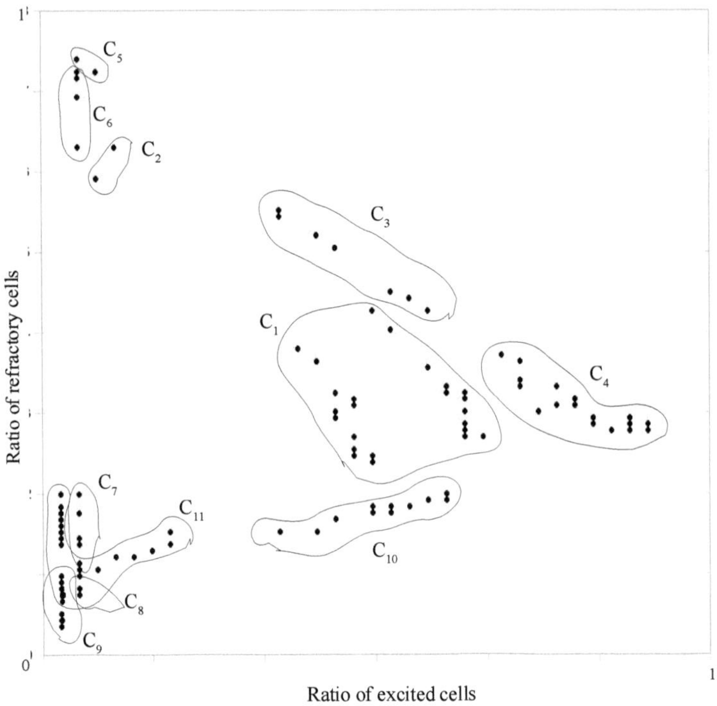

Fig. 14.2 Ratio of excited states to resting states vs ratio of refractory states to resting states in configurations developed by a cellular automat with retained refractoriness with *D*-stimulation

Class C_2

D-stimulation leads to propagating quasi-chaotic pattern with domination of refractory states. The pattern fills the automaton array (Fig. 14.1b). The pattern's propagating front is comprised of travelling localizations, which branch periodically. Excited localizations travelling along north-west-south-east and north-east-south-west axis leave a distinctive trail of refractory states (Fig. 14.1b). There is a tail of gradually extinct excitations. These spatially extended excitations collapse into localised oscillating localizations, most of which extinguish with time (Fig. 14.2, C_2). The class C_2 has three functions $\mathscr{M}(1122)$, $\mathscr{M}(1211)$, $\mathscr{M}(1222)$.

Class C_3

Initial random stimulation leads to formation of a single circular excitation wave, followed by trains of excitation waves travelling north-west, north-east, south-west and south-east. The wave-trains are followed by discoidal growing quasi-random pattern of excited and refractory states (Fig. 14.1c). North, south, east and west domains of cellular array lying between the growing patterns are filled entirely with refractory states. These refractory domains contribute towards slight prevalence of refractory states (Fig. 14.2, C_3). The class C_3 has eight functions $\mathscr{M}(1a1b)$, $a = 3,\ldots,6$, $b = 1,2$.

Class C_4

The D-stimulated automata response is characterised by a single circular excitation wave, which encapsulates wave trains propagating north-west, north-east, south-west and south-east, and envelopes of wave-fronts travelling north, east, south and west (Fig. 14.1d)). The wave envelopes in C_4 occupying the same space domains are refractory domains in C_3. This is why a cluster of C_4 functions is positioned symmetrically to cluster of C_3 functions with respect to the diagonal of equal ratios of excited and refractory states in Fig. 14.2. The class C_4 has 48 functions $\mathscr{M}(1abc)$, $a = 3,\ldots,6$, $b = 1,2$, $c = 3,\ldots,8$.

Classes C_5 and C_6

In response D-stimulation automata from these classes generate circular waves of excitation (Fig. 14.3ab). The domain D of initial perturbation is a random pattern of refractory and resting states. No excitation is observed inside D in automata from class C_6 and usually just few localised oscillating excitations in automata from C_5. The circular wave-front leaves behind an almost uniform field of refractory cells with radial traces of resting states towards north-west, north-east, south-west and south-east directions (Fig. 14.3ab). Functions from classes C_5 and C_6 are positioned close to class C_2 in Fig. 14.2 due to high ratio of refractory states in configurations they generate. The class C_5 has four functions $\mathscr{M}(1a22)$, $a = 3,\ldots,6$ and the class C_6 has six functions $\mathscr{M}(1abc)$, $a = 7,8$, $b = 1,2$, $b \leq c \leq 2$.

Classes C_7 and C_8

Automata from these classes generate a single circular excitation wave while perturbed by D and a domain of refractory and resting states inside boundaries of D. In automata from C_8 a propagating excitation wave leaves a trail of refractory states along north-west-south-east and north-east-south-west axis, the rest of cells remain in the resting state. Propagating wave-front leaves no traces in automata from C_8 (Fig. 14.3cd). Class C_7 has 16 functions $\mathscr{M}(1a1b)$, $a = 7,8$, $b = 3,4$; $\mathscr{M}(1a2b)$,

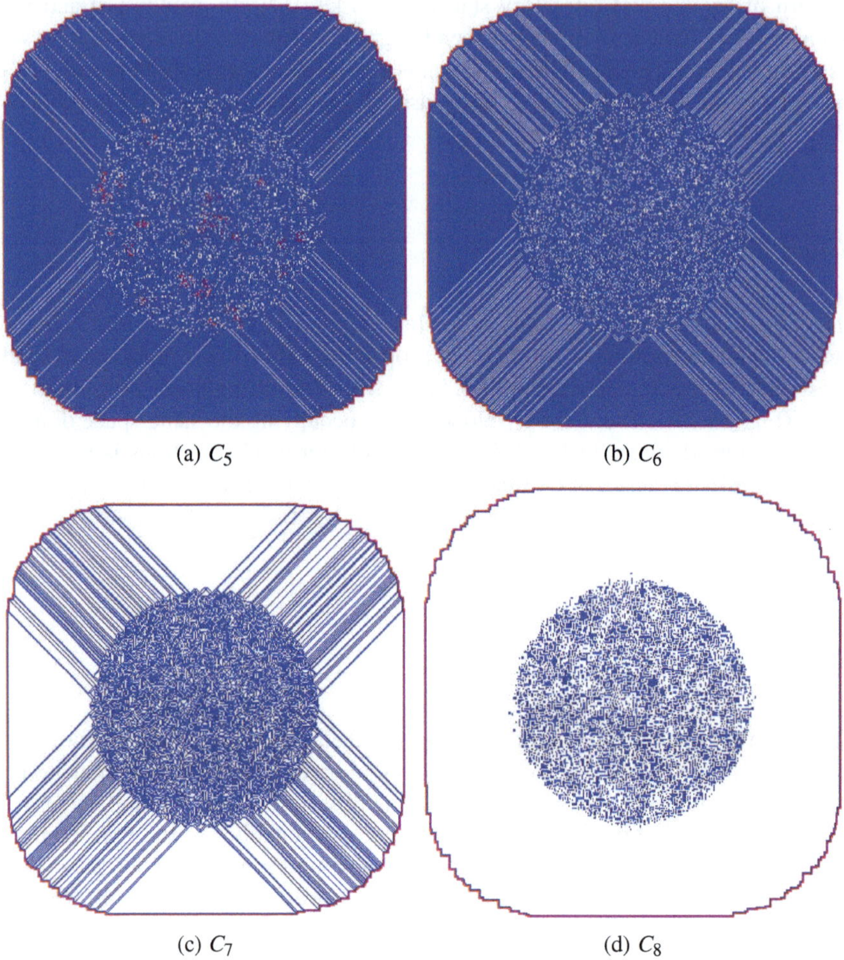

Fig. 14.3 Exemplars of configurations generated by functions from classes $C_5 \dots C_8$ with D-stimulation. Excited cells are red (light grey), refractory cells are blue (dark grey) and resting cells are white.

$a = 7,8$, $b = 3,\dots,8$; and, class C_8 has eight functions $\mathscr{M}(1a1b)$, $a = 7,8$ and $b = 5,\dots,8$.

Class C_9

D-stimulation of an automaton from C_9 produces a domain of refractory and resting states (inside boundaries of D) and localizations travelling along north-south and west-east axis (Fig. 14.4a). The localizations leave traces of refractory states.

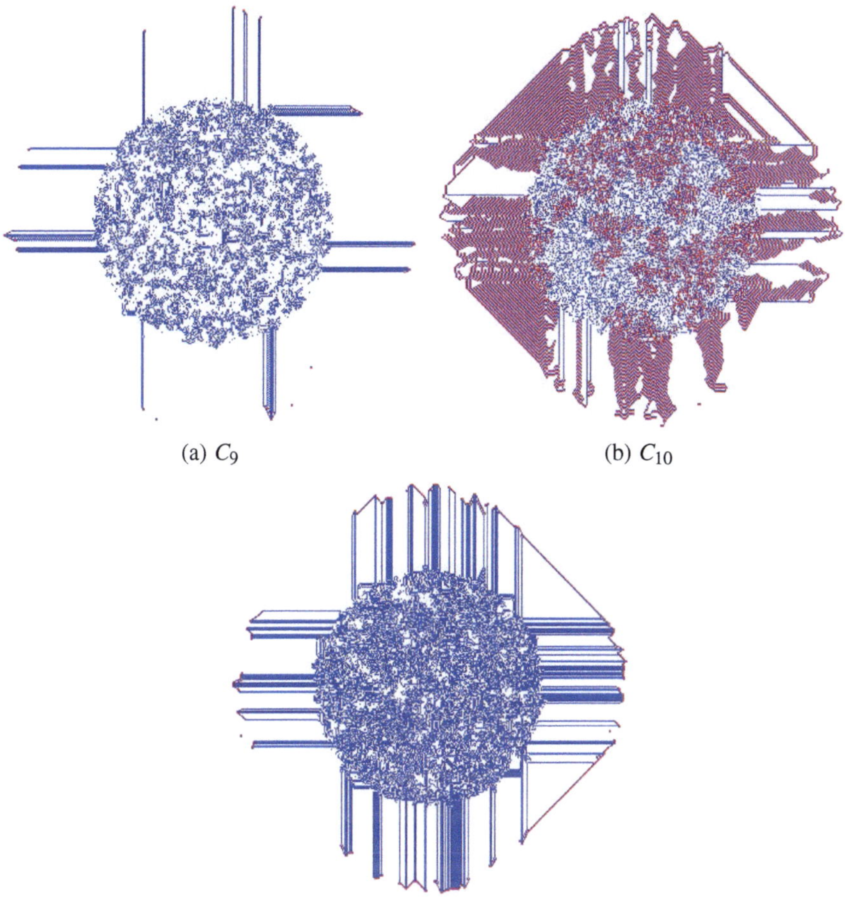

(a) C_9 (b) C_{10}

(c) C_{11}

Fig. 14.4 Exemplars of configurations generated by functions from classes $C_9 \ldots C_{11}$ with D-stimulation. Excited cells are red (light grey), refractory cells are blue (dark grey) and resting cells are white.

Configurations produced by functions from C_9 has the lowest (amongst all other classes) ratio of excited states (Fig. 14.2, C_9). The class C_9 has 17 functions $\mathcal{M}(22ab)$, $a = 1,2$, $a \leq b \leq 8$; $\mathcal{M}(2a11)$, $a = 3,4$.

Class C_{10}

Phenomenology of automata's behaviour is richest amongst all classes. D-stimulation leads to formation of trains of excitation wave-fragments, propagating in north-west,

Class	Shannon entropy	Simpson's index	Space-filling ratio	Expressiveness
C_1	7.01	1.0	0.94	7.41
C_2	5.06	0.95	0.95	5.35
C_3	5.99	0.96	0.96	6.24
C_4	5.55	0.98	0.96	5.8
C_5	1.79	0.53	0.96	1.87
C_6	2.15	0.65	0.96	2.23
C_7	3.95	0.96	0.42	9.5
C_8	4.56	0.99	0.15	31.21
C_9	3.89	0.95	0.14	27.69
C_{10}	5.20	0.98	0.60	8.7
C_{11}	4.27	0.97	0.27	15.71

Fig. 14.5 Average values of Shannon entropy, Simpson's index, space-filling ratio and expressiveness with D-stimulaton

north-east, south-west and south-east directions; domains of localised excitations inside boundaries of D, travelling localizations with and without traces of refractory states (Fig. 14.4b). Typically excited cells are in slight majority in configurations generated by automata form C_{10} (Fig. 14.2, C_{10}). The class C_{10} includes 61 functions $\mathscr{M}(2a1b)$, $a,b=3,\dots,8$; $\mathscr{M}(2a2b)$, $a,b=4,\dots,8$.

Class C_{11}

Behaviour of automata from class C_{11} is somewhat similar to that of class C_9 but sometimes localizations travelling along north-south and west-east axis are 'linked' by segments from excitation wave-fronts (Fig. 14.4c). The class C_{11} has 27 functions $\mathscr{M}(2312)$; $\mathscr{M}(232a)$, $a=2,\dots,8$; $\mathscr{M}(2412)$; $\mathscr{M}(242a)$, $a=2,3$; $\mathscr{M}(2abc)$, $a=5,\dots,8$, $b=1,2$, $b\leq c\leq 3$.

14.3 Hierarchies

Finding 42. *Classes of excitable cellular automata with retained refractoriness obey the following power hierarchy:*

$$C_4 \rhd C_{10} \rhd \{C_1,C_{11}\} \rhd C_7 \rhd C_9 \rhd \{C_3,C_8\} \rhd C_6 \rhd C_5 \rhd C_2.$$

This is a direct consequence of our joining functions into classes.

Finding 43. *Classes of excitable cellular automata with retained refractoriness obey the following hierarchy of space-filling ratio:*

$$\{C_1,C_2,C_3,C_5,C_6\} \rhd C_{10} \rhd C_7 \rhd C_{11} \rhd \{C_8,C_9\}$$

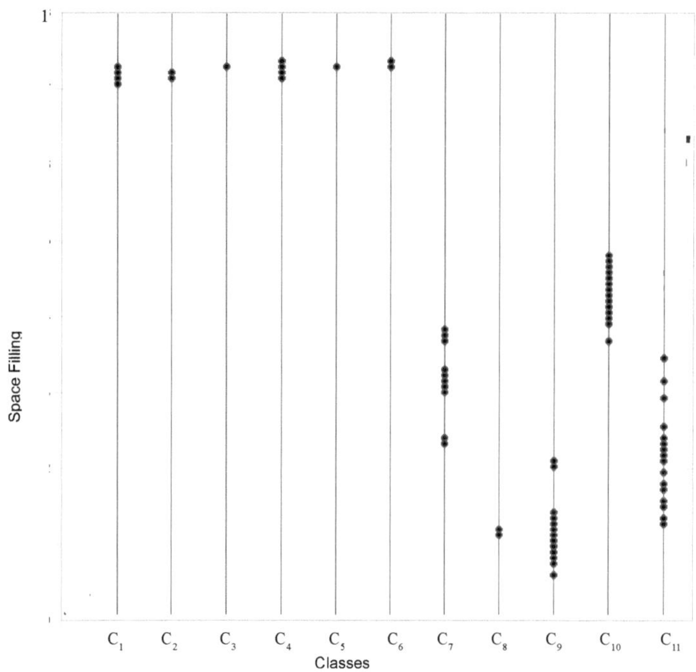

Fig. 14.6 Space-filling ratio values for classes of retained refractoriness rules. Classes are indicated on horizontal axis. Space-filling is shown on vertical axis. All 225 functions are mapped to show distribution of space-filling ratio inside each class. There is less discs on the chart than functions because some functions from the same class have the same space-filling ratio values.

Space-filling ratio values averaged amongst each class are shown in Fig. 14.5 and the distribution of functions inside each class in Fig. 14.6. Classes C_1 to C_6, and C_8 show small variance in space-filling ratio while class C_{11} shows largest variance. Roughly the classes can be split into two space-filling groups: high space-filling ratio $(C_1,\dots,C_6)$ and low space-filling ratio $(C_7,\dots,C_{11})$.

Finding 44. *Classes of excitable cellular automata with retained refractoriness obey the following hierarchy of diversity:*

$$C_1 \rhd C_3 \rhd C_4 \rhd C_{10} \rhd C_2 \rhd C_8 \rhd C_{11} \rhd C_7 \rhd C_9 \rhd C_6 \rhd C_5.$$

We can draw an analogy between neighbourhood states in a given configuration of cellular automaton and living species in a population. Thus we can measure a diversity of a function by calculating Simpson's index and Shannon entropy of a population of neighbourhood states in a single configuration generated by the function (Fig. 14.7). Classes C_5 and C_6 exhibit the lowest diversity because initial random excitation pattern D extinguishes quickly and only single propagating wave-front

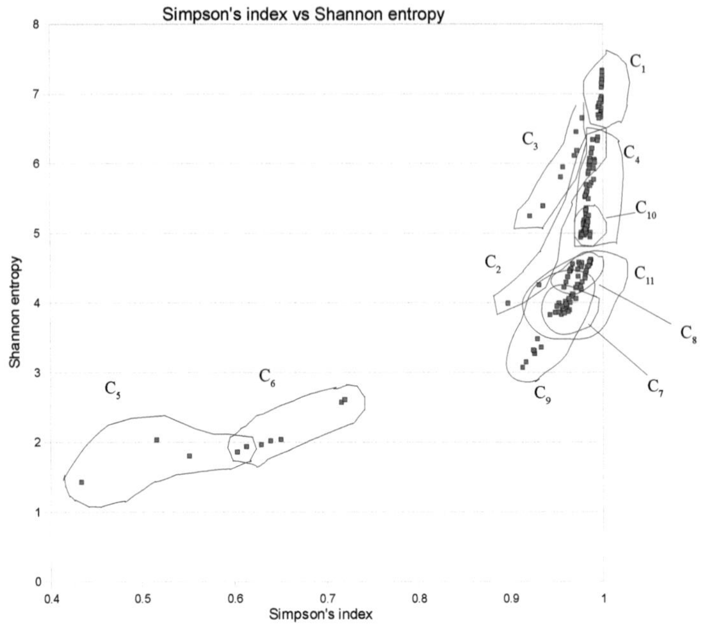

(a)

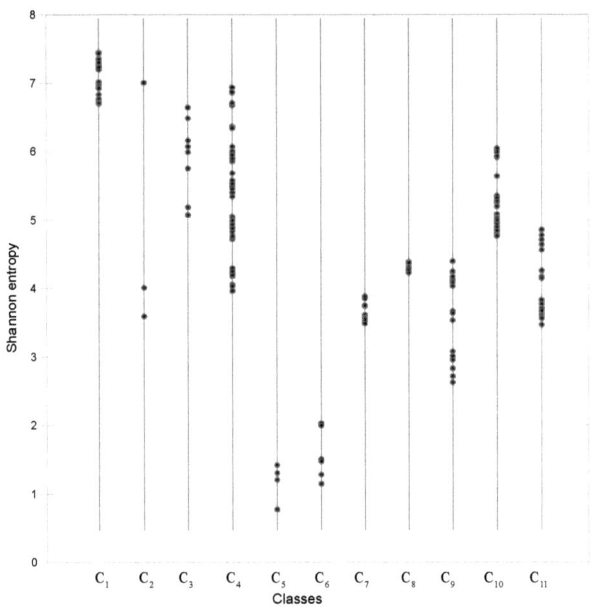

(b)

Fig. 14.7 Distribution of classes on morphological diversity. (a) Simpson's index vs Shannon entropy and (b) Shannon entropy with *D*-stimulation.

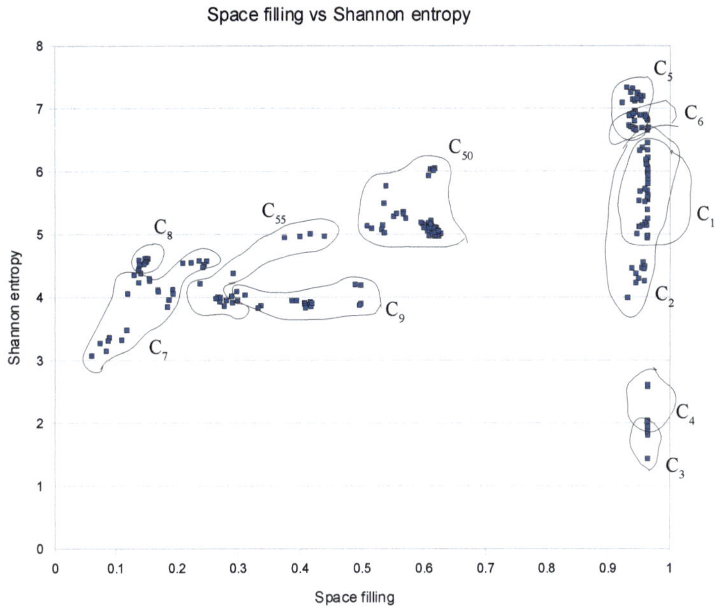

Fig. 14.8 Space-filling ratio vs Shannon entropy with *D*-stimulation

travels away from the site of original excitation. The front leaves behind an almost homogeneous domain of refractory states. Classes C_1, C_3 and C_4 hold top ranks in the diversity hierarchy because the initial random patterns expands into a growing quasi-random configuration of excited and refractory states. In terms of Shannon entropy class C_3 shows higher degree of diversity than class C_4 (Fig. 14.5), possibly, because space-time dynamics is biased towards formation of extended wave-fronts propagating north, south, east and west, which interact with fragmentary wave-patterns travelling north-west, north-east, south west and south-east. Class C_{10}, which demonstrates high variety of travelling localizations, and stays just above the middle of the diversity hierarchy.

Finding 45. *Classes of excitable cellular automata with retained refractoriness obey the following hierarchy of expressivenes:*

$$C_8 \rhd C_9 \rhd C_{11} \rhd C_7 \rhd C_{10} \rhd C_1 \rhd C_3 \rhd C_4 \rhd C_2 \rhd C_6 \rhd C_5.$$

Distribution of functions in a space filing versus Shannon entropy space is shown in Fig. 14.8. Despite having below average diversities classes C_8 and C_9 occupy top level of the expressiveness hierarchy. This is because their space filling demands are incredibly modest. Apart of a random refractory domain left after *D*-perturbation, we observe only propagating excitation wave-front in C_8 and trace leaving

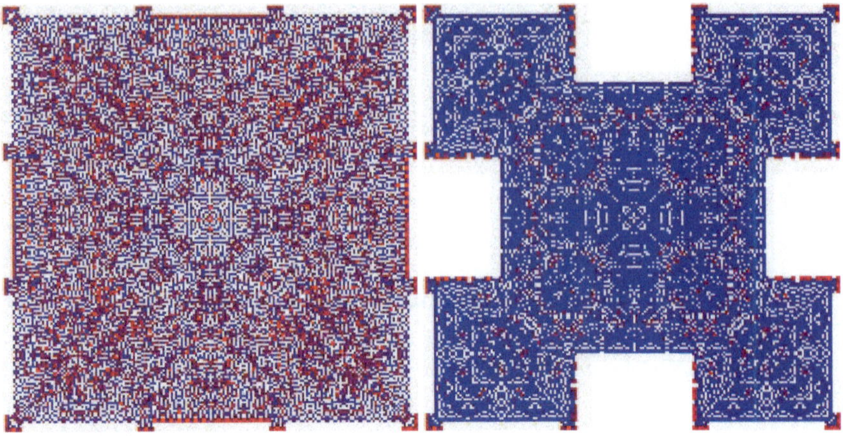

(a) $\mathscr{M}(1228) \in C_1$ (b) $\mathscr{M}(1222) \in C_2$

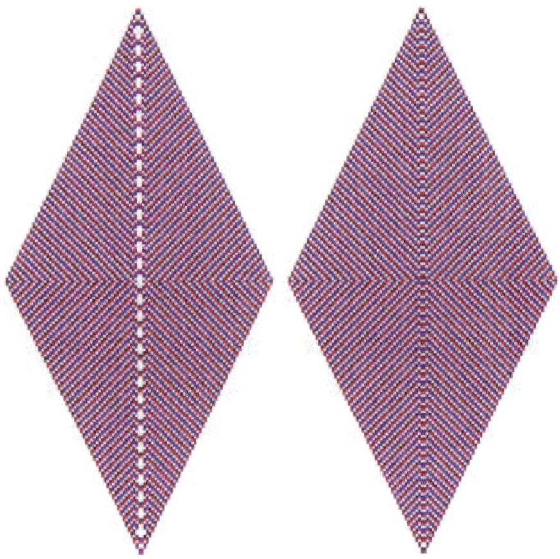

(c) $\mathscr{M}(2413) \in C_{10}$ (d) $\mathscr{M}(2426) \in C_{10}$

Fig. 14.9 Typical configurations, developed after point-wise excitation, generated by functions (a) C_1, (b) C_2, (cd) C_{10}. Names of functions are indicated in the sub-captions. Snapshots of configurations are taken 100 time steps after point-wise perturbation. Excited cells are red (light grey), refractory cells are blue (dark grey) and resting cells are white.

localizations in C_9. They are followed by C_{11}. Automata from C_{11} exhibit varieties of localised and spatially extended excitations.

Finding 46. *Classes of excitable cellular automata with retained refractoriness obey the following hierarchy of generative diversity:*

$$C_1 \rhd C_2 \rhd C_9 \rhd C_{11} \rhd C_{10} \rhd C_7 \rhd C_4 \rhd C_8 \rhd \{C_3, C_5, C_6\}.$$

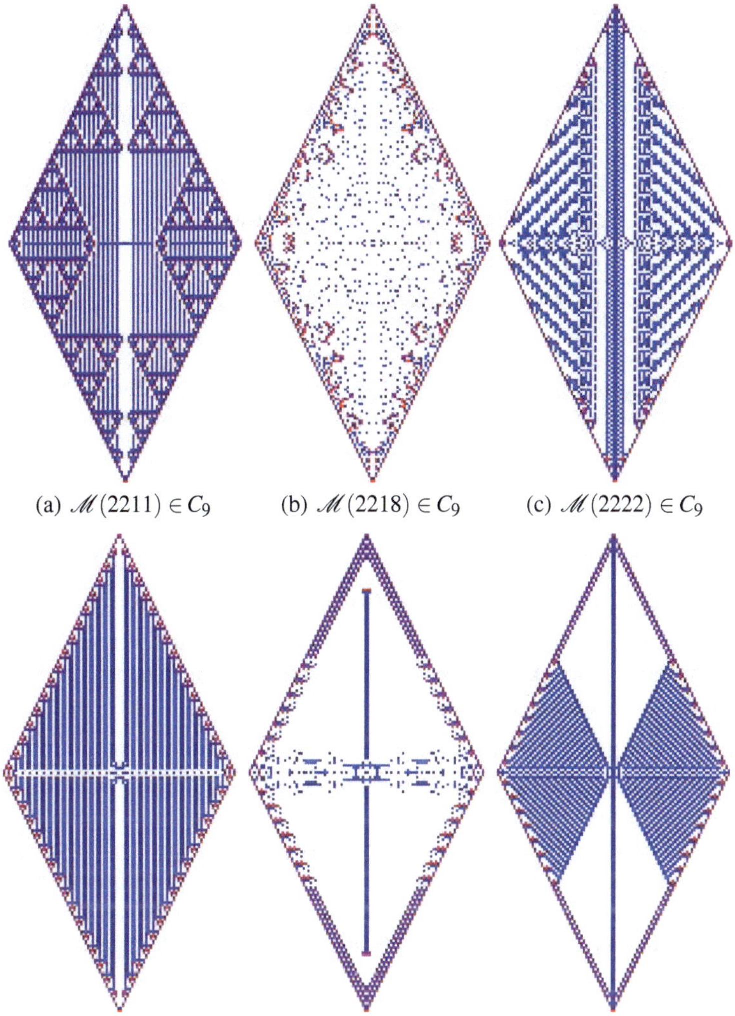

(a) $\mathcal{M}(2211) \in C_9$ (b) $\mathcal{M}(2218) \in C_9$ (c) $\mathcal{M}(2222) \in C_9$

(d) $\mathcal{M}(2223) \in C_9$ (e) $\mathcal{M}(2312) \in C_{11}$ (f) $\mathcal{M}(2322) \in C_{11}$

Fig. 14.10 Typical configurations, developed after point-wise excitation, generated by functions (a–d) C_9, and (ef) C_{11}. Exact functions are indicated in the sub-captions. Snapshots of configurations are taken 100 time steps after point-wise perturbation. Excited cells are red (light grey), refractory cells are blue (dark grey) and resting cells are white.

Class	Shannon entropy	Simpson's index	Space-filling ratio	Expressiveness
C_1	6.89	1.0	0.90	7.65
C_2	5.52	0.98	0.84	6.6
C_3	0.31	0.10	1.0	0.31
C_4	2.86	0.94	0.08	34.32
C_5	0.31	0.1	1.0	0.31
C_6	0.31	0.1	1.0	0.31
C_7	2.93	0.94	0.09	32.58
C_8	2.71	0.93	0.07	38.79
C_9	4.55	0.96	0.22	20.59
C_{10}	3.16	0.94	0.27	11.71
C_{11}	3.80	0.95	0.21	17.91

Fig. 14.11 Average values of Shannon entropy, Simpson's index, space-filling ratio and expressiveness calculated on configurations generated by a point-wise excitation

Average values of Shannon entropy of configurations generated by point-wise excitation are shown in Fig. 14.11. Classes C_1 and C_2 are at the top of generative diversity hierarchy. Typical configurations developed after a point-wise excitation include almost-rectangular patterns (Fig. 14.9a) and rectangular patterns with 2/3 diamond corners (Fig. 14.9b). These patterns are symmetric yet sophisticated in terms of visual complexity patterns of excited and refractory states.

In the descending order of generative diversity, classes C_1 and C_2 are followed by classes C_9 and C_{11}. Patterns generated in the result of point-wise stimulation include

- Sierpinski carpets expanding north and south from longest sides of two-cell point-wise excitation (Fig. 14.10a);
- rhomboid excitation wave fronts, followed by travelling localizations and leaving behind a domain of disordered sparsely distributed refractory states (Fig. 14.10b);
- rhomboid wave-fronts forming ordered strips of refractory states (Fig. 14.10cd);
- and combinations of localised and spatially-extended refractory-state domains (Fig. 14.10ef).

Class C_{10} occupies a mid-range of generative diversity. A point-wise excitation generates rhomboid patterns of wave packets, each excitation front is followed by a refractory tail (Fig. 14.9cd). Some functions of the class also produce a narrow band of resting states, surrounded by excitation domain (Fig. 14.9c). Classes C_3, C_5, C_4, C_6 C_8 have lowest generative diversity. A point-wise excitation leads to formation of a circular excitation wave, which leaves a trail of refractory states.

Finding 47. *Classes of excitable cellular automata with retained refractoriness obey the following hierarchy of generative expressiveness:*

$$C_8 \rhd C_4 \rhd C_7 \rhd C_9 \rhd C_{11} \rhd C_{10} \rhd C_1 \rhd C_2 \rhd \{C_3, C_5, C_6\}.$$

See detailed structure of the hierarchy in Fig. 14.12.

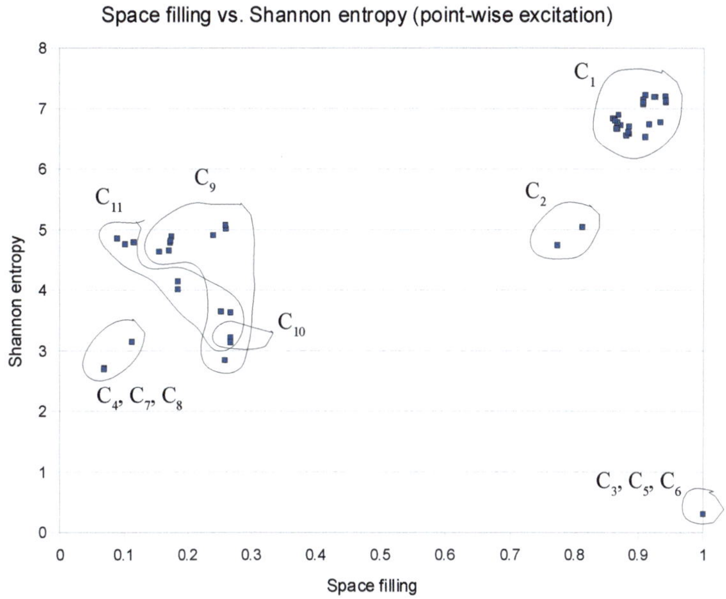

Fig. 14.12 Space-filling ratio vs Shannon entropy. Point-wise excitation

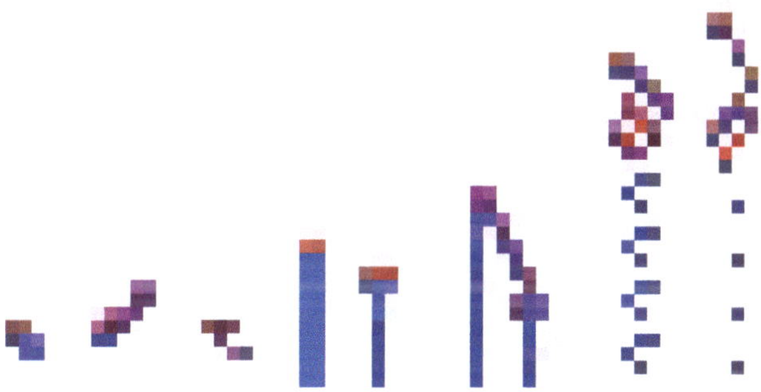

Fig. 14.13 Examples of travelling localizations found in configurations of automata with retained refractoriness. (a–e) Gliders, (f–h) Trains. Excited cells are red (light grey), refractory cells are blue (dark grey) and resting cells are white.

14.4 Travelling Localisations

Travelling localised excitations is a unique phenomenon of sub-excitable media, widely used in theoretical and experimental laboratory implementations of novel computing architectures, see e.g. [14]. Most common localizations found in experiments with excitable automata with retained refractoriness are gliders, trains, glider guns, and spaceships. These structures are well-known in Conway's Game-of-Life

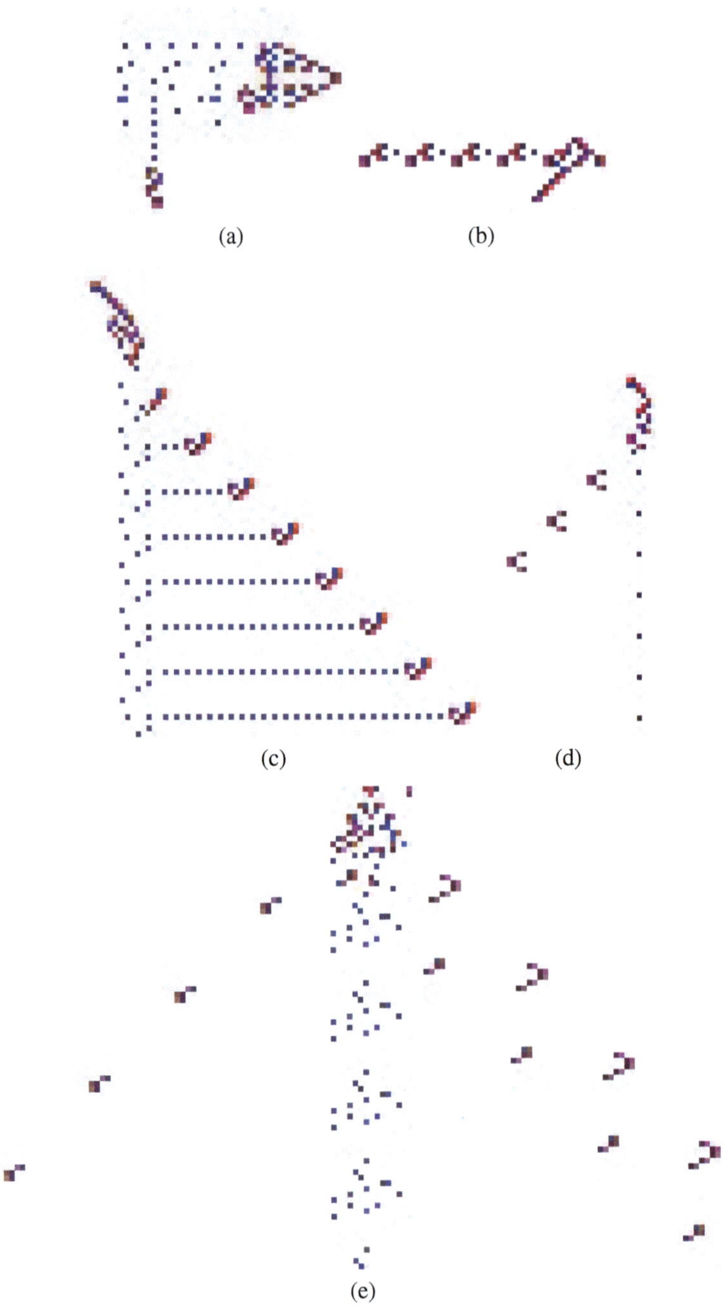

(a) (b) (c) (d) (e)

Fig. 14.14 Examples of mobile glider guns found in development of automata with retained refractoriness. Excited cells are red (light grey), refractory cells are blue (dark grey) and resting cells are white.

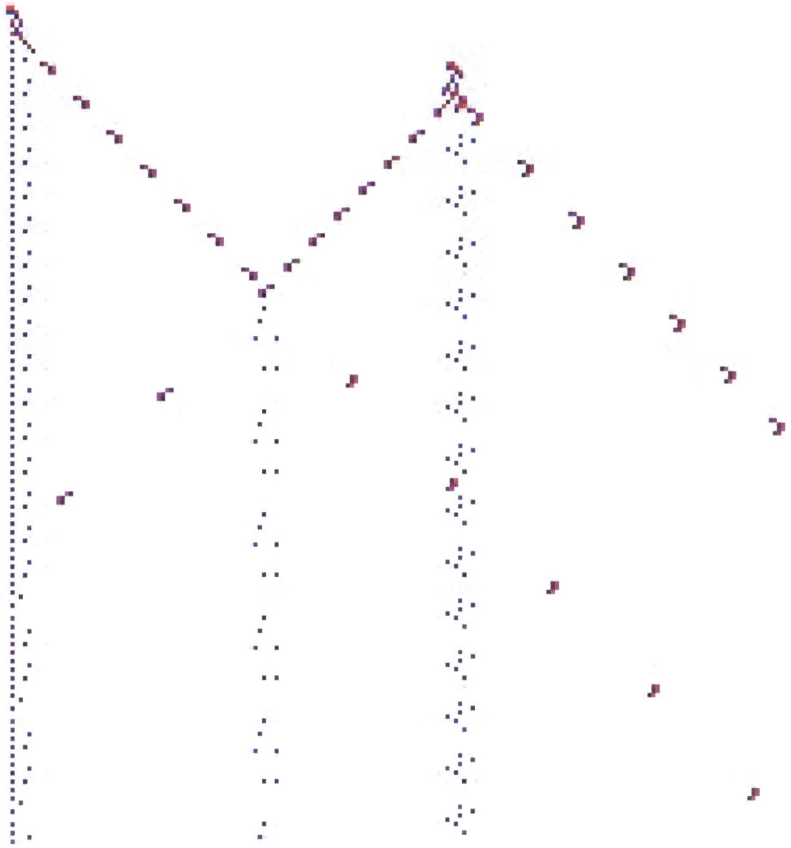

Fig. 14.15 Example of interactions between travelling excitations in automata with retained refractoriness. Excited cells are red (light grey), refractory cells are blue (dark grey) and resting cells are white.

cellular automata (and we are using terminology of the Conway's Game of Life) however they display a range of uncommon properties when inhabiting discrete excitable medium.

Examples of gliders (localizations which do not leave traces) and trains (localizations which leave traces made of stationary refractory patterns) are shown in Fig. 14.13; all localizations shown there travel north. A glider shown in Fig. 14.13a consists of three excited and states. The small gliders can be joined together to make propagating wave-fragments: in Fig. 14.13b we see a glider made of three excited and three refractory states joined with a localisation of two excited and two refractory states. The gliders move along columns and rows of a cellular array. Similarly to our previous models of sub-excitable discrete media, e.g. [14], we found only one excited localisation which travels along diagonals of the cellular array. The diagonal

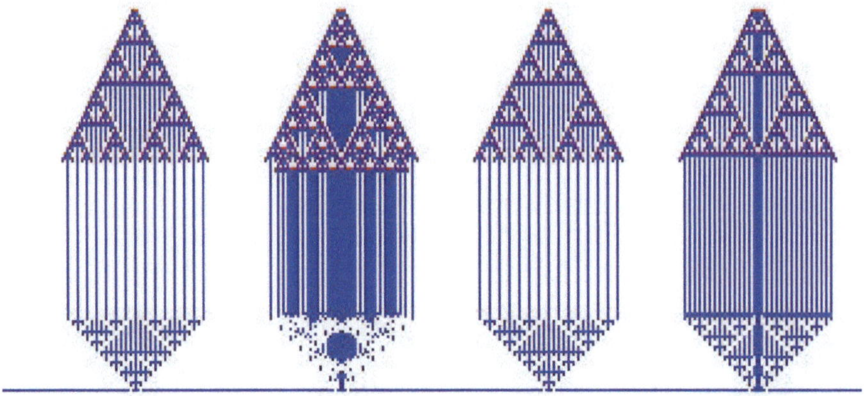

Fig. 14.16 'Flying' Sierpinski carpets, generated by automata of class C_9. Carpets travel north. Patterns (from the left to the right) are generated by lines of two, three, four and five excited cells. Launching pad is constructed by a glider, which travels east and leaves a refractory trail. The configurations are generated by function $\mathscr{M}(2311)$. Excited cells are red (light grey), refractory cells are blue (dark grey).

glider consists of three excited and three refractory states, an exemplar snapshot is shown in Fig. 14.13c.

Gliders which leave traces of stationary patterns behind are called trains. The trains are very common in classes C_9, C_{10} and C_{11}. A minimal train consists of a pair of excited states leaving a column/row of refractory pairs behind (Fig. 14.13d). In terms of trace minimality, the train shown in Fig. 14.13e is minimal localisation of three excited states which leaves a trace of refractory states one-cell wide. Trains can be linked with gliders and other trains, an example is shown in (Fig. 14.13f). Trains leaving localised traces of refractory states are shown in (Fig. 14.13gh).

Travelling localizations which generate streams of other localizations, e.g. gliders, are called mobile guns (Fig. 14.14). A mobile gun can emit localisation in any directions, subject to the lattice's discreteness, apart of the direction of the gun's motion. Here we illustrate just few examples: gun travels east and emits trains travelling south (Fig. 14.14a), gun travels east and emits gliders travelling west (Fig. 14.14b), gun travels north and emits trains travelling east (Fig. 14.14c), gun travels north and emits gliders travelling west (Fig. 14.14d). A complex gun is shown in (Fig. 14.14e): the gun travels north and emits three streams of gliders, one stream travels west and two streams travel east.

Interactions between gliders is pretty rich, gliders and trains can reflect or annihilate in the result of the collision, also refractory stationary patterns can be formed in the results of such collisions. An example is shown in Fig. 14.15. Two mobile guns travel north. They both leave a trace of refractory states, as trains do. Westward gun emits a single stream of gliders travelling east. Eastward gun emits two streams of

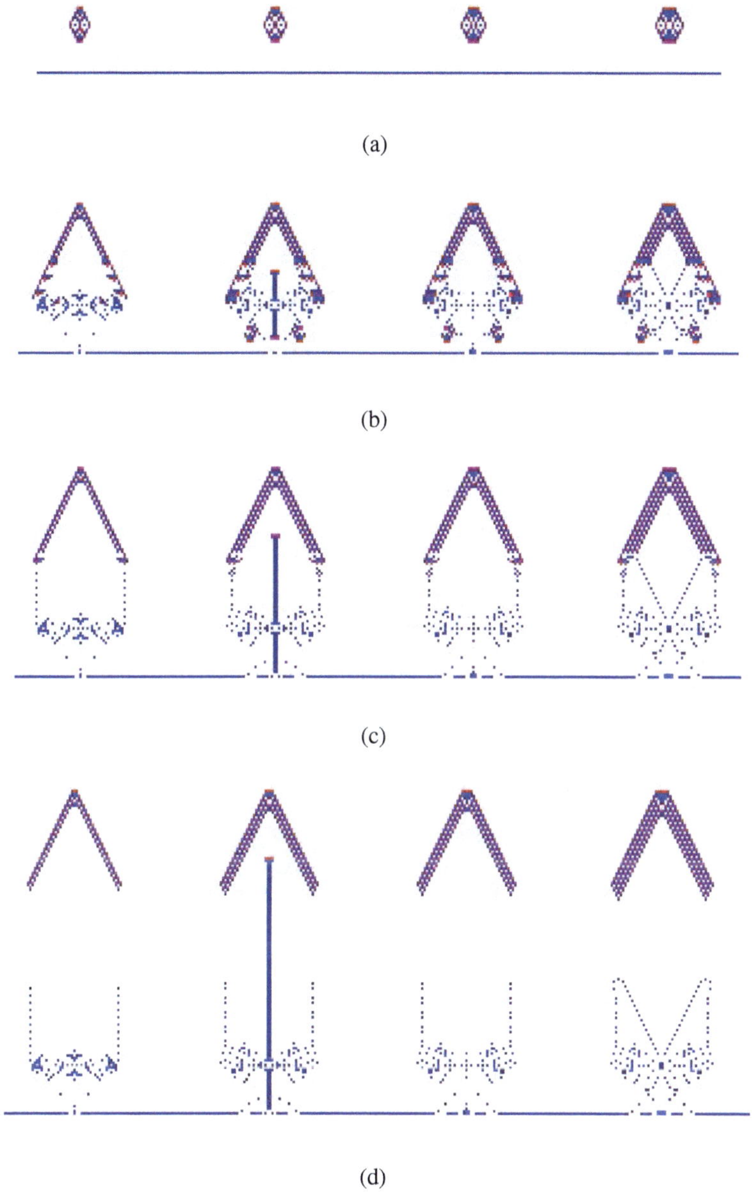

Fig. 14.17 Formation and propagation of spaceships in automata of class C_{11}. Patterns (from the left to the right) are generated by lines of two, three, four and five excited cells. Launching pad is constructed by glider travelling east and leaving refractory trail. (a) Generation stage. (bc) Take-off stage. (d) Flying. The configurations are generated by function $\mathscr{M}(2312)$. Excited cells are red (light grey), refractory cells are blue (dark grey).

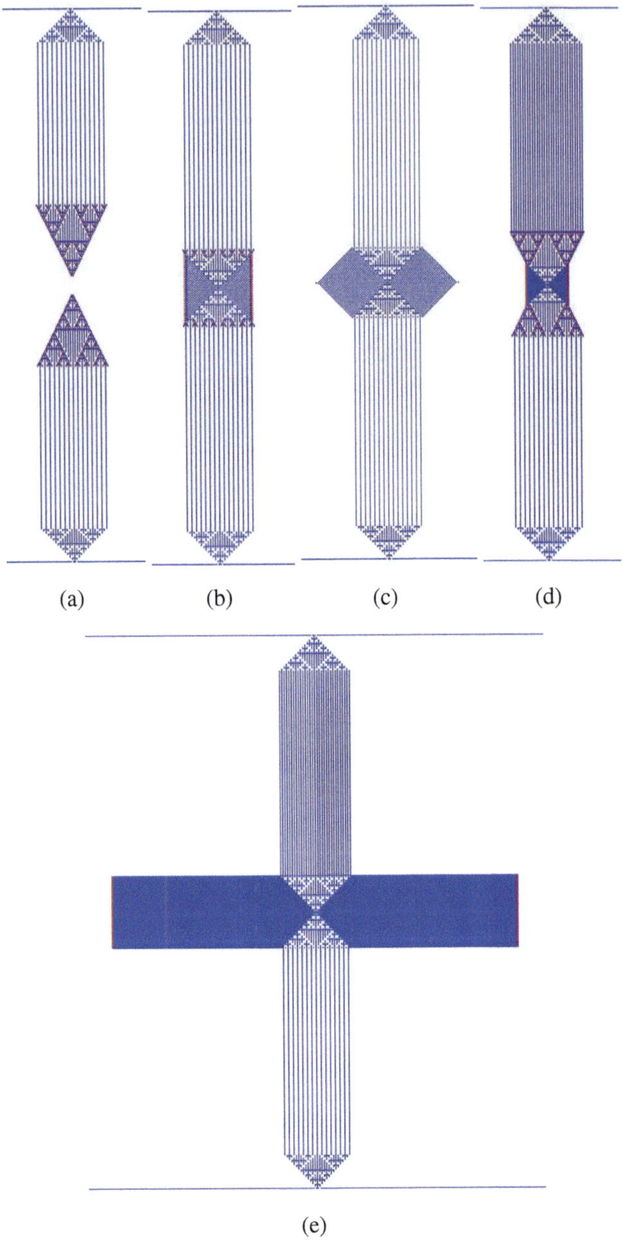

Fig. 14.18 Interaction of flying Sierpinski carpets in automata of class C_{11}. (a) Carpets approach each other. (bc) Even distance: (b) full body impact and (c) formation of debris/annihilation. (de) Odd distance: (d) full body impact and (e) formation of new travelling localizations. The configurations are generated by function $\mathscr{M}(2811)$; seed's size is two cells; distance of a seed from southern refractory strip (launchpad) is 50 cells. Excited cells are red (light grey), refractory cells are blue (dark grey).

gliders. Eastward stream of westward gun collides with westward stream of eastward gun. Refractory patters are formed in the result of collision, and glider streams are pruned: only every fifth glider of westward stream survives collision and only every fourth glider of eastward stream survives the collision (Fig. 14.15).

Functions $\mathcal{M}(2a1b)$, where $a = 3, \ldots, 8$ and $b = 1, 2$, generate spaceships ($b = 1$) and Sierpinski flying carpets ($b = 2$). Sierpinski carpets and spaceships observed in many types of cellular automata including Conway's Game of Life [229, 259]. However, we believe, travelling Sierpinski carpets have never been observed before in cellular automata. Examples of Sierpinski carpets generated by function $\mathcal{M}(2311)$ from seeds of different sizes are shown in Fig. 14.16. The carpets always leave a trail of refractory states. Therefore they can be considered as analogies of puffer trains [212] in the Game of Life cellular automata.

Spaceships do not leave refractory debris behind. Several stages of spaceships development are illustrated in Fig. 14.17. A strip of excited states, e.g. '+', "++', "+++' etc., leads to formation of self-similar growing rhomb (Fig. 14.17a and Fig. 14.10ae). When the south part of the rhomb gets in touch with a refractory strip it becomes 'infected' with refractory states, or frozen, thus only north part of the rhomb continues travelling (Fig. 14.17bc). At some stage a spaceship becomes disconnected from refractory debris it left and gets 'airborne' (Fig. 14.17d).

Both Sierpinski flying carpets and spaceships observed are rather close relatives of 'grayships' [229], because they are comprised of still patterns which can be extended to arbitrary size. The size of each carpet or spaceship is determined by the distance of its seed from the launchpad (a chain of refractory states, which prevents the carpet/spaceship from expanding in one direction).

Collisions between the flying carpets and spaceships develop along 'classical' scenarios of collisions between travelling localizations: annihilation and reflection, possibly with emergence of new localizations (Fig. 14.18).

14.5 Summary

Aiming to design a cellular automaton model of a discrete memristive medium we developed a model of excitable cellular automaton with retained refractoriness. The model is ideologically similar to our previous model of cellular automata with retained excitation [18] but the focus shifted to retained refractoriness. An excited cell takes refractory state unconditionally, i.e. independently of the number of its excited neighbours, however the refractory cell can remain in its refractory state depending on the number of its excited neighbours. A propagation of excitation in a medium can be considered as an analog of electrical current. A refractory state is an automaton analog of high-resistance state of a memristive element, and a switching form the refractory state to resting state, controlled by a density of local excitation, is an analog of switching the memristive element to its state of low-resistance. The automata studied are semi-memristive because switching from low-resistance (resting) state to high-resistance (refractory) state is not controlled by electrical current (density of local excitation).

Classical automaton models of excitable media are usually based on the principle of threshold excitation: a resting cell excites if the number of excited neighbours exceeds certain threshold. This threshold-view on excitation dynamics was the main reason why mobile localizations have not been discovered for so long time. In [9] we introduced an interval-based excitation: resting cell takes excited state if the number of its excited neighbours belongs to some specified interval. The approach proved to be productive — a range of non-classical computing devices [14], based on the rich phenomenology of excitation in interval-excitation-based models, was designed. Therefore we employed interval-based excitation and also interval-based recovery, i.e. transition from refractory to resting state, to imitate semi-memristive automata.

Based on functions' behavioural grouping, visual categorisation of configurations generated in a response to random perturbation, ordering functions on their morphological diversity and expressiveness we selected eleven classes of functions: $C_1, \dots, C_{11}$.

The functions of classes $C_1, \dots, C_6$ show high degree of space-filling ratios, while classes $C_7, \dots, C_{11}$ mid-range and low space-filling ratios. Low space-filling classes can be further sub-divided into classes with extinct excitation, namely C_7 and C_8, where only a single centripetal wave-front is formed, and classes that support travelling localizations: C_9, C_{10}, C_{11}. The low space-filling classes $C_7, \dots, C_{11}$ also demonstrate higher ratio of excited states over refractory states in configuration functions of the classes generate.

Top four classes in the hierarchy of morphological diversity are C_1, C_3, C_4 and C_{10}. Automata from C_1, C_3 and C_4 show rather quasi-chaotic response to spatially extended random excitation, and this is the reason for their high morphological diversity. Class C_{10} is more interesting. A spatially extended random perturbation leaves a random configuration of refractory states filled with non-expanding, sometimes breathing, domains of localised excitations and a combination of propagating wave-fronts, wave-fragments and travelling localizations.

Top four classes in the hierarchy of generative diversity are C_1, C_2, C_9 and $C_1 1$. Classes C_1 and C_2 show quasi-chaotic dynamics and even a point-wise excitation of a resting lattice leads to formation of a complex growing pattern. Class C_9 got to the top four of generative diversity because a point-wise excitations give birth to expanding Sierpinski carpets, and disordered and highly-ordered arrangements of refractory states.

To refine our classification even further we introduced a measure of expressiveness, which relates morphological diversity to space-filling ratio, and thus reflects an 'economy of diversity' in cellular automaton configurations. As you can see in Fig. 14.8, classes $C_9, \dots C_{11}$ show higher degrees of expressiveness. Thus we can propose that functions which support travelling localizations may be at the mid-range levels of morphological diversity but top levels of expressiveness.

We briefly characterised a range of travelling localizations which emerged in the cellular automata studied. We demonstrated that many localizations are capable for laying sophisticated patterns of refractory, high-resistive in memristor's terminology, states. Such patterns could be used in further studies to compartmentalise otherwise homogeneous spatially extended memristive media. We found outcomes of collisions between propagating localisation are quite rich and therefore all scenarios of collision-based computing [10] can be implemented in semi-memristive cellular automata.

Chapter 15
Structural Dynamics: Memristive Excitable Automata

A memristive automaton is a structurally-dynamic excitable cellular automaton where a link connecting two cells is removed or added if one of the cells is in excited state and another cell is in refractory state.

A cellular automaton $\mathscr{A}$ is an orthogonal array of uniform finite-state machines, or cells. Each cell takes finite number of states and updates its states in discrete time depending on states of its closest neighbours. All cells update their states simultaneously by the same rule. We consider eight-cell neighbourhood and three cell-states: resting $\circ$, excited $+$, and refractory $-$. Let $u(x) = \{y : |x-y|_{L\infty} = 1\}$ be a neighbourhood of cell x. A cell x has a set of incoming links $\{l_{xy} : y \in u(x)\}$ which take states 0 and 1. A link l_{xy} is a link of excitation transfer from cell y to cell x. A link in state 0 is considered to be high-resistant, or non-conductive, and link in state 1 low resistant, or conductive. A link-state l^t_{xy} is updated depending on states of cells x and y at time step t: $l^t_{xy} = f(x^t, y^t)$. Resting state gives little indication of cell's previous history, therefore we will consider only non-resting cells contributing to a link state updates. When cells x and y are in the same state (bother cells are in state $+$ or both are in state $-$) no 'current' can flow between the cells, therefore scenarios $x^t = y^t$ are not taken into account. Thus we assume that the only situations when $x^t, y^t \in \{+,-\}$ and $x^t \neq y^t$ may lead to changes in links conductivity:

$$l^{t+1}_{xy} = \begin{cases} a, x^t = + \text{ and } y^t = - \\ b, x^t = - \text{ and } y^t = + \\ l^t_{xy}, \text{ otherwise} \end{cases} \tag{15.1}$$

where $a \neq b$ and $a, b \in \{0,1\}$. Thus we consider two types of automata $\mathscr{A}^{ab}$: $\mathscr{A}^{01}$ and $\mathscr{A}^{10}$.

A resting cell excites ($x^t = \circ \rightarrow x^t = +$ transition) depending on number of excited neighbours.

There are two ways to calculate a weighted sum of number of excited neighbours:

1. $\Sigma_+ = \sum_{y \in u(x)} \chi(y^t, +)$
2. $\sigma_+ = \sum_{y \in u(x)} l_{xy} \chi(y^t, +)$,

where $\chi(y^t, +) = 1$ if $y^t = +$ and $\chi(y^t, +) = 0$ otherwise.

A. Adamatzky: Reaction-Diffusion Automata, ECC 1, pp. 287–310.
springerlink.com

Thus, we consider two types of memristive automata. In automaton $\mathscr{A}_1$ a resting cell excites if $\sigma_+ > 0$. In automaton $\mathscr{A}_2$ a resting cell excites if $\Sigma_+ > 1$ or $\sigma_+ > 0$.

Note 15.1. Automaton $\mathscr{A}_1$ is a pre-memristive cellular automaton because it imitates only polarity (links update (15.1)) not voltage; automaton $\mathscr{A}_2$ is a memristive cellular automaton because it imitates both polarity (links update (15.1)) and current intensity (use of Σ_+ and σ_+).

A polarity of a current is imitated by excitable cellular automaton using excited $+$ and $-$ refractory states. If a cell x is in state $-$ and a cell y is in state $+$ then cell y symbolises an anode, and cell x a cathode. And we can say that a current flows from y to x. Indeed, such an abstraction is at the edge of physical reality, however this is the only way to develop a *minimal* discrete model of a memristive network. In automaton $\mathscr{A}_2$ the condition $\Sigma_+ > 1$ symbolises propagation of a high intensity current along all links, including links non-conductive for a low intensity current. This high intensity current in $\mathscr{A}_2$ resets conductivity of the links and also states of cells. The condition $\sigma_+ > 0$ reflects propagation of a low intensity current along conductive links. The current of low intensity does not affect states of links but only states of cells.

We experiment with 300×300 cell automaton arrays, with non-periodic absorbing boundaries. We conduct experiments for two initial conditions on links' 'conductivity' —

- L_1-condition: all links are conductive (for every cell x and its neighbour y $l^0_{xy} = 1$), and
- L_0-condition: all links are non-conductive (for every cell x and its neighbour y $l^0_{xy} = 0$).

While testing automata's response to external excitation we use point-wise and spatially extended stimulations. By point-wise stimulation we mean excitation of a single cell ($\mathscr{A}_1$) or a couple of cells ($\mathscr{A}_2$) of resting automata. These are minimal excitations to start propagating activity.

Let D-disc be a set of cells which lie at distance not more than r from the array centre $(n/2, n/2)$, $n = 300, r = n/3)$. When undertaking D-stimulation we assign excited state $+$ to a cell of D with probability 0.05. In some cases we apply D-stimulation twice as follows. An automaton starts in L_1- or L_2-condition, we apply D-stimulation first time (we call it E_1-excitation) and wait till excitation waves propagate beyond boundaries of the array or a quasi-stationary structure is formed. After this transient period we apply D-stimulation again (E_2-excitation) without resetting links states.

Space-time dynamics of automata is illustrated by configurations of excitations and dynamics of link conductivity is shown either explicitly by arrows (in small configuration) or via grey-scale representation of cells' in-degrees: cell x is represented by pixel with grey-value $32 \cdot \Sigma_{y \in u(x)}$. Despite not representing exact configuration of local links, in-degrees give us a rough indicator of spatial distribution of conductivity in the medium. The higher is the in-degree at a given point, the higher is the conductivity at this point.

15.1 Phenomenology

Finding 48. *A point-wise stimulation of automaton $\mathscr{A}_2$ leads to a persistent excitation, while automaton $\mathscr{A}_1$ returns to a resting state.*

A single-cell excitation of resting automaton $\mathscr{A}_1$ (Fig. 15.1) or two-cell excitation of resting automaton $\mathscr{A}_2$ (Fig. 15.2) in L_1 initial conditions lead to formation of a 'classical' excitation wave-front. The wave-front propagates omni-directionally away from the initial perturbation site and updates states of links it is passing through.

Links leading from cells to the neighbours they excited are made non-conductive in development of $\mathscr{A}_i^{01}$ (Figs. 15.1 and 15.2); or we can say that links corresponding to normal vectors of propagating wave-front are made non-conductive. Cell excited at time step $t = 1$ becomes isolated. The situation is similar in development of automata $\mathscr{A}_1^{10}$ and $\mathscr{A}_2^{10}$ with the only difference that links connecting cells which are excited at time step t to cells they have been excited by are removed.

In summary, in automata $\mathscr{A}_i^{01}$ links associated with forward propagation of perturbation are made non-conductive, and in $\mathscr{A}_i^{10}$ links associated with backward propagation are made non-conductive. Excitation wave-front travelling from a single stimulation site forms a domain of co-aligned links. Excitation waves initiated in different cells collide and merge. Boundaries between domains formed by different fronts are represented by distinctive configurations of links (Fig. 15.3).

Automaton $\mathscr{A}_1$ is non-excitable in L_0-conditions because no excitation can propagate along non-conductive links. An outcome of two-site excitation of resting automaton $\mathscr{A}_2$ in L_0-condition depends on configuration of the initial excitation.

Let two diagonal neighbours (north-west and south-east) be excited (Fig. 15.4a) at time step t. These two cells transfer excitation to their two neighbours (north-east and south-west) as shown in Fig. 15.4b. Only links from north-west cell to north-east and south-west, and south-east to south-west and north-east are formed (Fig. 15.4c) and excitation becomes extinguished (Fig. 15.4d).

Let excited cells be neighbours at the same row of cells (Fig. 15.5a). If they are excited at time step t then their north and south neighbours are excited due to condition $\Sigma_+ > 1$ taking place (Fig. 15.5b). The localised (two-cell size) excitations propagate north and south (Fig. 15.5c) and make links pointing backwards (towards source of initial excitation) conductive. At the second iteration cell lying east and west of initially perturbed cells becomes excited (Fig. 15.5c). By that initially perturbed cells return to resting state (Fig. 15.5d) and thus they become excited again. A growing pattern of recurrent excitation fills the lattice (Fig. 15.5de).

Finding 49. *Repeated stimulation of memristive automata in a spatially-extended domain leads to formation of either disorganised activity domain emitting target waves of excitation or a sparse configuration of stationary oscillating localizations.*

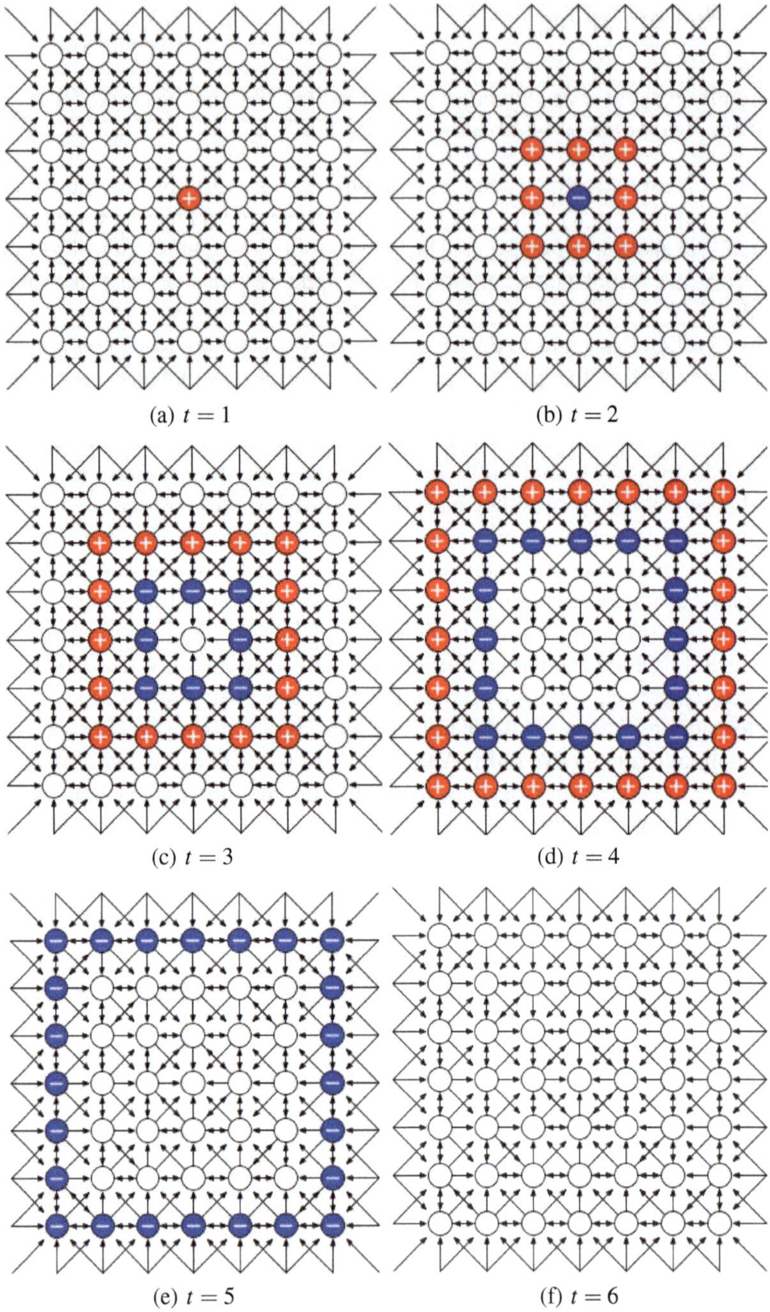

(a) $t = 1$ (b) $t = 2$

(c) $t = 3$ (d) $t = 4$

(e) $t = 5$ (f) $t = 6$

Fig. 15.1 Snapshots of excitation and links dynamics in automaton $\mathscr{A}_1^{01}$. In initial configuration ($t = 1$) one cell is excited others are resting. Arrow from cell x to cell y means $l_{xy} = 1$. Excited cells are shown by red discs with sign '+', refractory cells by blue discs with sign '-', resting cells are blank. Links at the edges of cellular array corresponds to an absorbing boundary cells, which are always in resting state independently on states of their neighbours.

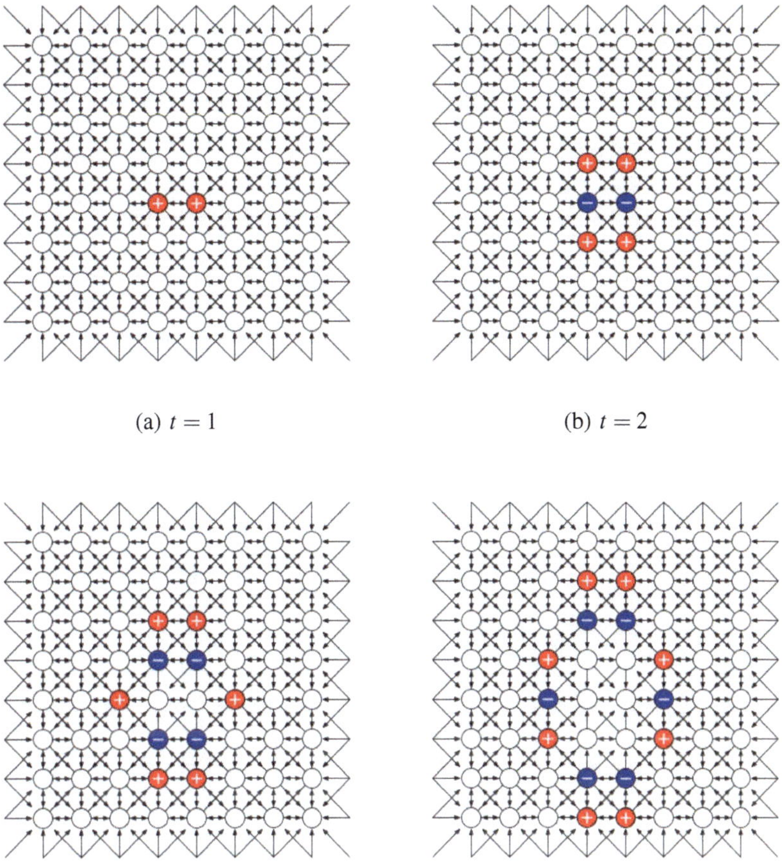

(a) $t = 1$ (b) $t = 2$

(c) $t = 3$ (d) $t = 4$

Fig. 15.2 Snapshots of excitation and links dynamics in automaton $\mathscr{A}_2^{01}$. In initial configuration two cells (minimum size of non-dying excitation for $\mathscr{A}_2$) are excited others are resting. Arrow from cell x to cell y means $l_{xy} = 1$. Excited cells are shown by red discs with sign '+', refractory cells by blue discs with sign '-', resting cells are blank.

Examples of excitation dynamics in response to E_1- and E_2-excitations are shown in Fig. 15.6. Each singular perturbation in the first stimulation of automata being in L_1-condition leads to propagation of excitation. The waves merge into a single wave and disappear beyond lattice boundary. Merging of waves outside D into a single wave is reflected in cells' in-degree distribution in Fig. 15.19.

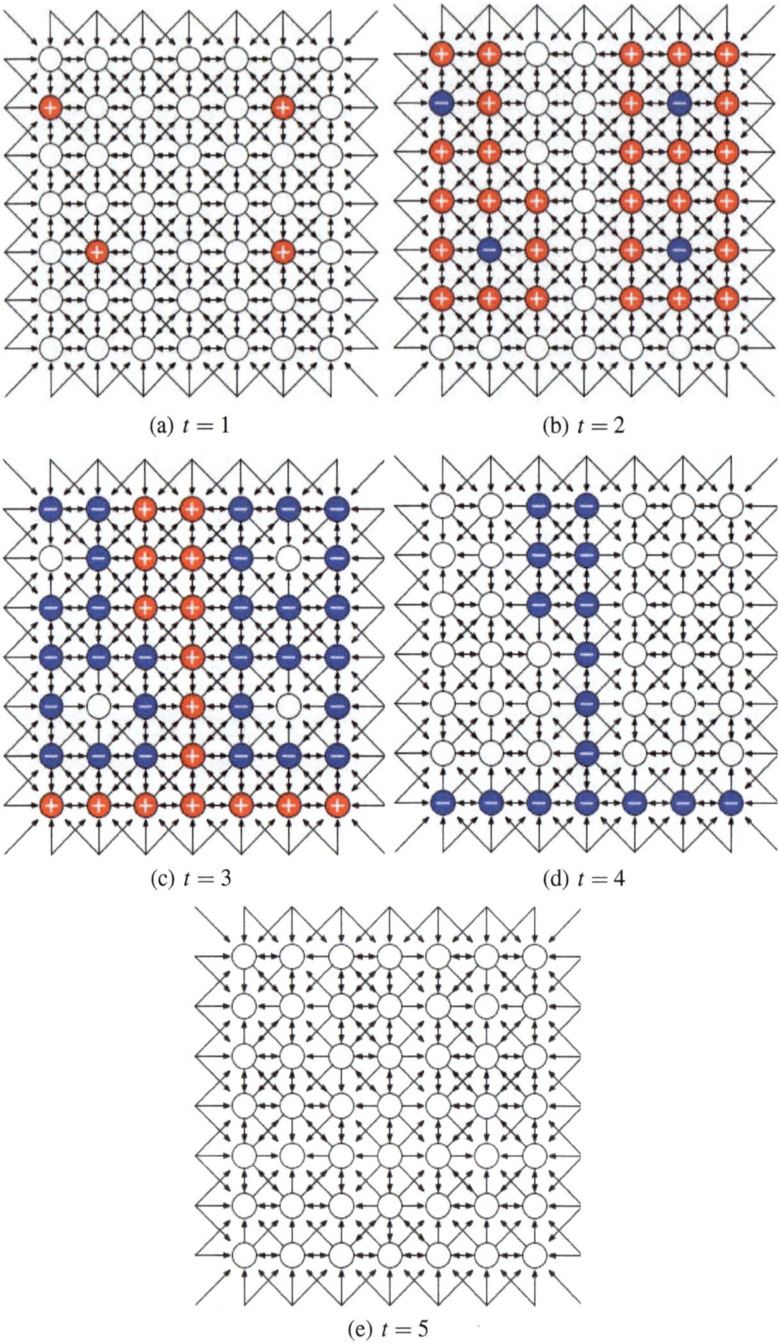

Fig. 15.3 Snapshots of excitation and links dynamics in automaton $\mathscr{A}_1^{01}$. In initial configuration several cells (those in '+'-state at $t = 1$) are excited others are resting. Arrow from cell x to cell y means $l_{xy} = 1$.

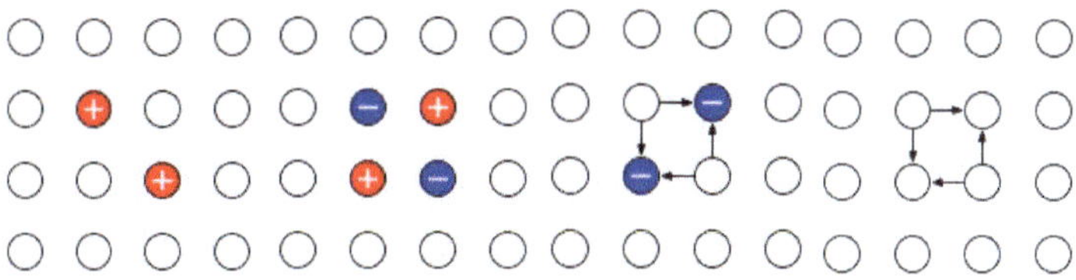

Fig. 15.4 Snapshots of excitation and links dynamics in automaton $\mathscr{A}_2^{01}$, L_0-start. In initial configuration two cells (minimum size of non-dying excitation in $\mathscr{A}$) are excited others are resting. Arrow from cell x to cell y means $l_{xy} = 1$.

Automata' behaviour become different after second D-stimulation (E_2-excitation). They all exhibit quasi-chaotic dynamic inside boundaries of disc D (shown by circle in Fig. 15.6); cell in-degrees' distributions of the automata are similar (Fig. 15.18). However the excitation dynamics in automaton $\mathscr{A}_1^{01}$ is reduced (after long transient period) to stationary oscillating localizations while in automata $\mathscr{A}_1^{10}$, $\mathscr{A}_2^{01}$, and $\mathscr{A}_2^{10}$ excitations outside boundaries of initial stimulation merge into target waves (Fig. 15.6 and Fig. 15.19). Oscillating localizations developed in $\mathscr{A}_1^{01}$ stay inside the boundaries of D.

Random excitation is extinguished immediately in automata $\mathscr{A}_1$ being in L_0-condition of total non-conductivity. When automaton $\mathscr{A}_2$ in L_0-condition is excited, the quasi-random excitation activity persists inside boundaries of D while omnidirectional waves are formed outside D. Patterns of activity are not changed significantly after second random perturbation, E_1-excitation (Fig. 15.6).

If there are both excited and refractory states in the external stimulation domain the developments are almost the same but $\mathscr{A}_1^{10}$ and $\mathscr{A}_2^{10}$ shows persistent excitation activity already at the first stimulation.

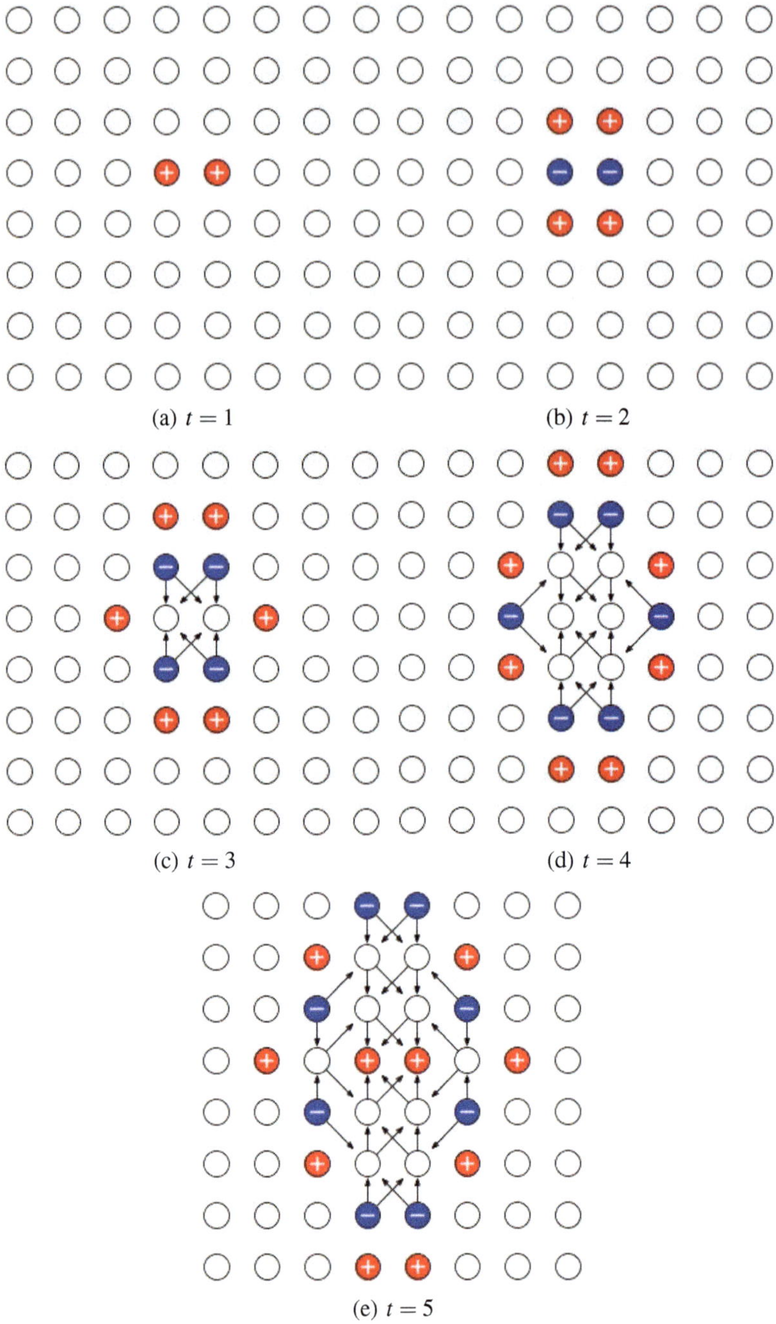

Fig. 15.5 Snapshots of excitation and links dynamics in automaton $\mathscr{A}_2^{10}$, L_0-start. In initial configuration two cells (minimum size of non-dying excitation) are excited others are resting. Arrow from cell x to cell y means $l_{xy} = 1$.

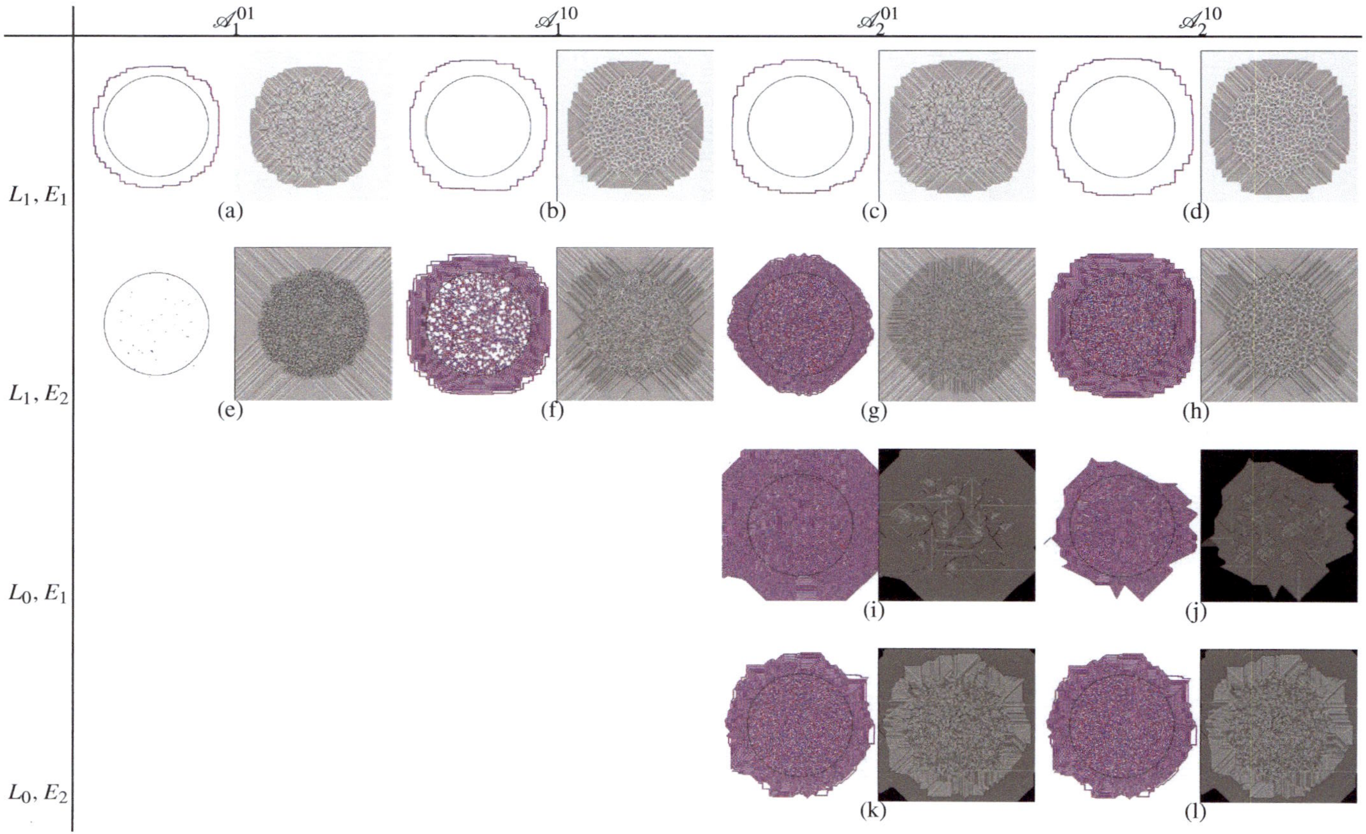

Fig. 15.6 Example of automaton behaviour in response to D-stimulation. Boundaries of disc D are shown as circles. Topology of links is represented by grey-levels of cells' in-degrees.

15.2 Oscillating Localisations

Finding 50. *E_2-excitation of $\mathscr{A}_1^{01}$ leads to formation of excitation wave-fragments trapped in a structurally defined domains.*

E_2-excitation of $\mathscr{A}_1^{01}$ leads to formation of sparsely distributed localised oscillating excitations, or oscillators (Fig. 15.7). An oscillating localisation (oscillator) usually consist of one or two mobile localizations which shuffle inside a small compact domain of the automaton array. This micro-wave is updating states of links and thus influencing its own behaviour. $\mathscr{A}_1^{01}$, after E_2-stimulation and formation of oscillating localizations, is characterised by a smooth balanced distribution of cells in-degrees (Fig. 15.18, table entry L_1, E_2).

Examples of two most commonly found oscillators — O_1 and O_2 — are shown in Figs. 15.8 and 15.9, and their characteristics in Fig. 15.10. Both oscillators have exactly the same minimum and maximum masses (measured as a sum of cells in excited and refractory states). Oscillator O_2 has much longer period than O_1 and larger maximum density. Oscillator O_1 spans large space during its transformation cycle, it occupies a sub-array of 6×6 cells when in its largest form.

In its minimal form O_1 consists of two cells: one cell is in state $+$ another in state $-$ (Fig. 15.8, $t = 0$). At next two steps of O_1's transformations a small localised excitation is formed. It propagates north (Fig. 15.8, $t = 1,2,3$). At fourth step of oscillator O_1's transformation the travelling localised excitation splits into two excitations: one travels north-west, another south (Fig. 15.8, $t = 4,5$). The localisation travelling south returns to exact position of the first step excitation, just with swapped excited and refractory states (Fig. 15.8, compare $t = 6$ with $t = 0$) and then repeats a cycle of transformations $t = 1,2$ (compare $t = 1$ with $t = 7$ an $t = 2$ with $t = 8$). The excitation travelling north-west become extinguished ($t = 8,9$).

A couple $(-+)$ is a minimal configuration of oscillator O_2 (Fig. 15.9, $t = 0$). When the localisation starts it is development in configuration $(-+)$, and configuration of links' conductivity as shown in Fig. 15.9, it is transformed into excitation wave-fragment propagating east and north-east (Fig. 15.9, $t = 1,2,3$). At fourth step of oscillator's transformation the wave-fragment shrinks and by step $t = 6$ the oscillator repeats its original state $(-+)$ yet rotated by 90° clockwise. New excitation wave-fragment emerges and propagates west and south-west (Fig. 15.9, $t = 7 \dots 10$). The fragment contracts to configuration $(+-)$ by step $t = 12$. Then development and transformation of excitation wave-fragments is repeated (Fig. 15.9, $t = 13, \cdots, 23$). The localisation returns to its original configuration $(-+)$ at $t = 26$.

Most oscillating localizations observed in experiments with $\mathscr{A}_1^{01}$ have very long periods. This is because the oscillators' behaviour is determined not only by cell-state transitions rules, as in classical cellular automata, but also by topology of links modified by repeated random stimulation and dynamics of the links affected by oscillators themselves.

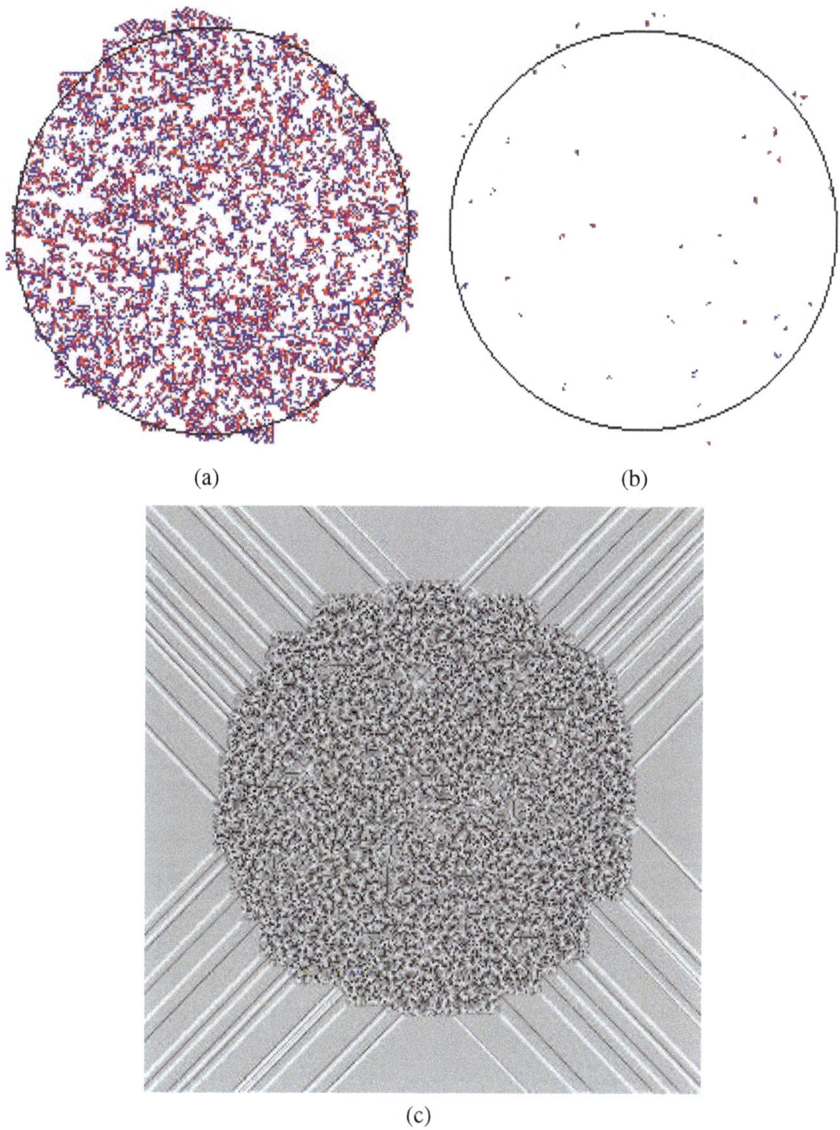

Fig. 15.7 Formation of localised excitations in $\mathscr{A}_1^{01}$, L_1-start. A localised excitation occupies a fixed size domain, the excitation may change its size periodically but never expands more then the certain fixed size. (a) configuration of automaton after E_2-excitation, boundary of D is shown by black circle; (b) the same automaton after transient period, only few oscillating localizations sustain, (c) links' conductivity presented via grey-values of cells' in-degrees.

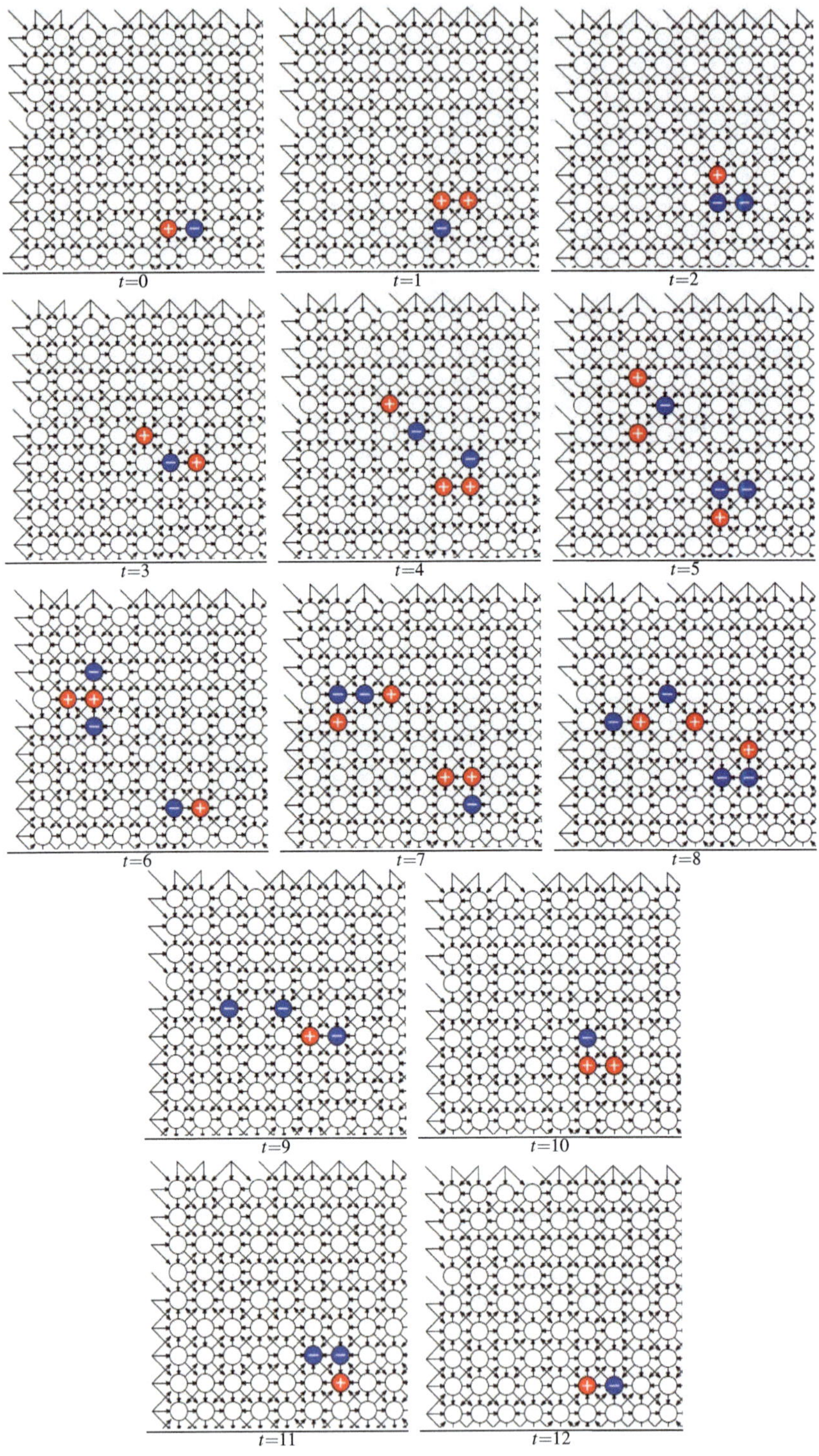

Fig. 15.8 Example of oscillator O_1 in $\mathscr{A}_1^{01}$

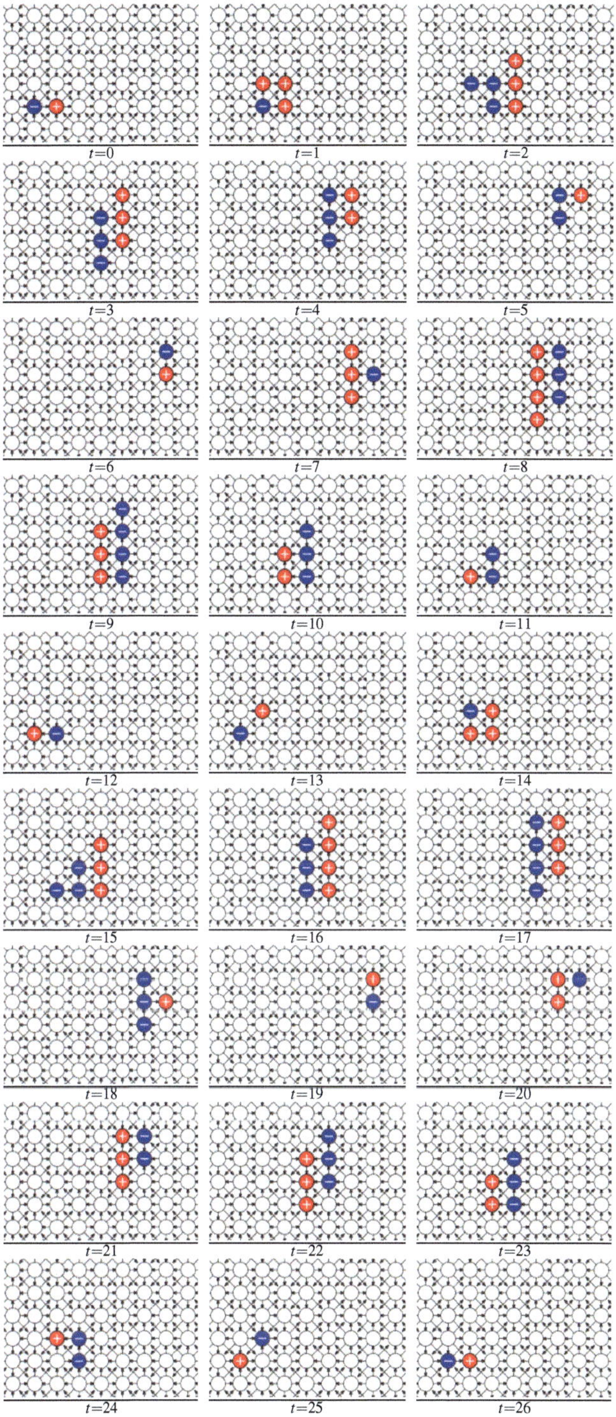

Fig. 15.9 Example of oscillator O_2 in $\mathscr{A}_1^{01}$

Characteristic	O_1	O_2
Period	12	26
Minimum mass	2	2
Maximum mass	7	7
Minimum size	2	2
Maximum size	36	9
Minimum density	1	1
Maximum density	$\frac{1}{6}$	$\frac{1}{2}$

Fig. 15.10 Characteristics of commonly found oscillators O_1 and O_2 in $\mathscr{A}_1^{01}$

15.3 Dynamics of Excitation on Interfaces

Finding 51. *Automaton $\mathscr{A}_1^{01}$ exhibits localizations propagating along the boundary of disordered and ordered conductivity domains.*

As previously discussed, after E_2-excitation of $\mathscr{A}_1^{01}$ the automaton becomes separated into two domains. One domain has a disorganised appearance, it lies inside boundaries of D. This domain is characterised by quasi-random discoidal configuration of cell in-degrees (Fig. 15.7c). Second domain, cells lying outside D, represents ordered configurations of links (Fig. 15.7c). The links are co-alined during target wave propagation outward D. See Figs. 15.18 and 15.19 for distribution of cell in-degrees in disordered and ordered domaines.

In automaton $\mathscr{A}_1^{01}$ a link from cell y to cell x becomes non-conductive at time step $t+1$ if $x^t = +$ and $y^t = -$. Thus target waves travel outside of D form non-conductive pathways. If we stimulate automaton in these domains of ordered links no distal propagation of excitation will be observed, however, a worm of disordered excitations is formed at the boundary between disordered and ordered configurations of links.

When configuration of oscillating localizations becomes stationary — in a sense that the localizations do not travel outside their fixed domains — we apply a few-cells wide excitation at the boundary of D (Fig. 15.11a), north-north-west localised excitation, which exceeds in size existing oscillating localizations). An extended cluster, or a worm (a group of excited and refractory states with a preferable direction of growth), of excited and refractory cells is formed (Fig. 15.11b). It expands along the boundary between domains. Two heads of the emerging worm — one head moves clockwise and another anticlockwise — are clearly visible in (Fig. 15.11). At some point the worm undergoes sub-division (Fig. 15.11c) and it breaks into two independent worms (Fig. 15.11d). Each new worm has an extended head and gradually thinning tail. Thickness of a worm, measured in a number of non-resting cells along the worm body, is shown in Fig. 15.12.

The worms move along boundary of D until they collide (Fig. 15.11g) and annihilate in the result of the collision (Fig. 15.11h). The configuration of oscillating localizations then returns to its original state (Fig. 15.11i).

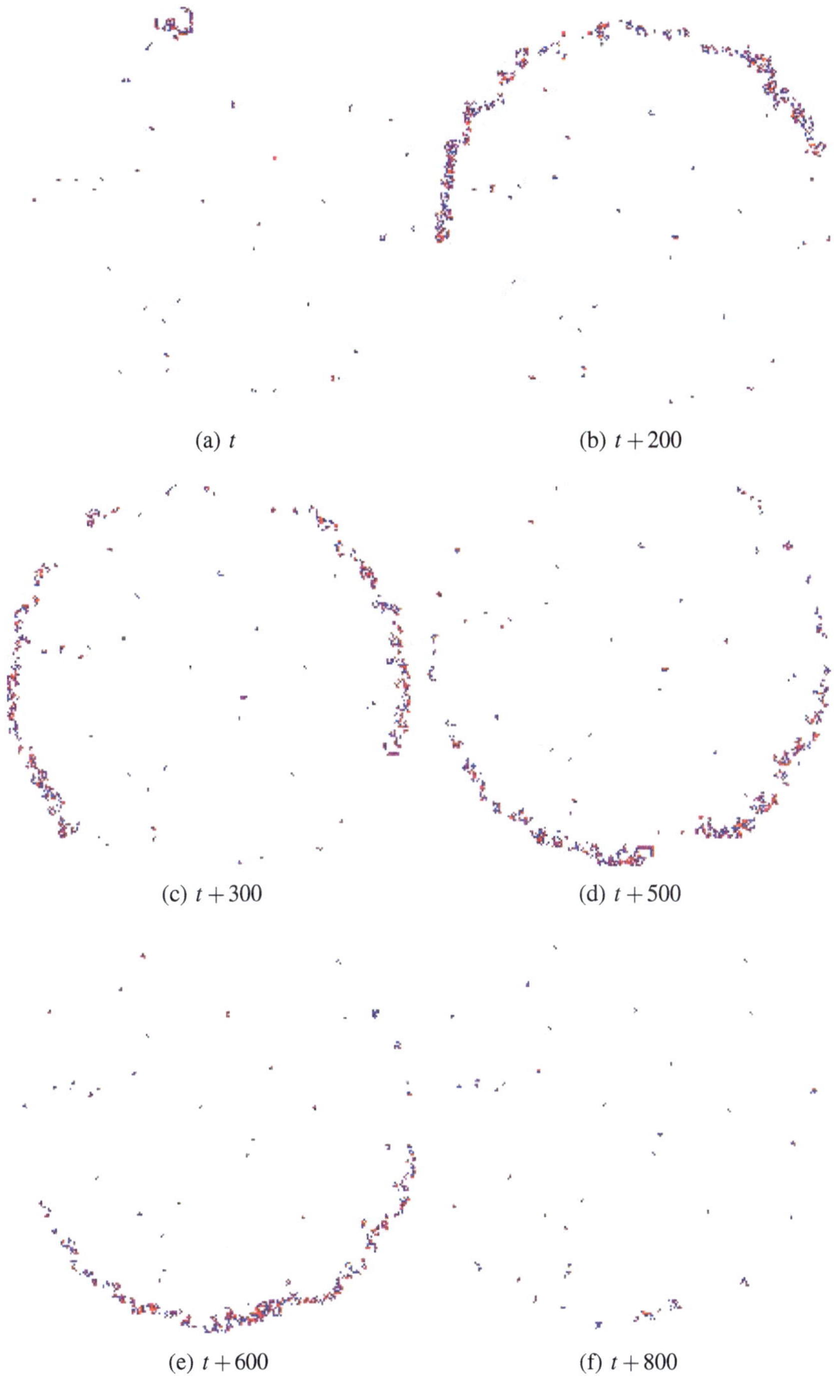

Fig. 15.11 Propagation of localised excitations at the interface between disordered and ordered conductivity domains, $\mathscr{A}_1^{01}$ automaton, L_1-start

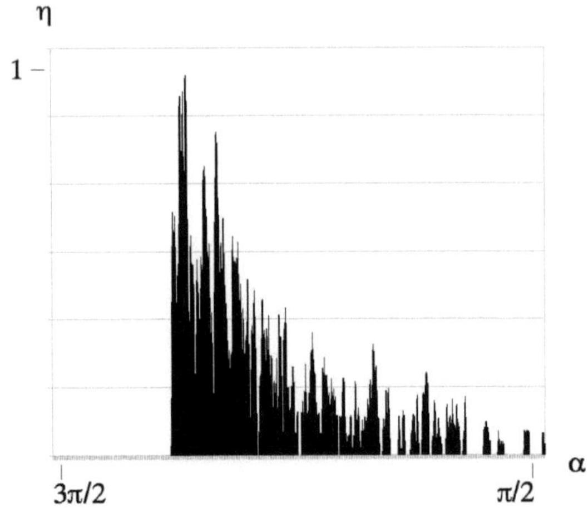

Fig. 15.12 Illustrative profile of a worm excitation propagating anti-clockwise in Fig. 15.11. $\eta(\alpha)$ is a normalised number of cells in non-resting states lying at distance $[n/3-10, n/3+10]$ and angle α from the centre of disc D.

15.4 Building Conductive Pathways

Finding 52. *By exploring collisions between excitation wave-fragments travelling in $\mathscr{A}_2$ one can built information transmission pathways in an initially non-conductive medium. Routing primitives realised include signal splitting, signal echoing and signal turning.*

Localizations travelling in $\mathscr{A}_2$, L_0-condition, can form pathways conductive for low-strength excitations. For example, a localisation of two excited and two refractory states propagates in $\mathscr{A}_2$ in the direction of its excited 'head'. The localisation forms a chain of conductive links oriented opposite to the localisation's velocity vector in case of $\mathscr{A}_2^{01}$, and in the direction of the localisation's propagation in case of $\mathscr{A}_2^{10}$ (Fig. 15.13).

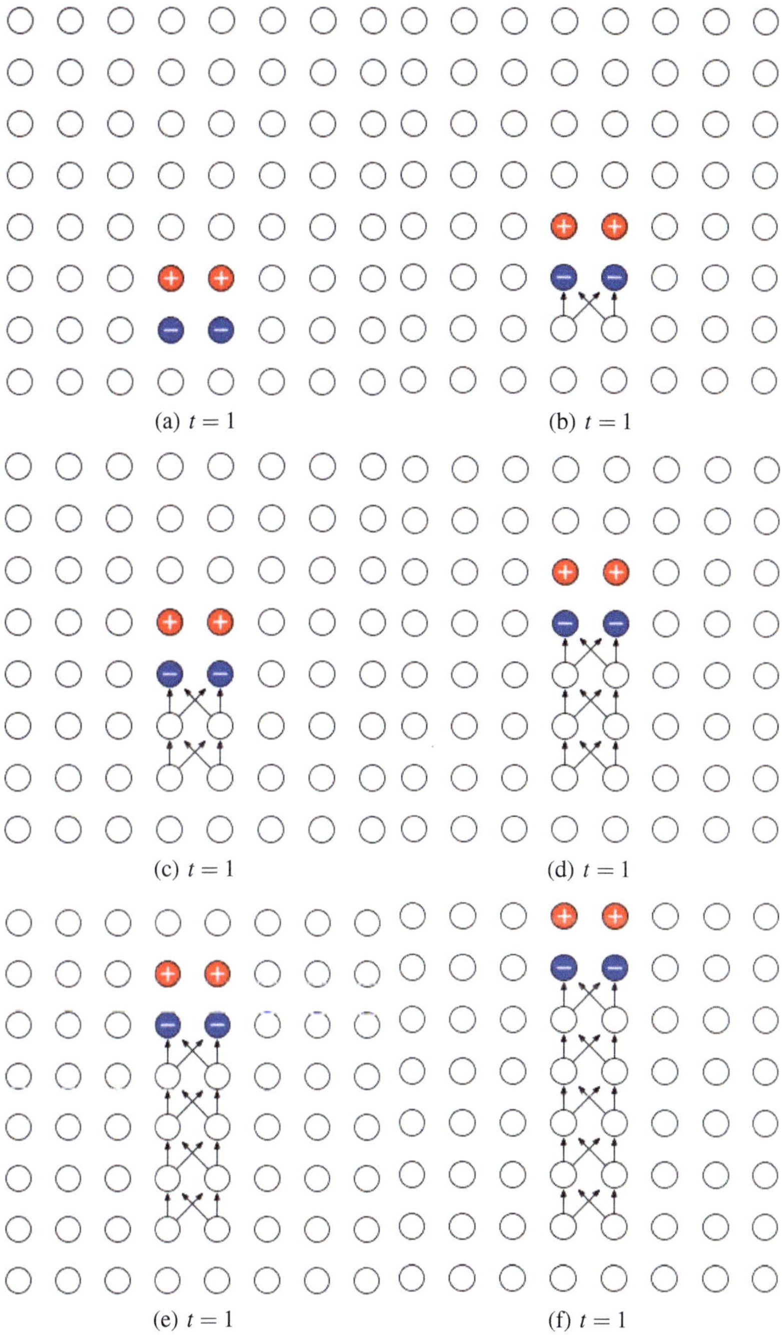

Fig. 15.13 Snapshots of excitation and links dynamics in automaton $\mathscr{A}_2^{10}$, L_0-start. In initial configuration two cells are excited, two cells are in refractory state, others are resting. Arrow from cell x to cell y means $l_{xy} = 1$.

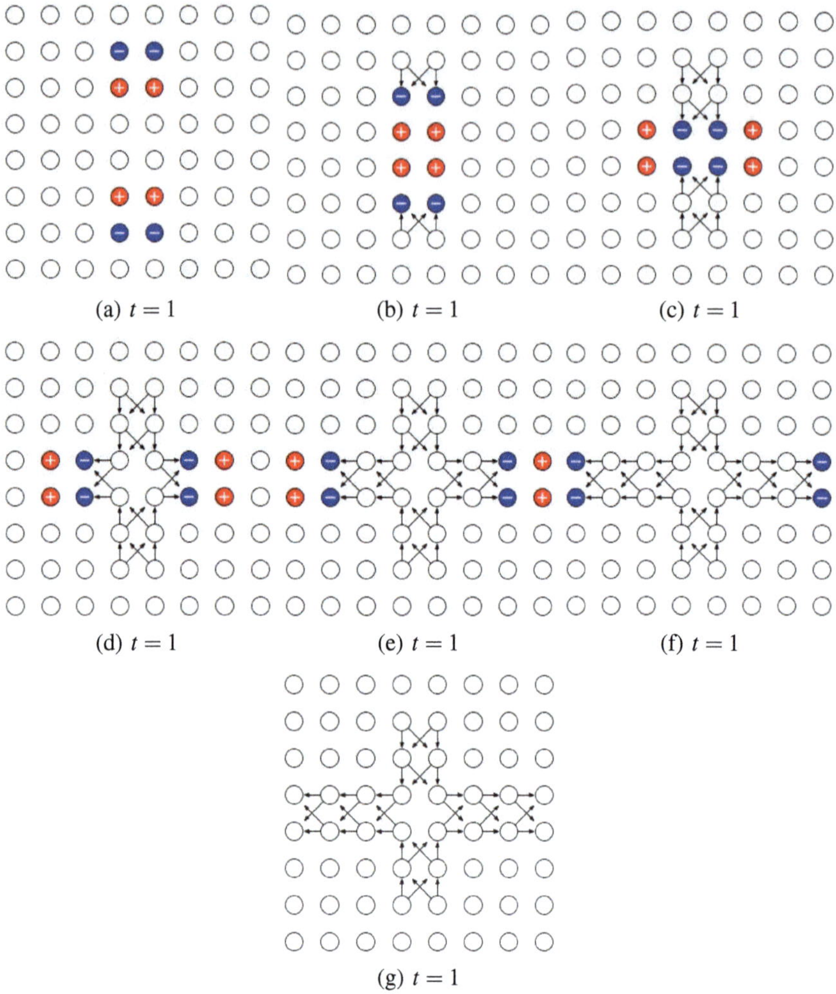

Fig. 15.14 Snapshots of excitation and link dynamics in automaton $\mathscr{A}_2^{10}$, L_0-start. In initial configuration two 2^+-particles are introduced in the medium. Four cells are excited, four cells are in refractory state, others are resting. Arrow from cell x to cell y means $l_{xy} = 1$.

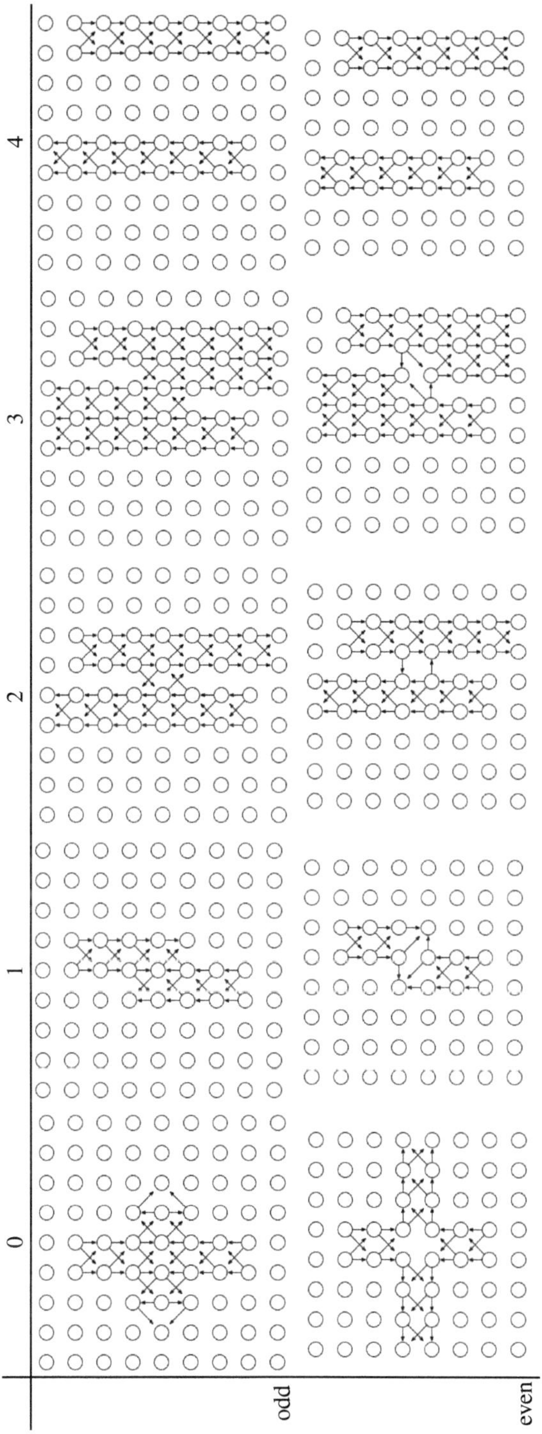

Fig. 15.15 Intersection of 'wires' build by propagating and colliding 2^+-particles. Automaton $\mathscr{A}_2^{10}$.

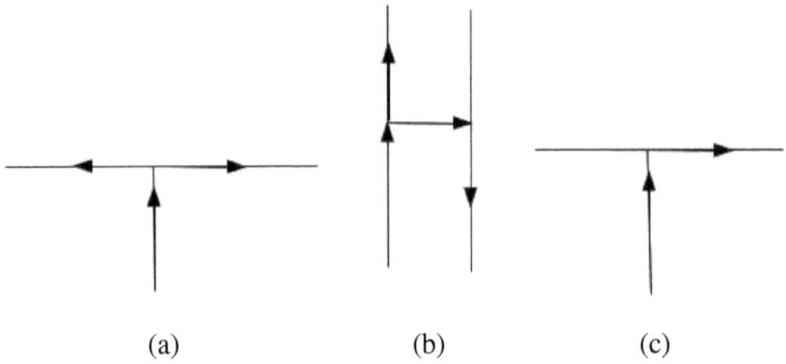

Fig. 15.16 Types of information transfer pathways laid out by interacting travelling localisations: (a) T-branching, (b) echo, (c) turning

A layout of conductive pathways, or 'wires', is determined by outcomes of collisions between path-laying particles. A detailed example of such collision-determined pathway building is shown in Fig. 15.14. Two travelling localizations, 2^+-particles[1], are initiated in $\mathscr{A}_2^{10}$ facing each other with their excited heads (Fig. 15.14a). One particle propagates south another north (Fig. 15.14a). When the particles collide they undergo elastic-like collision, in the result of which two 2^+-particles are formed: one travels east another west (Fig. 15.14c–f). When a weak excitation rule — a cell is excited if at least one neighbour is excited and no links are updated — is imposed on the automaton the pathways formed by these two colliding particles becomes selective. If an excitation is initiated in the southern or northern channel the excitation propagates till cross-junction and then branches into eastern and northern channels.

Pathways built by two 2^+-particles undergoing head-on collision being in different phases (odd and even distance between their start positions) and lateral offsets are shown in Fig. 15.15. Most T-bone collisions between 2^+-particles lead to formation of omni-directionally growing excitation patterns. Only in situations when one particle hits a tail of another particle no uncontrollable growth occurs. The particle hitting a tail of another particle extinguishes and the other particle continues its journey undisturbed.

There are primitives of information routing implementable in collisions between 2^+-particles (Fig. 15.16). T-branching, or signal splitting, (Fig. 15.16a) is built by particles colliding with zero lateral shift and even number of cells between their initial positions (e.g. Fig. 15.15, entry (even, 0)). When signal travelling north reaches cross-junctions it splits into two signals — one travel west and another travels east; no signal continues straight propagation across junction.

Echo primitive (Fig. 15.16b) is constructed in a head-on collision between 2^+-particles with lateral shift two or three cells (see (e.g. Fig. 15.15, entries (odd, 2), (odd, 3), (even, 2), (even, 3)). The echo primitive consists of two anti-aligned (e.g.

[1] 2^+-particles are localizations consisting of two excited and two refractory states, which move along rows or columns of an excitable cellular array [2, 3, 9].

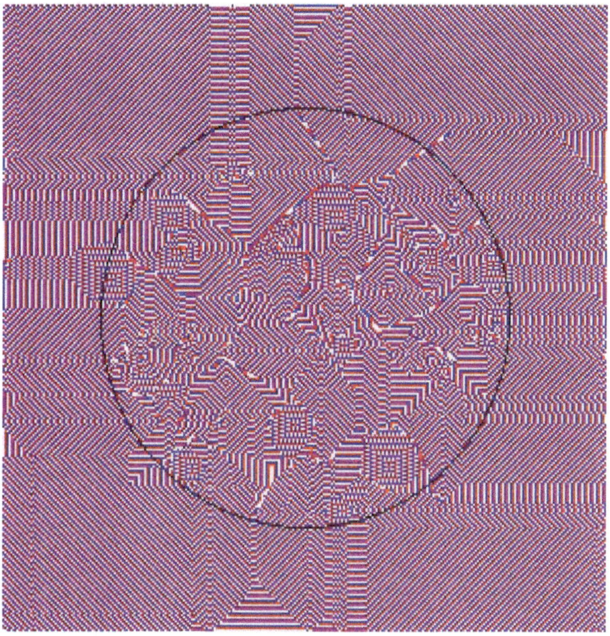

(a)

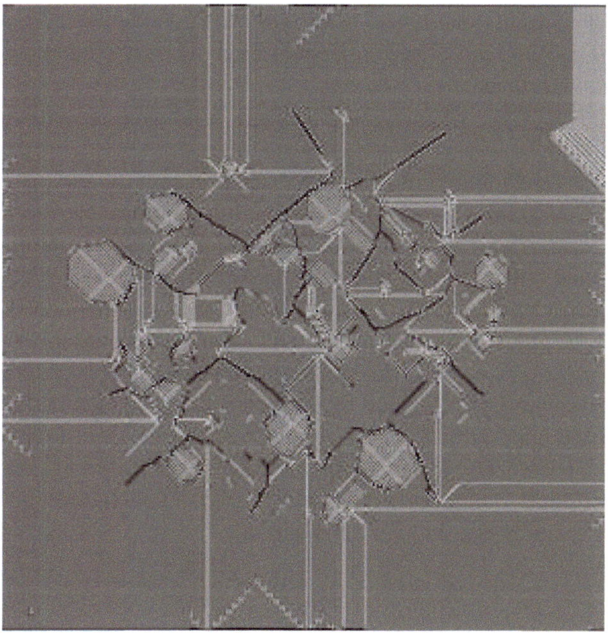

(b)

Fig. 15.17 Configuration of target waves (a) and configuration of cell in-degrees (b) in $\mathscr{A}_2^{01}$, L_0-start, D-stimulation, E_1-excitation. Boundary of D is shown by circle in (a).

one propagates information northward and another southward) parallel information pathways. There is a bridge between the pathways. When signal propagating along one pathway reaches the bridge, it splits, daughter signal enters second pathway and propagates in the direction of mother signal's origination. In (Fig. 15.16b) mother-signal propagates north and daughter-signal south.

Turn primitive (Fig. 15.16c) is implemented in T-bone collision between two 2^+-particles. A signal generated at either of the pathways propagates in the direction of 2^+-particle which trajectory was undisturbed during collision.

15.5 Summary

A minimalistic automaton model of a two-dimensional discrete memristive medium is a hybrid between classical excitable cellular automata [122] and classical structurally-dynamic cellular automata [143]. Every site of such automaton takes three states, and a binary conductivity of links is updated depending on states of sites the links connect. A memristive automaton with binary cell-states would give us even more elegant model however by using binary cell-states we could not easily detect source and sink of simulated 'currents' unless several step of historical development are taken into account. Excitable cellular automata provide us with all necessary tools to imitate current polarity and to control local conductivity. From topology of excitation wave-fronts and wave-fragments we can reconstruct relative location of a source of initiated current.

We defined two type of memristive cellular automata and characterised their space-time dynamics in response to point-wise and spatially extended perturbations. We classified several regimes of automata excitation activity, and provided detailed accounts of most common types of oscillating localizations. We did not undertake any systematic search for minimal oscillators though but just exemplified two most commonly found after random spatially-extended stimulation.

We demonstrated that few possible versions of growing conductive pathways. Opportunities to grow 'wires' in memristive automata are virtually unlimited. For example, in $\mathscr{A}_2^{01}$, L_0-start, after first D-stimulation (Fig. 15.17) generators of spiral and target waves are formed inside D. Boundaries between the generators — they provide a partial approximation of a discrete Voronoi diagram over centres of the generators — are comprised of cells with high degrees of incoming conductive links. Such chains of high in-degree cells can play a role of conductive pathways even if we increase excitation threshold of the medium.

15.6 Appendix

Distributions of cell in-degrees for automata $\mathscr{A}_1$ and $\mathscr{A}_2$ after E_1- and E_2-excitations are shown in Figs. 15.18 and 15.19. Calculations are done separately for cells lyings inside stimulation disc D (Fig. 15.18) and outside the disc (Fig. 15.19). Each chart represents distribution of cell in-degrees, where horizontal axis is a number of incoming links and vertical axis is a ratio of cells with given in-degree to a total number of cells in the analysed domain.

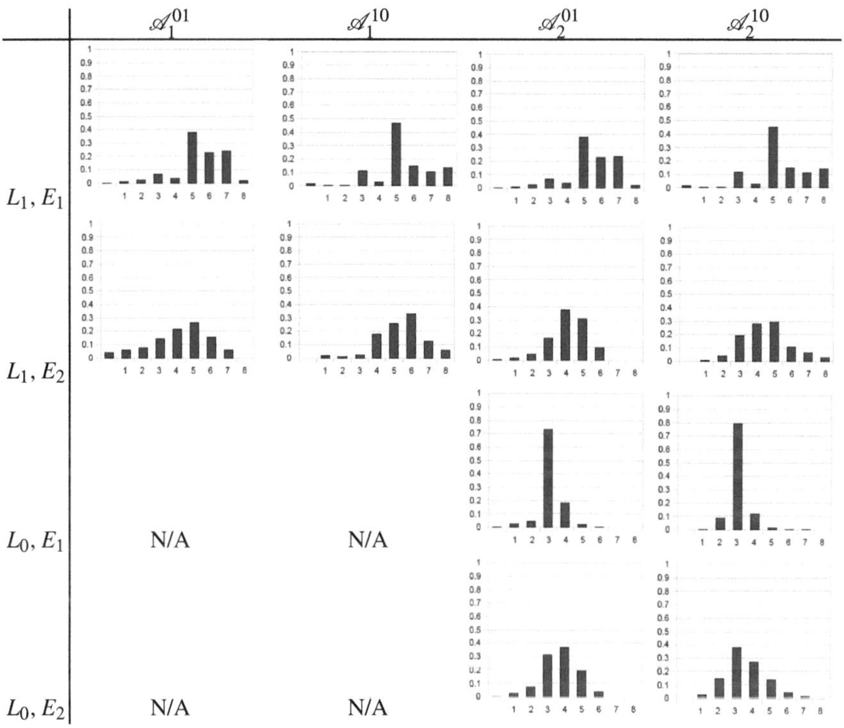

Fig. 15.18 Distributions of cells' in-degrees inside disc D

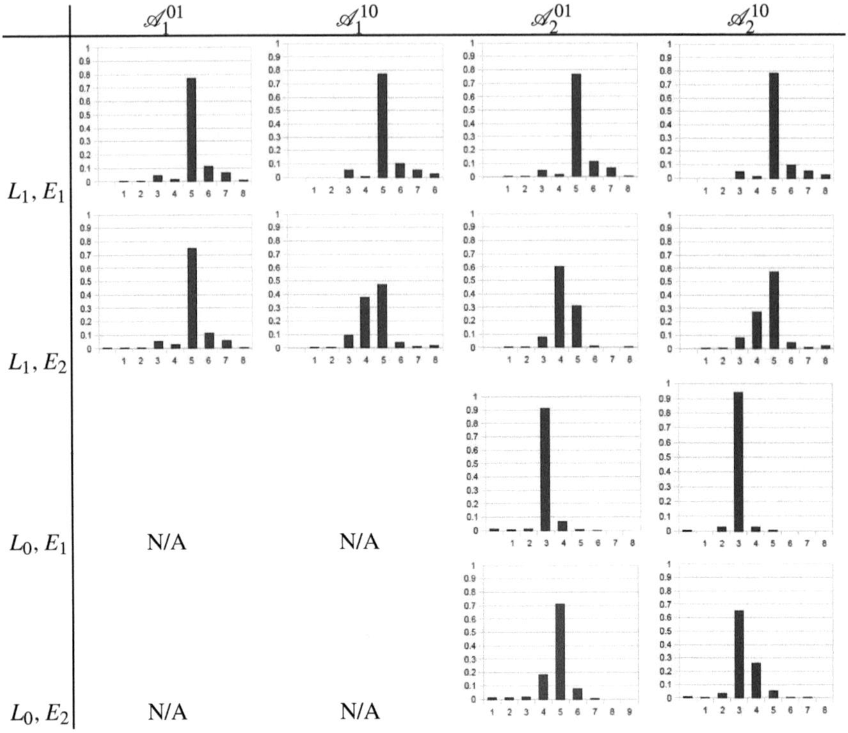

Fig. 15.19 Distributions of cells' in-degrees outside disc D

Epilogue

Automaton networks with just two or three cell-states produce very rich spectra of reaction-diffusion and excitation patterns. A control space-time dynamics of automata —- quasi-chaotic, waves, localisations — is controlled by changing lower and upper boundaries of intervals determining cell-state transitions. In the course of the book we explored some unusual automaton models of excitation and reaction-diffusion, e.g. where excited or refractory cells retain their states, refractory and excited states mutualistically contribute to excitation dynamics, and non-regular automaton networks, e.g. β-skeletons, Delaunay triangulation and Voronoi diagram. We uncovered complexity of interspecies interactions in automaton models of population dynamics: amensalism, commensalism, competition, parasitism, mutualism, and provided evidences for travelling waves, localisations and switching patterns in these systems. We demonstrated efficiency of automaton models of reaction-diffusion and excitable media in unconventional solutions of some problems of computational geometry and binary arithmetics, and implemented two automaton models of memristor networks. The automaton models provided are in no way exhaustive but are rather aimed to act as encouragement for those interested in spatially extended non-linear systems, automaton networks and unconventional computing to develop their own paradigms, architectures and implementations.

References

1. Abrams, P.A., Matsuda, H.: Positive indirect effects between prey species that share predators. Ecology 77, 610–616 (1996)
2. Adamatzky, A.: Identification of Cellular Automata. Taylor & Francis, London (1994)
3. Adamatzky, A.: Controllable transmission of information in the excitable media: the 2^+ medium. Advanced Materials for Optics and Electronics 5, 145–155 (1995)
4. Adamatzky, A.: Computation of discrete convex hull in homogeneous automata networks. Neural Network World 3, 241–254 (1995)
5. Adamatzky, A.: On the particle like waves in the discrete model of excitable media. Neural Networks World 1, 3–10 (1996)
6. Adamatzky, A.: Cellular automaton labyrinths and solution finding. Computers & Graphics 21, 519–522 (1997)
7. Adamatzky, A., Holland, O.: Phenomenology of excitation in 2D cellular automata and swarm systems. Chaos, Solitons & Fractals 3, 1233–1265 (1998)
8. Adamatzky, A.: Universal dynamical computation in multi-dimensional excitable lattices. International Journal of Theoretical Physics 37, 3069–3108 (1998)
9. Adamatzky, A.: Computing in Nonlinear Media and Automata Collectives. IoP Publishing, London (2001)
10. Adamatzky, A. (ed.): Collision-Based Computing. Springer, London (2002)
11. Adamatzky, A., Holland, O.: Reaction-diffusion and ant-based load balancing of communication networks. Kybernetes 31, 667–681 (2002)
12. Adamatzky, A.: Collision-based computing in Belousov-Zhabotinsky medium. Chaos, Solitons & Fractals 21, 1259–1264 (2004)
13. Adamatzky, A.: Computing with waves in chemical media: Massively parallel reaction-diffusion processors. IEICE Trans. Electron. E87-C, 1748–1756 (2004)
14. Adamatzky, A., De Lacy Costello, B., Asai, T.: Reaction-Diffusion Computers. Elsevier, Amsterdam (2005)
15. Adamatzky, A., Wuensche, A., De Lacy Costello, B.: Glider-based computing in reaction-diffusion hexagonal cellular automata. Chaos, Solitons & Fractals 27, 287–295 (2006)
16. Adamatzky, A., Martinez, G.J., Seck Tuoh Mora, J.C.: Phenomenology of reaction-diffusion binary-state cellular automata. Int. J. Bifurcation & Chaos 16, 2985–3005 (2006)
17. Adamatzky, A., Wuensche, A.: Computing in spiral rule reaction-diffusion hexagonal cellular automaton. Complex Systems 16, 277–297 (2007)

18. Adamatzky, A.: Phenomenology of retained excitation. Int. J. Bifurcation and Chaos 17, 3985–4014 (2007)
19. Adamatzky, A.: Developing proximity graphs by Physarum Polycephalum: Does the plasmodium follow Toussaint hierarchy? Parallel Processing Letters 19, 105–127 (2008)
20. Adamatzky, A., Bull, L., Collet, P., Sapin, E.: Evolving localisations in reaction-diffusion cellular automata. Int. J. Modern Physics C 19, 557–568 (2009)
21. Adamatzky, A., Grube, M.: Minimal model of two-species interactions on hexagonal lattice. Int. J. Bifurcation and Chaos 19, 2885–2897 (2009)
22. Adamatzky, A.: Game of Life Cellular Automata. Springer, London (2010)
23. Adamatzky, A., Jones, J.: Road planning with slime mould: If Physarum built motorways it would route M6/M74 through Newcastle. Int. J. Bifurcation and Chaos 20, 3065–3084 (2010)
24. Adamatzky, A.: On excitable beta-skeletons. J. Computational Sci. 1, 175–186 (2010)
25. Adamatzky, A. (ed.): Game of Life Cellular Automata. Springer (2010)
26. Adamatzky, A., Holley, J., Bull, L., De Lacy Costello, B.: On computing in fine-grained compartmentalised Belousov-Zhabotinsky medium arXiv:1006.1900v1 [nlin.PS]
27. Adamatzky, A., De Lacy Costello, B., Bull, L.: On polymorphic logical gates in sub-excitable chemical medium. arXiv:1007.0034v1 [nlin.PS]
28. Adamatzky, A.: On excitable Delaunay automata. Kybernetes 40, 719–735 (2011)
29. Adamatzky, A., Chua, L.: Memristive excitable cellular automata. Int. J. Bifurcation Chaos 21, 3083–3102 (2011)
30. Adamatzky, A., Chua, L.: Phenomenology of retained refractoriness: On semi-memristive discrete media (2011) arXiv:1111.3525v1 [nlin.CG]
31. Adamatzky, A., Kitson, S., De Lacy Costello, B., Matranga, M.A., Younger, D.: Computing with liquid crystal fingers: Models of geometric and logical computation. Phys. Rev. E 84, 061702 (2011)
32. Adamatzky, A., Holley, J., Dittrich, P., Gorecki, J., De Lacy Costello, B., Zauner, K.-P., Bull, L.: On architectures of circuits implemented in simulated Belousov-Zhabotinsky droplets. Biosystems (in press, 2012)
33. Alonso-Sanz, R.: A structurally dynamic cellular automaton with memory. Chaos, Solitons and Fractals 32, 1285–1295 (2006)
34. Alonso-Sanz, R., Martin, M.: Elementary cellular Automata with elementary memory rules in cells: the case of linear rules. J. Cellular Automata 1, 70–86 (2006)
35. Alonso-Sanz, R.: Cellular Automata with Memory. Old City Publishing (2009)
36. Alonso-Sanz, R., Adamatzky, A.: On beta-skeleton automata with memory. J. Computational Sci. 2, 57–66 (2010)
37. Anikeenko, A.V., Alinchenko, M.G., Voloshin, V.P., Medvedev, N.N., Gavrilova, M.L., Jedlovszky, P.: Implementation of the Voronoi-Delaunay Method for Analysis of Intermolecular Voids. In: Laganá, A., Gavrilova, M.L., Kumar, V., Mun, Y., Tan, C.J.K., Gervasi, O. (eds.) ICCSA 2004, Part III. LNCS, vol. 3045, pp. 217–226. Springer, Heidelberg (2004)
38. Arii, K., Parrott, L.: Examining the colonization process of exotic species varying in competitive abilities using a cellular automaton model. Ecological Modelling 199, 219–228 (2006)
39. Asai, T. (ed.): Single-electron circuits. Special Issue of the Int. J. of Nanotechnology and Molecular Computation 1(2), 1–92 (2009)
40. Astrov Yu, A.: Semiconductor gas-discharge planar structures as devices for unconventional computing. Int. J. Unconventional Computing 6, 33–73 (2010)

41. Attali, D., Montanvert, A.: Computing and simplifying 2D and 3D continuous skeletons. Computer Vision and Image Understanding 67, 261–273 (1997)
42. Aubry, S.: Breathers in nonlinear lattices: Existence, linear stability and quantization. Physica D 103, 201–250 (1997)
43. Aubry, S.: Discrete Breathers: Localization and transfer of energy in discrete Hamiltonian nonlinear systems. Physica D 216, 1–30 (2006)
44. Aziza, M., Amrane, A.: Commensalism during submerged mixed culture of Geotrichum candidum and Penicillium camembertii on glutamate and lactate. Process Biochemistry 41, 2452–2457 (2006)
45. Balzter, H., Braun, P.W., Köhler, P.: Cellular automata models for vegetation dynamics. Ecological Modelling 107, 113–125 (1998)
46. Bandyopadhyay, A., Acharya, S.: A 16-bit parallel processing in a molecular assembly. Proc. Natl. Acad. Sci. 105, 3668–3672 (2008)
47. Bandyopadhyay, A., Pati, R., Sahu, S., Peper, F., Fujita, D.: Massively parallel computing on an organic molecular layer. Nature Phys. 6, 369–375 (2010)
48. Bär, M., Brusch, L.: Breakup of spiral waves caused by radial dynamics: Eckhaus and finite wavenumber instabilities. New J. of Physics 6, 5 (2004)
49. Bell, D.I.: HighLife — An interesting variant of Life (1994), `http://www.tip.net.au/~dbell/articles/HighLife.zip`
50. Berlekamp, E., Conway, J., Guy, R.: Winning Ways, vol. 2. Academic Press, New York (1982)
51. Bernal, J.D., Mason, J.: Packing of spheres: co-ordination of randomly packed spheres. Nature 188, 910–911 (1960)
52. Billiot, J.M., Corset, F., Fontenas, E.: Continuum percolation in the relative neighbourhood graph (2010), arXiv:1004.5292
53. Blum, H.: A transformation for extracting new descriptors of shape. In: Wathen-Dunn, W. (ed.) Models for the Perception of Speech and Visual Form. MIT Press (1967)
54. Blum, H.: Biological shape and visual science. J. Theor. Biol. 38, 205–287 (1973)
55. Boccara, N.: Automata network models of interacting population. In: Goles, E., Martinez, S. (eds.) Cellular Automata, Dynamical Systems and Neural Networks. Springer (1994)
56. Boccara, N.: Modeling Complex Systems. Springer (2010)
57. Bode, M., Liehr, A.W., Schenk, C.P., Purwins, H.-G.: Interaction of dissipative solitons: Particle-Like behaviour of localized structures in a three-component reaction-diffusion system. Physica D 161, 45–66 (2002)
58. Bondavalli, C., Ulanowicz, R.E.: Unexpected effects of predators upon their prey: the case of the American Alligator. Ecosystems 2, 49–63 (1999)
59. Boucher, D.H.: The idea of mutualism, past and future. In: Boucher, D.H. (ed.) The Biology of Mutualism, pp. 1–28. Oxford University Press, New York (1985)
60. Boucher, D.H.: The Biology of Mutualism: Ecology and Evolution. Oxford University Press (1988)
61. Boza, G., Scheuring, I.: Environmental heterogeneity and the evolution of mutualism. Ecological Complexity 1, 329–339 (2004)
62. Calabi, L., Hartnett, W.E.: Shape recognition, prairie fires, convex deficiencies and skeletons. Am. Math. Mon. 75, 335–342 (1968)
63. Camazine, S.: Self-organizing pattern formation on the combs of honey bee colonies. Behav. Ecol. Sociobiol. 28, 61–76 (1991)

64. Caswell, H., Etter, R.: Cellular automaton models for competition in patchy environments: Facilitation, inhibition, and tolerance. Bulletin of Mathematical Biology 61, 625–649 (1999)
65. Chazottes, J.-R., Fernandez, B. (eds.): Dynamics of Coupled Map Lattices and of Related Spatially. Springer (2005)
66. Chen, Q., Mynett, A.E., Minns, A.W.: Application of cellular automata to modelling competitive growths of two underwater species. Chara Aspera and Potamogeton Pectinatus in Lake Veluwe Ecological Modelling 147, 253–265 (2002)
67. Chen, Q., Mynett, A.E.: Effects of cell size and configuration in cellular automata based prey-predator modelling. Simulation Modelling Practice and Theory 11, 609–625 (2003)
68. Chen, Z., Martin, H., Eugenieva, E.D., Xu, J., Yang, J.: Formation of discrete solitons in light-induced photonic lattices. Optics Express 13, 1816 (2005)
69. Chen, Q., Mynett, A.E.: Modelling algal blooms in the Dutch coastal waters by integrated numerical and fuzzy cellular automata approaches. Ecological Modelling 199, 73–81 (2006)
70. Cushing, J.M., Henson, S.M., Desharnais, R.A., Dennis, B., Costantino, R.F., King, A.: A chaotic attractor in ecology: theory and experimental data. Chaos, Solitons & Fractals 12, 219–234 (2001)
71. Comins, H.N., Hassell, M.P., May, R.M.: The spatial dynamics of host-parasitoid systems. J. Animal Ecology 61, 735–748 (1992)
72. Christodoulides, D., Lederer, F., Silberberg, Y.: Discretizing light behaviour in linear and nonlinear waveguide lattices. Nature 424, 817–823 (2003)
73. Cannas, S.A., Páez, S.A., Marco, D.E.: Modeling plant spread in forest ecology using cellular automata. Computer Physics Communications 121/122, 131–135 (1999)
74. Cannas, S.A., Marco, D.E., Páez, S.A.: Modelling biological invasions: species traits, species interactions, and habitat heterogeneity. Mathematical Biosciences 183, 93–110 (2003)
75. Cantrell, R.S., Cosner, C.: Spatial Ecology via Reaction-Diffusion Equations. Wiley (2003)
76. Chen, F., You, M.: Permanence for an integro-differential model of mutualism. Applied Mathematics and Computation 186, 30–34 (2007)
77. Chopard, B., Droz, M.: Cellular Automata Modeling of Physical Systems. Cambridge University Press (2005)
78. Chua, L.O.: Memristor — the missing circuit element. IEEE Trans. Circuit Theory 18, 507–519 (1971)
79. Chua, L.O., Kang, S.M.: Memristive devices and systems. Proc. IEEE 64, 209–223 (1976)
80. Chua, L.O.: Device modeling via non-linear circuit elements. IEEE Trans. Circuits Systems 27, 1014–1044 (1980)
81. Dale, M.R.T.: Spatial Analysis in Plant Ecology. Cambridge University Press (2000)
82. Dale, M.R.T., Dixon, P., Fortin, M.-J., Legendre, P., Myers, D.E., Rosenberg, M.S.: Conceptual and mathematical relationships among methods for spatial analysis. Ecography 25, 558–577 (2002)
83. Darwen, P.J., Green, D.G.: Viability of populations in a landscape. Ecological Modelling 85, 165–171 (1996)
84. Dauby, P.C., Desaive, T., Croisier, H.: Standing waves in the FitzHugh-Nagumo model of cardiac electrical activity. Phys. Rev. E 73, 021908 (2006)
85. de Bary, A.: Die Erscheinungen der Symbiose Strassburg, Trubner: 30s (1879)

86. De Lacy Costello, B.P.J.: Constructive chemical processors — Experimental evidence that shows this class of programmable pattern forming reactions exist at the edge of a highly nonlinear region. Int. J. Bifurcation and Chaos 13, 1561–1564 (2003)
87. De Lacy Costello, B.P.J., Adamatzky, A.: On multitasking in parallel chemical processors: experimental results. Int. J. Bifurcation and Chaos 13, 521–533 (2003)
88. De Lacy Costello, B., Hantz, P., Ratcliffe, N.: Voronoi diagrams generated by regressing edges of precipitation fronts. J. Chem. Phys. 120, 2413 (2004)
89. De Lacy Costello, B., Adamatzky, A.: Experimental implementation of collision-based gates in Belousov-Zhabotinsky medium. Chaos, Solitons & Fractals 25, 535–544 (2005)
90. De Lacy Costello, B., Toth, R., Stone, C., Adamatzky, A., Bull, L.: Implementation of glider guns in the light-sensitive Belousov-Zhabotinsky medium. Phys. Rev. E 79, 26114 (2009)
91. De Lacy Costello, B.P.J., Jahan, I., Adamatzky, A., Ratcliffe, N.M.: Chemical tesselations. Int. J. Bifurcation and Chaos 19, 619–622 (2009)
92. Delaunay, B.: Sur la sphère vide, Izvestia Akademii Nauk SSSR. Otdelenie Matematicheskikh i Estestvennykh Nauk 7, 793–800 (1934)
93. de Cardozo, G., de Silvestre, D., Colato, A.: Periodical cicadas: A minimal automaton model. Physica A 382, 439–444 (2007)
94. Delgado, M., López-Goómez, J., Suárez, A.: On the symbiotic Lotka-Volterra model with diffusion and transport effects. Journal of Differential Equations 160, 175–262 (2000)
95. Deutsch, A., Dormann, S.: Cellular Automaton Modeling of Biological Pattern Formation. Birkhäuser, Boston (2004)
96. Dewdney, A.K.: Armchair Universe: An Exploration of Computer Worlds. Freeman and Co. (1988)
97. Dieckmann, U., Law, R., Metz, J.A.J.: The Geometry of Ecological Interactions: Simplifying Spatial Complexity. Cambridge University Press (2000)
98. Dockery, J.D., Keener, J.P.: Diffusive effects on dispersion in excitable media. SIAM J. on Applied Mathematics 49, 539–566 (1989)
99. Droz, M., Pekalski, A.: Dynamics of Populations in Extended Systems. In: Bandini, S., Chopard, B., Tomassini, M. (eds.) ACRI 2002. LNCS, vol. 2493. Springer, Heidelberg (2002)
100. Duryea, M., Caraco, T., Gardner, G., Maniatty, W., Szymanski, B.K.: Population dispersion and equilibrium infection frequency in a spatial epidemic. Physica D 132, 511–519 (1999)
101. Elettreby, M.F.: Two-prey one-predator model. Chaos, Solitons & Fractals 36, 1097–1107 (2007)
102. El Naschie, M.S.: A very brief history of localization. Chaos, Solitons & Fractals 11, 1479–1480 (2000)
103. El Naschie, M.S., Kapitaniak, T.: Soliton chaos models for mechanical and biological elastic chains. Phys. Lett. A 147, 275–281 (1990)
104. Ermentrout, G.B., Edelstein-Keshet, L.: Cellular automata approaches to biological modeling. J. Theor. Biology 160, 97–133 (1993)
105. Erokhin, V., Fontana M.T.: Electrochemically controlled polymeric device: a memristors (and more) found two years ago (2008) arXiv:0807.0333v1 [cond-mat.soft]
106. Fath, B.D.: Network mutualism: Positive community-level relations in ecosystems. Ecological Modelling 208, 56–67 (2007)

107. Filatovs, G.J.: Delaunay-Subgraph Statistics of Disc Packings. Materials Characterization 40, 27–35 (1998)
108. Flach, S., Willis, C.R.: Discrete breathers. Physics Reports 295, 181 (1998)
109. Fredkin, E., Toffoli, T.: Conservative logic. Int. J. Theor. Phys. 21, 219–253 (1982)
110. Frey-Klett, P., Garbaye, J., Tarkka, M.: The mycorrhiza helper bacteria revisited. New Phytologist 176, 22–36 (2007)
111. Fortune, S.: A sweepline algorithm for Voronoi diagrams. In: Proc. 2nd Annual Symp. on Comput. Geom., pp. 313–322 (1986)
112. Gabriel, K.R., Sokal, R.R.: A new statistical approach to geographic variation analysis. Systematic Zoology 18, 259–270 (1969)
113. Galam, S., Chopard, B., Droz, M.: Killer geometries in competing species dynamics. Physica A 314, 256–263 (2002)
114. Gardner, M.: The fantastic combinations of John H. Conway's new solitaire game Life. Scientific American 223, 120–123 (1970)
115. Gause, G.F., Witt, A.A.: Behavior of mixed populations and the problem of natural selection. The American Naturalist 69, 596–609 (1935)
116. Gerhardt, M., Schuster, H., Tyson, J.J.: A cellular excitable media. Physica D 46, 392–415 (1990)
117. Gervois, A., Annic, C., Lemaitre, J., Ammi, M., Oger, L., Troadec, J.-P.: Arrangement of discs in 2D binary assemblies. Physica A 218, 403–418 (1995)
118. Ghergu, M., Rdulescu, V.D.: Nonlinear PDEs: Mathematical Models in Biology, Chemistry and Population Genetics. Springer (2011)
119. Grattepanche, F., Audet, P., Lacroi, C.: Milk fermentation by functional mixed culture producing nisin Z and exopolysaccharides in a fresh cheese model. Int. Dairy Journal 17, 123–132 (2007)
120. Graves, V.G., Peckham, B.B., Pastor, J.: A 2D differential equations model for mutualism. Department of Mathematics and Statistics. Technical Report TR 2006-2. University of Minnesota Duluth (2006)
121. Gray, R.A., Mornev, O.A., Jalife, J., Aslanidi, O.V., Pertsov, A.M.: Standing excitation waves in the heart induced by strong alternating electric fields. Phys. Rev. Lett. 87, 168104 (2001)
122. Greenberg, J.M., Hastings, S.P.: Spatial patterns for discrete models of diffusion in excitable media. SIAM Journal on Applied Mathematics 34, 515–523 (1978)
123. Griffeath, D., Moore, C. (eds.): New Constructions in Cellular Automata. Oxford University Press, Oxford (2003)
124. Grimm, V., Frank, K., Jeltsch, F., Brandl, R., Uchmanski, J., Wissel, C.: Pattern-oriented modelling in population ecology. Sci. of The Total Environment 183, 151–166 (1996)
125. Gomes, A.A., Manica, E., Varriale, M.C.: Applications of chaos control techniques to a three-species food chain. Chaos, Solitons & Fractals 36, 1097–1107 (2008)
126. Górecki, J., Górecka, J.N.: Information processing with chemical excitations — from instant machines to an artificial chemical brain. Int. J. Unconv. Comput. 2, 321–336 (2006)
127. Górecka, J.N., Górecki, J., Igarashi, Y.: On the simplest chemical signal diodes constructed with an excitable medium. Int. J. Unconventional Computing 5, 129–143 (2009)
128. Gotoh, K., Finney, J.L.: Statistical geometrical approach to random packing density of equal spheres. Nature 252, 202–205 (1974)
129. G'unther, R., Schapiro, B., Wagner, P.: Complex systems, complexity measures, grammars and model-inferring. Chaos, Solitons & Fractals 4, 635–651 (1994)

130. Halpern, P., Caltagirone, G.: Behavior of topological cellular automata. Complex Systems 4, 623–651 (1990)
131. Hartman, H., Tamayo, P.: Reversible cellular automata and chemical turbulence. Physica D 45, 293–306 (1990)
132. Hassell, M.P., Pacala, S.W., May, R.M., Chesson, P.L.: Chesson The persistence of host-parasitoid associations in patchy environments. I. A general criterion. American Naturalist 138, 568–583 (1991)
133. Hasslacher, B., Meyer, D.A.: Modelling dynamical geometry with lattice gas automata. Int. J. Modern Physics C 9, 1597–1605 (1998)
134. He, M., Lin, J., Jiang, H., Liu, X.: The two populations cellular automata model with predation based on the Penna model. Physica A 312, 243–250 (2002)
135. Heard, S.B.: Pitcher-plant midges and mosquitoes: a processing chain commensalism. Ecology 75, 1647–1660 (1994)
136. Heudin, J.-K.: A new candidate rule for the game of two-dimensional life. Complex Systems 10, 367–381 (1996)
137. Hillman, D.: Combinatorial spacetimes (Ph.D. dissertation) 234 pp. arXiv:hep-th/9805066v1
138. Hobbs, W.L.: Network topology in aperiodic networks. J. of Non-Crystalline Solids 192/193, 79–91 (1995)
139. Holland, J.N., DeAngelis, D.L., Bronstein, J.L.: Population dynamics and mutualism: functional responses of benefits and costs. American Naturalist 159, 231–244 (2002)
140. Holley, J., Adamatzky, A., Bull, L., De Lacy Costello, B., Jahan, I.: Modalities of Belousov-Zhabotinsky encapsulated vesicles. Nanocommunication Networks J. 2, 50–61 (2011)
141. Holley, J., Jahan, I., de Lacy Costello, B., Bull, L., Adamatzky, A.: Logical and arithmetic circuits in Belousov-Zhabotinsky encapsulated discs. Physical Review E, APS 84, 56110 (2011)
142. Hui, C., McGeoch, M.A.: Evolution of body size, range size, and food composition in a predator-prey metapopulation. Ecological Complexity 3, 148–159 (2006)
143. Ilachinsky, A., Halpern, P.: Structurally dynamic cellular automata. Complex Systems 1, 503–527 (1987)
144. Ilachinski, A.: Cellular Automata: A Discrete University. World Scientific, Singapore (2001)
145. Ilachinsky, A.: Structurally Dynamic Cellular Automata Encyclopedia of Complexity and Systems Science, part 19, pp. 8815–8850 (2009)
146. Isokawa, T., Kowada, S., Takada, Y., Peper, F., Kamiura, N., Matsui, N.: Defect-tolerance in cellular nanocomputers. New Generation Computing 25, 171–199 (2007)
147. Isokawa, T., Abo, F., Peper, F., Adachi, S., Lee, J., Matsui, N., Mashiko, S.: Fault-tolerant nanocomputers based on asynchronpous cellular automata. Int. J. Modern Physics C 15, 893–915 (2004)
148. Itoh, M., Chua, L.: Memristor cellular automata and memristor discrete-time cellular neural networks. Int. J. Bifurcation and Chaos 19, 3605–3656 (2009)
149. Jakubith, S., Rotermund, H.H., Engel, W., von Oertzen, A., Ertl, G.: Spatiotemporal concentration patterns in a surface reaction: Propagating and standing waves, rotating spirals, and turbulence. Phys. Rev. Lett. 65, 3013–3016 (1990)
150. Jaromczyk, J.W., Toussaint, G.T.: Relative neighbourhood graphs and their relatives. Proc. IEEE 80, 1502–1517 (1992)
151. Jeon, K.W.: Development of cellular dependence on infective organisms: micrsurgical studies in amoebas. Science 176, 1122–1123 (1972)

152. Jombart, T., Devillard, S., Dufour, A.-B., Pontier, D.: Revelaing cryptic spatial patterns in genetic variability by a new multivariate method. Heredity 101, 92–103 (2008)
153. Kaminaga, A., Vanag, V., Epstein, I.: Black spots in a surfactant-rich Belousov-Zhabotinsky reaction dispersed in a water-in-oil microemulsion system. J. Chem. Phys. 122, 174706–174711 (2005)
154. Kaminaga, A., Vanag, V.K., Epstein, I.R.: A reaction-diffusion memory device. Angew. Chem. Int. Ed. 45, 3087–3089 (2006)
155. Karim, M.F.A., Ismail, M.F.A., Ching, H.B.: Cellular automata modelling of hantarvirus infection. Chaos, Solitons & Fractals 41, 2847–2853 (2008)
156. Karst, J., Marczak, L., Jones, M.D., Turkington, R.: The mutualism-parasitism continuum in ectomycorrhizas: a quantitative assessment using meta-analysis. Ecology 89, 1032–1042 (2008)
157. Kawaguchi, I., Sasaki, A.: The wave speed of intergradation zone in two-species lattice Müllerian mimicry model. J. Theor. Biology 243, 594–603 (2006)
158. Kendall, B.E.: Cycles, chaos, and noise in predator-prey dynamics. Chaos, Solitons & Fractals 12, 321–332 (2001)
159. King, P.H., Corsi, J.C., Pan, B.-H., Morgan, H., de Planque, M.R., Zauner, K.-P.: Towards molecular computing: Co-development of microfluidic devices and chemical reaction media. BioSystems (in press, 2012)
160. Kirkpatrick, D.G., Radke, J.D.: A framework for computational morphology. In: Toussaint, G.T. (ed.) Computational Geometry, pp. 217–248. North-Holland (1985)
161. Kizaki, S., Katori, M.: A stochastic lattice model for locust outbreak. Physica A 266, 339–342 (1999)
162. Klein, R.: Concrete and Abstract Voronoi Diagrams. Springer, Berlin (1990)
163. van de Koppel, J., Gascoigne, J.C., Theraulaz, G., Rietkerk, M., Mooij, W.M., Herman, P.M.J.: Experimental evidence for spatial self-organization and its emergent effects in mussel bed ecosystems. Science 322, 937–942 (2008)
164. Kuo, C.-H., Corby-Harris, V., Promislow, D.E.L.: The unavoidable costs and unexpected benefits of parasitism: Population and metapopulation models of parasite-mediated competition. J. Theor. Biology 250, 244–256 (2008)
165. van der Laan, J.D., Bradbury, R.H.: Futures for the Great Barrier Reef ecosystem. Mathematical and Computer Modelling 14, 705–709 (1990)
166. van der Laan, J.D., Lhotka, L., Hogeweg, P.: Sequential predation: a multi-model study. Journal of Theoretical Biology 174, 149–167 (1995)
167. Langton, C.G.: Self-Reproduction in cellular automata. Physica D 10, 135–144 (1984)
168. Lee, J., Eggert, S., Kim, H., Kahng, S.-J., Shinohara, H., Kuk, Y.: Real space imaging of one-dimensional standing waves: direct evidence for a Luttinger liquid. Phys. Rev. Letts. 93, 166403 (2004)
169. Lee, J., Peper, F.: On Brownian cellular automata. In: Adamatzky, A., et al. (eds.) AUTOMATA-2008: Theory and Applications of Cellular Automata, pp. 278–302. Luniver Press (2008)
170. Legendre, P., Fortin, M.-J.: Spatial pattern and ecological analysis. Vegetatio 80, 107–138 (1989)
171. Leymarie, F., Levine, M.D.: Simulating the grassfire transform using an active contour model. IEEE. Trans. on Pattern Analysis and Machine Intelligence 14, 56–75 (1992)
172. Li, X.-Y.: Application of computation geometry in wireless networks. In: Cheng, X., Huang, X., Du, D.-Z. (eds.) Ad Hoc Wireless Networking, pp. 197–264. Kluwer Academic Publishers (2004)

173. Liehr, A.W., Bode, M., Purwins, H.-G.: The generation of dissipative quasi-particles near Turing's bifurcation in three-dimensional reaction-diffusion-systems. In: Krause, E., Jüger, W. (eds.) High Performance Computing in Science and Engineering 2000. Transactions of the High Performance Computing Center, Stuttgart (HLRS) 2000, pp. 425–439. Springer (2001)
174. Little, A.E.M., Currie, C.R.: Symbiotic complexity: discovery of a fifth symbiont in the attine antimicrobe symbiosis. Biology Letters 3, 501–504 (2007)
175. Little, A.E.M., Currie, C.R.: Black yeast symbionts compromise the efficiency of antibiotic defences in fungus-growing ants. Ecology 89, 1216–1222 (2008)
176. Liu, Z., Tan, R., Chen, Y., Chen, L.: On the stable periodic solutions of a delayed two-species model of facultative mutualism. Applied Mathematics and Computation 196, 105–117 (2008)
177. Lochmann, K., Oger, L., Stoyan, D.: Statistical analysis of random sphere packings with variable radius distribution. Solid State Sciences 8, 1397–1413 (2006)
178. Lozano, J., Virgós, E., Cabezas-Díaz, S., Mangas, J.G.: Increase of large game species in Mediterranean areas: Is the European wildcat (Felis silvestris) facing a new threat? Biological Conservation 138, 321–329 (2007)
179. Luchnikov, V.A., Gavrilova, M.L., Medvedev, N.N., Voloshin, V.P.: The Voronoi-Delaunay approach for the free volume analysis of a packing of balls in a cylindrical container. Future Generation Computer Systems 18, 673–679 (2002)
180. Lv, S., Fang, Z.: The dynamic complexity of a host-parasitoid model with a Beddington–De Angelis functional response. Chaos, Solitons & Fractals (2008) (in press)
181. Maccari, A.: Chaos, solitons and fractals in hidden symmetry models. Chaos, Solitons & Fractals 27, 363–376 (2006)
182. Magnier, M., Lattaud, C., Heudin, J.-C.: Complexity classes in the two-dimensional Life cellular automata subspace. Complex Systems 11, 419–436 (1997)
183. Magwene, P.W.: Using correlation proximity graphs to study phenotypic integration. Evolutionary Biology 35, 191–198 (2008)
184. Margolus, N.: Physics–like models of computation. Physica D 10, 81–95 (1984)
185. Martinez, G., Adamatzky, A., McIntosh, H.: Localization dynamic in binary two-dimensional cellular automaton: Diffusion Rule. J. Cellular Automata 5, 289–313 (2010)
186. Manz, N., Steinbock, O.: Propagation failures, breathing pulses, and backfiring in an excitable reaction-diffusion system. Chaos 16, 37112 (2006)
187. Markus, M., Hess, B.: Isotropic cellular automata for modelling excitable media. Nature 347, 56–58 (1990)
188. Maruno, K., Biondini, G.: Resonance and web structure in discrete soliton systems: the two-dimensional Toda lattice and its fully discrete and ultra-discrete analogues. J. Phys. A: Math. Gen. 37, 11819–11839 (2004)
189. Matsinos, Y.G., Troumbis, A.Y.: Modeling competition, dispersal and effects of disturbance in the dynamics of a grassland community using a cellular automaton model. Ecological Modelling 149, 71–83 (2002)
190. Matsuoka, C., Shimodoi, K., Iizuka, T., Hasegawa, T.: Reflection of active waves in reaction-diffusion media. Phys. Lett. A 243, 47–51 (1998)
191. Matula, D.W., Sokal, R.R.: Properties of Gabriel graphs relevant to geographic variation research and clustering of points in the plane. Geogr. Anal. 12, 205–222 (1980)
192. McIntosh, H.V.: One Dimensional Cellular Automata. Luniver Press (2009)

193. Mikhailov, A.S., Showalter, K.: Control of waves, patterns and turbulence in chemical systems. Physics Reports 425, 79–194 (2006)
194. Moran, N.: Symbiosis. Current Biology 871, R866–R871 (2006)
195. Morozov, A., Ruan, S., Li, B.-L.: Patterns of patchy spread in multi-species reaction-diffusion models. Ecological Complexity 5, 313–328 (2008)
196. Muhammad, R.B.: A distributed graph algorithm for geometric routing in ad hoc wireless networks. J. Networks 2, 49–57 (2007)
197. Murra, J.D.: Mathematical Biology II: Spatial Models and Biomedical Applications. Springer (2003)
198. Neuhauser, C., Fargione, J.E.: A mutualism-parasitism continuum model and its application to plant-mycorrhizae interactions. Ecological Modelling 177, 337–352 (2004)
199. von Neumann, J.: Theory of Self-reproducing Automata. Edited and completed by Burks, A.W. University of Illinois Press, Urbana and London (1966)
200. NeuNeu: Artificial Wet Neuronal Networks from Compartmentalised Excitable Chemical Media (2010), `http://neu-n.eu/`
201. Nowak, M.A.: Five rules for the evolution of cooperation. Science 314, 1560–1563 (2006)
202. Odum, E.P.: Fundamentals of Ecology. W B Saunders Co., New York (1971)
203. Odum, E.P.: Basic Ecology. Harcourt Brace College Publisher (1983)
204. Ouyang, Q., Swinney, H.L., Li, G.: Transition from spirals to defect-mediated turbulence driven by a doppler instability. Phys. Rev. Lett. 84, 1047–1050 (2000)
205. Overmann, J.: Phototrophic consortia. A tight cooperation between non-related eubacteria. In: Seckbach, J. (ed.) Symbiosis. Mechanisms and Model Systems, pp. 239–255. Kluwer, Dordrecht (2001)
206. Okabe, A., Boots, B., Sugihara, K., Chiu, S.N.: Spatial Tesselations: Concepts and Applications of Voronoi diagrams. John Wiley and Sons (2000)
207. Packard, N.H., Wolfram, S.: Two-dimensional cellular automata. J. Stat. Phys. 38, 901–946 (1985)
208. Pearce, A.R., Caelli, T., Sestito, S., Goss, S., Selvestrel, M., Murray, G.: Skeletonizing topographical regions for navigational path planning. Technical Report CVPRL and ARL, Australia (1993)
209. Pommier, S., Strehaiano, P., Délia, M.L.: Modelling the growth dynamics of interacting mixed cultures: a case of amensalism. Int. J. Food Microbiology 100, 131–139 (2005)
210. Poupon, A.: Voronoi and Voronoi-related tessellations in studies of protein structure and interaction. Current Opinion in Structural Biology 14, 233–241 (2004)
211. Preparata, F., Shamos, M.: Computational Geometry: An Introduction. Springer (1985)
212. Puffer train. LifeWiki (2010), `http://www.conwaylife.com/wiki/index.php?title=Puffer_train`
213. Requardt, M.: A geometric renormalization group in discrete quantum space time. J. Math. Phys. 44, 5588 (2003)
214. Renshaw, E.: Modelling Biological Populations in Space and Time. Cambridge University Press (1993)
215. Rosé, H., Hempel, H., Schimansky-Geier, L.: Stochastic dynamics of catalytic CO oxidation on Pt (100). Physica A 206, 421–440 (1994)
216. Rosenfeld, A., Pfaltz, J.L.: Distance functions on digital pictures. Pattern Recognition 1, 33–61 (1968)
217. Rotermund, H.H., Jakubith, S., von Oertzen, A., Ertl, G.: Solitons in a surface reaction. Physical Review Letters 66, 3083–3086 (1991)

218. Runions, A., Fuhrer, M., Lane, B., Federl, P., Rolland-Lagan, A.-G., Prusinkiewicz, P.: Modeling and visualization of leaf venation patterns. ACM Transactions on Graphics 24, 702–711 (2005)
219. Sediña-Nadal, I., Mihaliuk, E., Wang, J., Pérez-Muñuzuri, V. and Showalter, K.: Wave propagation in subexcitable media with periodically modulated excitability. Phys. Rev. Lett. 86, 1646–1649 (2001)
220. Salles, P., Bredeweg, B., Bensusan, N.: The ants' garden: Qualitative models of complex interactions between populations. Ecological Modelling 194, 90–101 (2006)
221. Santi, P.: Topology Control in Wireless Ad Hoc and Sensor Networks. Wiley (2005)
222. Sapin, E., Bailleux, O., Chabrier, J.J.: Research of a Cellular Automaton Simulating Logic Gates by Evolutionary Algorithms. In: Ryan, C., Soule, T., Keijzer, M., Tsang, E.P.K., Poli, R., Costa, E. (eds.) EuroGP 2003. LNCS, vol. 2610, pp. 414–423. Springer, Heidelberg (2003)
223. Sapin, E., Bailleux, O., Chabrier, J.-J.: Research of Complex Forms in Cellular Automata by Evolutionary Algorithms. In: Liardet, P., Collet, P., Fonlupt, C., Lutton, E., Schoenauer, M. (eds.) EA 2003. LNCS, vol. 2936, pp. 357–367. Springer, Heidelberg (2004)
224. Sapin, E., Bailleux, O., Chabrier, J.-J., Collet, P.: A New Universal Cellular Automaton Discovered by Evolutionary Algorithms. In: Deb, K., Tari, Z. (eds.) GECCO 2004. LNCS, vol. 3102, pp. 175–187. Springer, Heidelberg (2004)
225. Sapin, E., Bailleux, O., Chabrier, J.J., Collet, P.: Demonstration of the universality of a new cellular automaton. Int. J. Unconventional Computing 3, 79–103 (2007)
226. Sokal, R.R., Oden, N.L.: Spatial autocorrelation in biology 1. Methodology. Biological Journal of the Linnean Society 10, 199–228 (2008)
227. Song, W.-Z., Wang, Y., Li, X.-Y.: Localized algorithms for energy efficient topology in wireless ad hoc networks. In: Proc. MobiHoc 2004, Roppongi, Japan, May 24-26 (2004)
228. Soukoulis, C.M. (ed.): Photonic Crystals and Light Localization in the 21st Century: Proceedings of the NATO Advanced Study Institute on Photonic Crystals and Light Localization. Kluwer Academic Publishers (2001)
229. Spaceship. LifeWiki (2010), `http://www.conwaylife.com/wiki/index.php?title=Spaceship`
230. Sridharan, M., Ramasamy, A.M.S.: Gabriel graph of geomagnetic Sq variations. Acta Geophysica (2010), doi:10.2478/s11600-010-0004-y
231. Stadler, B., Dixon, A.F.G.: Mutualism: Ants and their Insect Partners. Cambridge University Press (2008)
232. Strukov, D.B., Snider, G.S., Stewart, D.R., Williams, R.S.: The missing memristor found. Nature 453, 80–83 (2008)
233. Summers, D., Cranford, J.G., Healey, B.P.: Chaos in periodically forced discrete-time ecosystem models. Chaos, Solitons & Fractals 11, 2331–2342 (2000)
234. Szaran, T.: Spatiotemporal Models of Population and Community Dynamics. Springer (1997)
235. Sun, C., Loreau, M.: Dynamics of a three-species food chain model with adaptive traits. Chaos, Solitons & Fractals 41, 2812–2819 (2008)
236. Tainaka, K., Terazawa, N., Yoshida, N., Nakagiri, N., Takeuchi, Y.: Spatial pattern formation in a model ecosystem: exchange between symbiosis and competition. Physics Letters A 282, 373–379 (2001)
237. Takada, Y., Isokawa, T., Peper, F., Matsui, N.: Construction universality in purely asynchronous cellular automata. J. Computer System Sci. 72, 1368–1385 (2006)

238. Takagi, S., Pumir, A., Kramer, L., Krinsky, V.: Mechanism of standing wave patterns in cardiac muscle. Phys. Rev. Lett. 90, 124101 (2003)
239. Teuscher, C., Adamatzky, A.: Unconventional Computing 2005: From Cellular Automata to WetWare. Luniver Press (2005)
240. Tilman, D., Kareiva, P. (eds.): Spatial Ecology. Princeton University Press (1997)
241. Toffoli, T., Margolus, N.: Cellular Automata Machines. The MIT Press, Cambridge (1987)
242. Margolus, N., Toffoli, T.: Cellular Automata Machines: A New Environment for Modeling. The MIT Press (1987)
243. Tolmachev, D., Adamatzky, A.: Chemical processor for computation of Voronoi diagram. Adv. Mater. Opt. Electron. 6, 191–196 (1996)
244. Tomita, K., Kurokawa, H., Murata, S.: Graph-rewriting automata as a natural extension of cellular automata. In: Understanding Complex Systems, pp. 291–309. Springer (2009)
245. Toroczkai, Z., Guclu, H.: Proximity networks and epidemics. Physica A 378, 68 (2007), arXiv:physics/0701255v1
246. Toth, R., Stone, C., Adamatzky, A., de Lacy Costello, B., Bull, L.: Experimental validation of binary collisions between wave fragments in the photosensitive Belousov-Zhabotinsky reaction. Chaos, Solitons & Fractals 41, 1605–1615 (2009)
247. Toth, T., Stone, C., de Lacy Costello, B., Adamatzky, A., Bull, L.: Simple collision-based chemical logic gates with adaptive computing. Int. J. Nanotechnology and Molecular Computation 1, 1–17 (2009)
248. Vakakis, A. (ed.): Normal Modes and Localization in Nonlinear Systems. Kluwer Academic Publishers (2001)
249. Vanag, V.K., Epstein, I.R.: Pattern formation in a tunable medium: The Belousov-Zhabotinsky reaction in an aerosol OT microemulsion. Phys. Rev. Lett. 87, 1–4 (2001)
250. Vanag, V.K., Epstein, I.R.: Segmented spiral waves in a reaction–diffusion system. Proc. Nat. Acad. Sci. USA 100, 14635–14638 (2003)
251. Viana, M.P., Travencolo, B., Tanck, E., da Costa, L.F.: Characterizing topological and dynamical properties of complex networks without border effects. Physica A 389, 1771–1778 (2010)
252. Volkenborn, N., Reise, K.: Lugworm exclusion experiment: Responses by deposit feeding worms to biogenic habitat transformations. J. Experimental Marine Biology and Ecology 330, 169–179 (2006)
253. Voronoi, G.: Nouvelles applications des paramétres continus à la théorie des formes quadratiques. Journal fur die Reine und Angewandte Mathematik 133, 97–178 (1907)
254. Wan, P.-J., Yi, C.-W.: On the longest edge of Gabriel Graphs in wireless ad hoc networks. IEEE Trans. on Parallel and Distributed Systems 18, 111–125 (2007)
255. Wang, J., Kada, S., Jung, P., Showalter, K.: Noise driven avalanche behaviour in subexitable media. Phys. Rev. Lett. 82, 855–858 (1999)
256. Watanabe, D.: A study on analyzing the road network pattern using proximity graphs. J. of the City Planning Institute of Japan 40, 133–138 (2005)
257. Watanabe, D.: Evaluating the configuration and the travel efficiency on proximity graphs as transportation networks. Forma 23, 81–87 (2008)
258. Weimar, J.R.: Simulation with Cellular Automata. Logos (2003)
259. Wisialowski, B.: Conways Game of Life: Lines to Fractals, `https://webfiles.uci.edu/bwisialo/www/gameoflife2.html`

260. Williams, J.D., McDermott, J.J.: Hermit crab biocoenoses: a worldwide review of the diversity and natural history of hermit crab associates. J. of Experimental Marine Biology and Ecology 305, 1–128 (2004)
261. Williams R. S. How we found the missing memristor. IEEE Spectrum (December 18, 2008)
262. Wolfram, S.: A New Kind of Science. Wolfram Media (2002)
263. Wootton, J.T.: Local interactions predict large-scale pattern in empirically derived cellular automata. Nature 413, 841–843 (2001)
264. Wright, D.H.: A simple, state model of mutualism incorporating handling time. American Naturalist 134, 664–667 (1989)
265. Wu, J., David, J.L.: A spatially explicit hierarchical approach to modeling complex ecological systems: theory and applications. Ecological Modelling 153, 7–26 (2002)
266. Wuensche, A., Lesser, M.: The Global Dynamics of Cellular Automata: An Atlas of Basin of Attraction Fields of One-Dimensional Cellular Automata. Perseus Books (1992)
267. Wuensche, A.: Self-reproduction by glider collisions: the beehive rule (A Life9 Proceedings), pp. 286–291. MIT Press (2004)
268. Wuensche, A.: The spiral glider-gun rule (2005), `http://www.cogs.susx.ac.uk/users/andywu/multi_value/spiral_rule.html`
269. Wuensche, A.: Glider dynamics in 3-value hexagonal cellular automata: the beehive rule. Int. J. Unconventional Computing 1, 375–398 (2005)
270. Wuensche, A.: Glider dynamics in 3-value hexagonal cellular automata: The beehive rule. Int. J. of Unconventional Computing 1, 375–398 (2005)
271. Wuensche, A., Adamatzky, A.: On spiral glider-guns in hexagonal cellular automata: activator-inhibitor paradigm. Int. J. Modern Physics C 17, 1009–1026 (2006)
272. Xavier, J.B., Foster, K.R.: Cooperation and conflict in microbial biofilms. PNAS 104, 876–881 (2006)
273. Yaguma, S., Odagir, K., Takatsuka, K.: Coupled-cellular-automata study on stochastic and pattern-formation dynamics under spatiotemporal fluctuation of temperature. Physica D 197, 34–62 (2004)
274. Yang, X.: Computational modelling of nonlinear calcium waves. Appl. Mathem. Modelling 30, 200–208 (2006)
275. Yang, X.: Pattern formation in enzyme inhibition and co-operativity with parallel cellular automata. Parallel Computing 30, 741–751 (2004)
276. Yang, J.J., Pickett, M.D., Li, X., Ohlberg, D.A.A., Stewart, D.R., Williams, R.S.: Memristive switching mechanism for metal-oxide-metal nanodevices. Nature Nano 3(7) (2008)
277. Yang, L., Zhabotinsky, A.M., Epstein, I.R.: Stable squares and other oscillatory Turing patterns in a reaction-diffusion model. Phys. Rev. Let. 92, 19833–01 (2004)
278. Young, D.: A local activator–inhibitor model of vertebrate skin patterns. Math. Biosci. 72, 51–58 (1984)
279. Yu, H., Zhao, M., Lv, S., Zhu, L.: Dynamic complexities in a parasitoid-host-parasitoid ecological model. Chaos, Solitons & Fractals 39, 39–48 (2009)
280. Zarzycki, J.: Structure of dense gels. J. of Non-Crystalline Solids 147/148, 176–182 (1992)
281. Zhang, L., Adamatzky, A.: Collision-based implementation of a two-bit adder in excitable cellular automaton. Chaos, Solitons & Fractals 41, 1191–1200 (2009)
282. Zhang, L., Adamatzky, A.: Towards arithmetical chips in sub-excitable media: cellular automaton models. Int. J. Nanotechnology and Molecular Computation 1, 1–19 (2009)

Index

MIX
Papier aus verantwortungsvollen Quellen
Paper from responsible sources
FSC® C105338

If you have any concerns about our products,
you can contact us on
ProductSafety@springernature.com

In case Publisher is established outside the EU,
the EU authorized representative is:
Springer Nature Customer Service Center GmbH
Europaplatz 3, 69115 Heidelberg, Germany

Printed by Libri Plureos GmbH
in Hamburg, Germany